Glycolipids, Glycoproteins, and Mucopolysaccharides of the Nervous System

ADVANCES IN EXPERIMENTAL MEDICINE AND BIOLOGY

Volume 1
THE RETICULOENDOTHELIAL SYSTEM AND ATHEROSCLEROSIS
Edited by N. R. Di Luzio and R. Paoletti • 1967

Volume 2
PHARMACOLOGY OF HORMONAL POLYPEPTIDES AND PROTEINS
Edited by N. Back, L. Martini, and R. Paoletti • 1968

Volume 3
GERM-FREE BIOLOGY: Experimental and Clinical Aspects
Edited by E. A. Mirand and N. Back • 1969

Volume 4
DRUGS AFFECTING LIPID METABOLISM
Edited by W. L. Holmes, L. A. Carlson, and R. Paoletti • 1969

Volume 5
LYMPHATIC TISSUE AND GERMINAL CENTERS IN IMMUNE RESPONSE
Edited by L. Fiore-Donati and M. G. Hanna, Jr. • 1969

Volume 6
RED CELL METABOLISM AND FUNCTION
Edited by George J. Brewer • 1970

Volume 7
SURFACE CHEMISTRY OF BIOLOGICAL SYSTEMS
Edited by Martin Blank • 1970

Volume 8
BRADYKININ AND RELATED KININS: Cardiovascular, Biochemical, and Neural Actions
Edited by F. Sicuteri, M. Rocha e Silva, and N. Back • 1970

Volume 9
SHOCK: Biochemical, Pharmacological, and Clinical Aspects
Edited by A. Bertelli and N. Back • 1970

Volume 10
THE HUMAN TESTIS
Edited by E. Rosemberg and C. A. Paulsen • 1970

Volume 11
MUSCLE METABOLISM DURING EXERCISE
Edited by B. Pernow and B. Saltin • 1971

Volume 12
MORPHOLOGICAL AND FUNCTIONAL ASPECTS OF IMMUNITY
Edited by K. Lindahl-Kiessling, G. Alm, and M. G. Hanna, Jr. • 1971

Volume 13
CHEMISTRY AND BRAIN DEVELOPMENT
Edited by R. Paoletti and A. N. Davison • 1971

Volume 14
MEMBRANE-BOUND ENZYMES
Edited by G. Porcellati and F. di Jeso • 1971

Volume 15
THE RETICULOENDOTHELIAL SYSTEM AND IMMUNE PHENOMENA
Edited by N. R. Di Luzio and K. Flemming • 1971

Volume 16A
THE ARTERY AND THE PROCESS OF ARTERIOSCLEROSIS: Pathogenesis
Edited by Stewart Wolf • 1971

Volume 16B
THE ARTERY AND THE PROCESS OF ARTERIOSCLEROSIS: Measurement and Modification
Edited by Stewart Wolf • 1971

Volume 17
CONTROL OF RENIN SECRETION
Edited by Tatiana A. Assaykeen • 1972

Volume 18
THE DYNAMICS OF MERISTEM CELL POPULATIONS
Edited by Morton W. Miller and Charles C. Kuehnert • 1972

Volume 19
SPHINGOLIPIDS, SPHINGOLIPIDOSES AND ALLIED DISORDERS
Edited by Bruno W. Volk and Stanley M. Aronson • 1972

Volume 20
DRUG ABUSE: Nonmedical Use of Dependence-Producing Drugs
Edited by Simon Btesh • 1972

Volume 21
VASOPEPTIDES: Chemistry, Pharmacology, and Pathophysiology
Edited by N. Back and F. Sicuteri • 1972

Volume 22
COMPARATIVE PATHOPHYSIOLOGY OF CIRCULATORY DISTURBANCES
Edited by Colin M. Bloor • 1972

Volume 23
THE FUNDAMENTAL MECHANISMS OF SHOCK
Edited by Lerner B. Hinshaw and Barbara G. Cox • 1972

Volume 24
THE VISUAL SYSTEM: Neurophysiology, Biophysics, and Their Clinical Applications
Edited by G. B. Arden • 1972

Volume 25
GLYCOLIPIDS, GLYCOPROTEINS, AND MUCOPOLYSACCHARIDES OF THE NERVOUS SYSTEM
Edited by Vittorio Zambotti, Guido Tettamanti, and Mariagrazia Arrigoni • 1972

Volume 26
PHARMACOLOGICAL CONTROL OF LIPID METABOLISM
Edited by William L. Holmes, Rodolfo Paoletti, and David Kritchevsky • 1972

Glycolipids, Glycoproteins, and Mucopolysaccharides of the Nervous System

Proceedings of the International Symposium on Glycolipids, Glycoproteins, and Mucopolysaccharides of the Nervous System: Chemical and Metabolic Correlations (Satellite Symposium of the XXV International Congress of Physiological Sciences — Munich, Germany) held in Milan, Italy, July 19-21, 1971

Edited by

Vittorio Zambotti
Symposium Chairman
Head, Institute of Biological Chemistry
Medical School
University of Milan
Milan, Italy

Guido Tettamanti
Professor of Biochemistry
Institute of Biological Chemistry
Medical School
University of Milan
Milan, Italy

and

Mariagrazia Arrigoni
Institute of Biological Chemistry
Medical School
University of Milan
Milan, Italy

PLENUM PRESS • NEW YORK - LONDON • 1972

Library of Congress Catalog Card Number 72-78628
ISBN 978-1-4684-3218-3 ISBN 978-1-4684-3216-9 (eBOOK)
DOI 10.1007/978-1-4684-3216-9

Softcover reprint of the hardcover 1st edition 1972
A Division of Plenum Publishing Corporation
227 West 17th Street, New York, N. Y. 10011

United Kingdom edition published by Plenum Press, London
A Division of Plenum Publishing Company, Ltd.
Davis House (4th Floor), 8 Scrubs Lane, Harlesden, London,
NW10 6SE, England

ACKNOWLEDGEMENTS

Acknowledgement is gratefully made to the following organizations who sponsored this Symposium:

ISTITUTO LOMBARDO ACCADEMIA
DI SCIENZE E LETTERE
Prof. Giuseppe Menotti de Francesco, President
Via Borgonuovo, 25
Milano, Italy

ISTITUTO NEUROLOGICO "C. BESTA"
Via Celoria, 11
Milano, Italy

LEPETIT S.p.A.
Via R. Lepetit, 8
Milano, Italy

ISTITUTO DE ANGELI S.p.A.
Via Serio, 15
Milano, Italy

The Editors are also most grateful to Miss Silvana Marai for her skilfulness in typing all the papers for the present volume.

Professor Vittorio Zambotti
President of the Symposium

CONTENTS

INTRODUCTION

SECTION I

CHEMISTRY AND METABOLISM OF GLYCOPROTEINS AND MUCOPOLYSACCHARIDES OF THE NERVOUS SYSTEM

SECTION II

CHEMISTRY AND METABOLISM OF GLYCOLIPIDS OF THE NERVOUS SYSTEM

SECTION III

SEPARATION AND PURIFICATION OF THE SUBCELLULAR COMPONENTS AND PLASMA MEMBRANES OF THE NERVOUS SYSTEM

SECTION IV

CHEMICAL PATHOLOGY AND DIAGNOSIS OF LIPID AND MUCOPOLYSACCHARIDE STORAGE DISEASES

INTRODUCTION

WHAT CAN WE EXPECT FROM THE STUDY OF NEUROLOGICAL MUTANTS

P. MANDEL

Centre de Neurochimie du CNRS and

Institut de Chimie Biologique, Faculté de Médecine

67 - Strasbourg, France

The great advances made in molecular biology by studying the properties of mutant microrganisms are well-known. It seems likely that similar advances can be made by studying mutants of higher organisms, in particular in unravelling the details of complex processes difficult to reproduce experimentally in simple but valid systems.

In the present paper we deal with two types of neurological mutants, first mutant mice extremely susceptible to audiogenic crises and then mutant mice in which myelination is impaired.

I. MUTANT MICE SUSCEPTIBLE TO AUDIOGENIC CRISES

Swiss white mice (strain Rb) when exposed to a sound stimulus (110 db, 8 Khz) pass through convulsive crises (1). There is first a running phase, a clonic phase and a tonic phase followed by a few clonic movements. The total crisis lasts for around 20 seconds. Mice of the same strain which are resistant to the sound stimulus can be used as controls in biochemical, and metabolic studies. Using these mice we can study the mechanism which is responsible for the crises, the biochemical events which are caused by the crises or factors which can be used to modify them.

We have studied the changes in high energy phosphate compounds during or after crises (2, 3) and the effect of n-dipropylacetate (nDPA) (4-6), which is, in certain respects, a GABA analogue.

A) Changes in High Energy Phosphate Compounds

Mice were placed in a cage, where the floor could be opened at will, dropping the mice into liquid nitrogen at determined stages of their crises. Seizures were induced by exposing mice to a loud-speaker emitting sound of controlled frequency and intensity. The effects of crises of around 20 sec duration on brain nucleoside triphosphates and creatine phosphate are shown in Table I and Fig. 1. There was a slight decrease in ATP which returned to normal within 20 sec, and a marked decrease in creatine phosphate (77 p. 100), both of which returned to normal values after 90 min. It is clear that these changes are a reflection of the equilibria between the utilization and biosynthesis of ATP and creatine phosphate, but it is interesting to note that the time necessary to regain normal creatine phosphate levels was long, relative to the duration of the crisis. This is in contrast to crises induced by electroshock where the high energy phosphate reserves return quickly to normal. Thus it seems that, depending upon the type of crisis, the degree of hypoxia and the loss of energy reserves, the time necessary to return to normal can vary greatly.

Table I - Nucleoside Triphosphates and Phosphocreatine in the Brains of Mice Sensitive to Sound Stimulation.

	ATP	GTP	UTP	Phosphocreatine
Unstimulated sensitive mice	186.3 ± 3.2	22 ± 2.1	21.13 ± 4.7	121.18 ± 4.18
Sensitive mice treated with nDPA (400 mg/kg)	214 ± 4.0	28.2 ± 1.7	21.5 ± 2.4	183.0 ± 8.30
Sensitive mice killed 10" after the beginning of the tonic phase	137.8 ± 0.75	21.7 ± 2.8	23.5 ± 8.7	28.0 ± 3.50

Results are expressed in µmole/100 g brain.

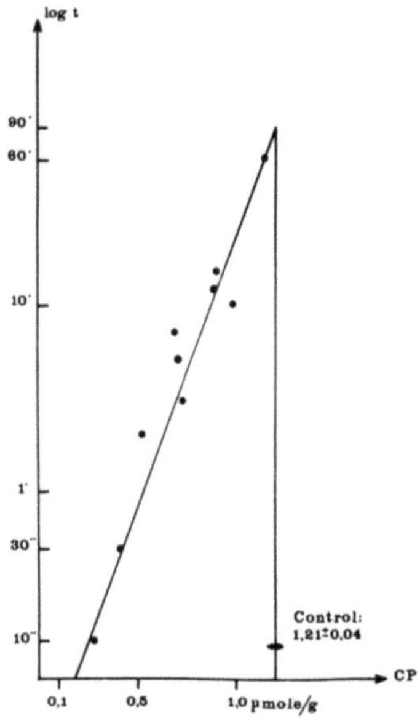

Fig. 1. Recovery of cerebral phosphocreatine after audiogenic crises. CP = creatine phosphate; log t = logarithm of the time after the beginning of the tonic phase.

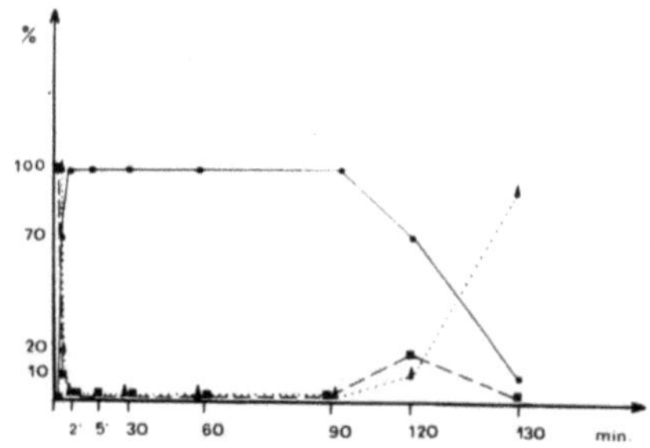

Fig. 2. Percent protection of sensitive mice against audiogenic crises after intramuscular injection of 400 mg/kg nDPA. ●—● total resistance; ■--■ partial resistance; ▲·····▲ clonic + tonic crisis.

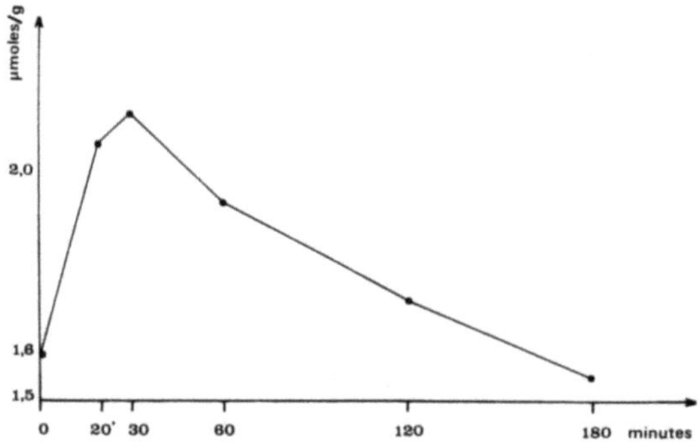

Fig. 3. Changes, as function of time, in the amount of GABA in the brains of mice sensitive to sound stimulation treated with 400 mg/kg nDPA. Results are expressed in µmole/g brain.

B) The Effect of n-Dipropylacetate (nDPA)

This compound protects the mice from audiogenic seizures. The time course after injection of 400 mg nDPA/kg is shown in Fig. 2. Protection was complete for two hours after injection. In parallel there was an increase in the level of GABA in the central nervous system which closely followed the protection observed (Fig. 3). These results confirm those obtained with a much more toxic compound, aminoxyacetic acid (7) and suggest that the protective effect may be due to the inhibitory action of GABA. Work currently in progress suggests that nDPA acts as a competitive inhibitor of GABA transaminase, thus reducing the degradation of GABA and increasing GABA levels (8).

II. MYELIN-DEFICIENT MUTANTS

Three myelin-deficient mutants have been studied by Sidman et al. (9) and by Meier (10). The Jimpy (9) and MSD (10) mutants are sex-linked and die after 25-30 days. The Quaking mutation is autosomal recessive and can survive for several months. In all cases tremors appear spontaneously around the 12th day. From 15 to 18 days, any shock and sometimes sounds can induce seizures.

A) Changes in Jimpy Mouse Brain Constituents During Brain Development

The most marked difference between control and Jimpy mice in the appearance of normal myelin constituents (phospholipids, cerebrosides, sulphatides and cholesterol) was the markedly lower amount of cerebrosides in Jimpy mouse brain (Fig. 4) compared to controls (11). However, brain cholesterol, which normally increases during myelination, did not increase in the mutants. Among the phospholipids in Jimpy mice there were slightly lower amounts of phosphatidyl ethanolamine and the plasmalogens (12), both of which are concentrated, although by no means exclusively, in myelin (Table II). Phospholipid synthesis however did not seem to be affected. Similar results were obtained for the gangliosides (13). Levels of most gangliosides were normal, except for GM_1 or G4, which, as it has been suggested by Suzuki et al. (14), is associated with myelin (Table III).

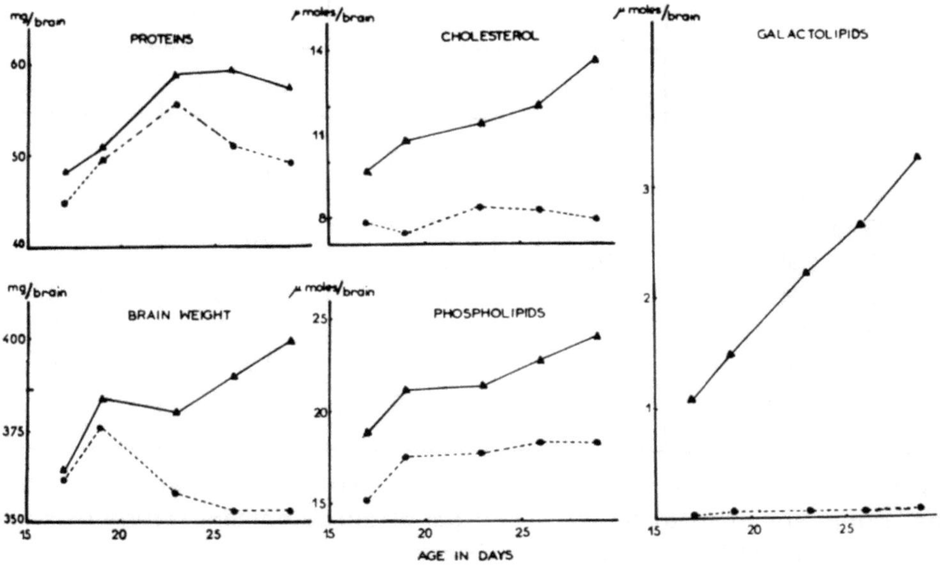

Fig. 4. Wet weight, protein and lipid contents of normal (——) and Jimpy (----) mouse brain as a function of age.

Table II - Distribution of Various Phosphatides in the Brains of Control and Jimpy Mice.

Phospholipids	Percent of total lipid P									
	17 days		19 days		23 days		26 days		29 days	
	C	Jp	C	Jp	C	Jp	C	Jp	C	Jp
Phosphatidyl choline	44.6	46.6	44.6	46.6	43.3	45.2	41.9	45.7	41.0	45.2
Choline plasmalogens	1.0	0.9	0.8	1.0	1.0	1.0	1.0	0.9	1.2	1.0
Phosphatidyl ethanolamine	25.2	25.0	26.8	26.1	24.6	25.1	21.7	25.7	23.2	25.4
Ethanolamine plasmalogens	10.3	8.8	9.6	8.1	11.8	8.8	14.4	8.8	14.1	8.4
Phosphatidyl serine	8.3	8.0	8.8	7.5	9.0	8.4	9.1	8.0	9.4	8.7
Sphingomyelin	3.9	3.6	4.4	4.1	4.3	3.9	4.8	4.3	5.2	4.2
Phosphatidyl inositol	3.7	4.0	4.2	3.8	3.5	3.6	3.9	4.6	3.7	3.9
Cardiolipin	2.4	2.6	1.9	2.5	2.1	2.9	2.4	3.1	2.1	2.8

The figures are the means of 2-4 experiments. The low amount of lipid P present at the origin has been taken into account in calculating the distributions. C = control mice; Jp = Jimpy mice.

Table III - Ganglioside Distribution in Control, Jimpy and Quaking Mouse Brain.

	28 days		60 days	
	Control	Jimpy	Control	Quaking
G_5 (GM_2)	4.6	4.2	14.3 +	10.2
G_4 (GM_1)	12.1 +	8.0		
G_{3a} (GD_3)	7.2	7.9	38.1	37.1
G_3 (GD_{1a})	30.6	30.3		
G_{2a} (GD_2)	10.2	11.4	10.5	10.4
G_2 (GD_{1b})	12.6	12.6	12.8	12.9
G_1 (GT_1)	15.3	15.4	15.3	14.9
G_0 (GQ_1)	4.9	4.6	4.9	5.0
Origin	2.7	3.4	3.6	4.1

$^+p < 0.01$. The results (average of 4-6 experiments) are expressed as a percent of total NANA. The ganglioside nomenclature of Korey and Gonatas (15) is used, with Svennerholm equivalents in brackets (16).

B) Cerebroside and Sulphatide Synthesis

Since the levels of cerebrosides were the most clearly decreased in the mutants, we have looked for the enzyme defect or defects responsible. For cerebroside synthesis there are two possible pathways: formation of psychosine followed by addition of a fatty acid or synthesis of ceramide followed by addition of galactose. Similarly there exist two pathways for sulphatide synthesis: synthesis of psychosine sulphate followed by addition of fatty acid or direct synthesis of cerebrosides followed by sulphation. In both Jimpy and Quaking mice, the activities of UDP-galactose:sphingosine--galactosyl transferase (17), UDP-galactose:ceramide-galactosyl transferase (18), PAPS:psychosine sulphotransferase (19) and PAPS:cerebroside sulphotransferase (20, 21) were all reduced both in total homogenates and in microsomal fractions (Fig. 5, 6). The enzyme activities, which in normal animals reached maxima during myelination around 19 days showed bell-shaped developmental curves both in mutants and controls, but the enzyme activities in the mutants were reduced.

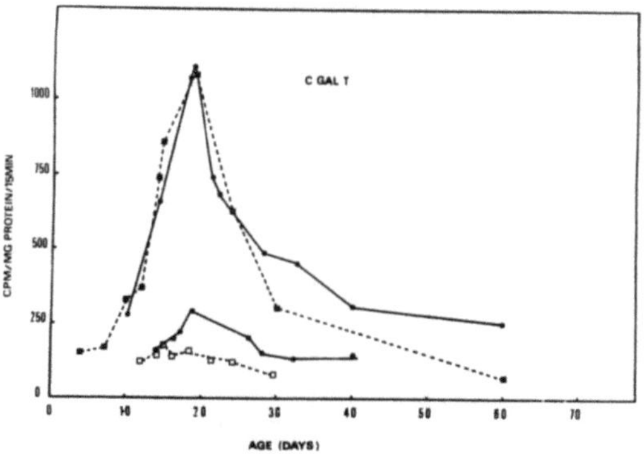

Fig. 5. UDP-galactose:ceramide-galactosyl transferase (C GAL T) in the brains of two mouse neurological mutants. □--□ Jimpy mice; ■--■ corresponding controls; ○—○ Quaking mice; ●—● corresponding controls.

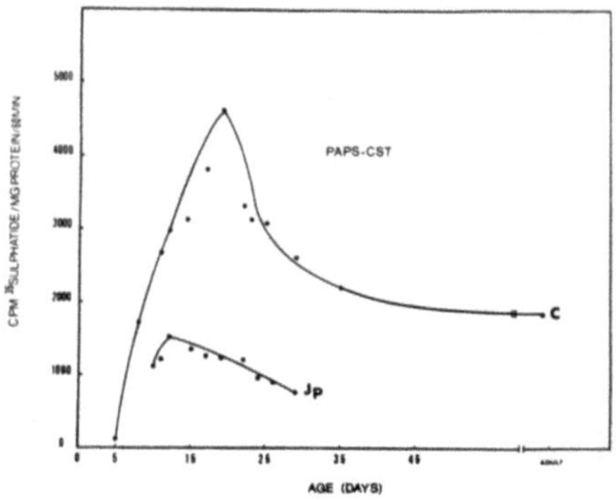

Fig. 6. 3'-phosphoadenosine 5'-phosphosulphate cerebroside-sulphotransferase (PAPS--CST) in brain. C = controls; Jp = Jimpy mice.

For the sulphotransferases there was little difference between control and mutant animals before myelination began, but clear differences were observed during myelination. A similar phenomenon has been observed for hydroxymethyl glutaryl CoA reductase, the rate-limiting enzyme in cholesterol synthesis, by Kandutsch (22). Thus, at least five enzymes are reduced in Jimpy and Quaking mice. There was also a marked reduction in the levels of long chain fatty acids in Jimpy (23), Quaking (24) and MSD mice (10). Work in progress has demonstrated similar deficiencies in UDP-galactose galactosyl transferases and PAPS: sulphotransferases in MSD mutants.

C) 2', 3'-Cyclic AMP 3'-Phosphohydrolase

This enzyme, first described by Drummond et al. (25) and found in high specific activities in myelin by Kurihara and Tsukada (26), was reduced in the central nervous system and spinal cord of Jimpy (Table IV) and Quaking mice. Enzyme activities were however normal in the peripheral nervous system, as were the UDP-galactose galactosyl transferase activities (27). These results are in accord with the localization of 2', 3'-cyclic AMP 3'-phosphohydrolase in myelin. The enzyme appears to be a good myelin marker, although, as might be expected, the enzyme is also present in glial cells (28).

Table IV - 2', 3'-Cyclic Nucleotide 3'-Phosphohydrolase Activity in Nervous Tissues of Jimpy and Control Mice.

Age days	Mice	Brain U/mg protein	Sciatic nerve U/mg protein	Spinal cord U/mg protein
17	Jimpy	0.36	0.98	1.7
	Control	2.57	4.24	1.7
28	Jimpy	0.33	0.70	1.8
	Control	3.54	5.18	1.9

D) Myelin Proteins

It is difficult to study myelin proteins in myelin fractions isolated from mutant mice since the quantity of myelin is very much reduced, and the fractions obtained are correspondingly impure. For this reason, myelin proteolipids have been studied on whole brain (29). The specific myelin proteolipids were identified among the total brain proteolipids of control and mutant mice. There was a marked reduction (about 80 p. 100) in the levels of the myelin-specific proteolipids in the Jimpy and Quaking mice from densitometric measurements (Fig. 7). The basic proteins have been studied in partially purified fractions. In Jimpy mouse "myelin" these proteins were hardly detectable but existed in normal amounts in Quaking mouse myelin. However relative to other myelin components, there was a marked decrease in the level of total basic proteins in Quaking mice.

DISCUSSION

Two types of information have been obtained from these studies.

1) Information on Normal Myelinogenesis

The parallel developmental changes in several enzyme activities, and the parallel reduced activities in mutants suggest that myelination is a coordinated phenomenon. This is true both for the increase of enzyme activities and their subsequent decrease. The parallel decrease in the two pathways for cerebroside synthesis:

sphingosine ⟶ psychosine ⟶ cerebroside

sphingosine ⟶ ceramide ⟶ cerebroside

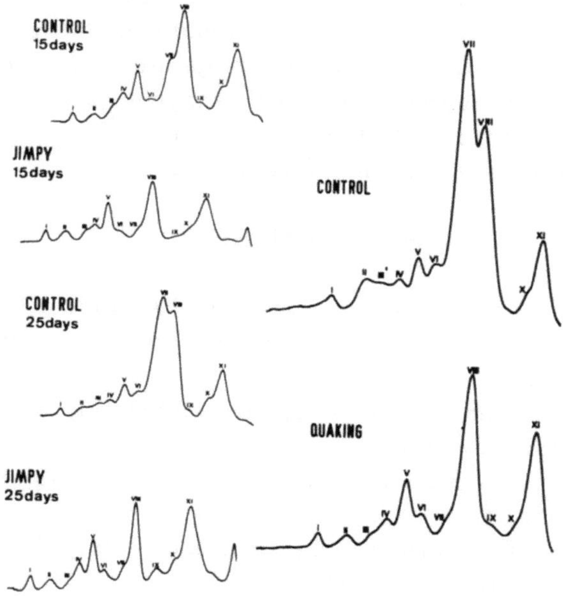

Fig. 7. Densitometric profiles in 12% polyacrylamide gels of whole brain proteolipids from Jimpy and control mice of the same age, and from adult Quaking and controls. Band VII corresponds to the specific myelin proteolipid.

suggests that both are operative in vivo. The enzyme 2', 3'-cyclic AMP 3'-phosphohydrolase appears to be genuinely associated with myelin and can be used as a myelin marker; its localization in myelin cannot be explained by an homogenization artifact since when there is a reduction in myelin, there is a reduced enzyme activity.

2) Information on the Mutants

The enzyme defect which characterizes these mutants is obiously multi-enzymatic. Given the coordinated appearance of the enzymes of cerebroside biosynthesis during normal development, and their coordinated reduction in the mutants, the basic defect is unlikely to be due to a mutation of a structural gene since this would imply a simultaneous mutation of several structural genes. It seems more likely that there is a mutation of a gene which regulates the enzyme levels. There could however be a sequential transcription of the enzymes involved in cerebroside synthesis where a mutation of the gene specifying one enzyme or protein affects the synthesis or transcription of subsequent enzymes or proteins. It is also possible to explain these results by a mutation which affects oligodendrocyte differentiation: a morphogenetic mutation. It is important to stress that the same multi-enzyme defects are found in mutations which are situated on two different chromosomes, on the X chromosome and on chromosome 20.

REFERENCES

(1) Lehmann, A. and Busnel, R.G., in "Acoustic Behaviour of Animals", Ed. R.G. Busnel, Elsevier Publ., Amsterdam, p. 244 (1964).
(2) Simler, S., Randrianarisoa, H., Lehmann, A. and Mandel, P., J. Physiol., in press.
(3) Simler, S., Lehmann, A. and Randrianarisoa, H., C.R. Soc. Biol., in press.
(4) Simler, S., Randrianarisoa, H., Lehmann, A. and Mandel, P., J. Physiol. 60: 547 (1968).
(5) Simler, S., Randrianarisoa, H., Godin, Y., Ciesielski, L., Maitre, M. and Lehmann, A., 3rd Int. Meeting Neurochem., Budapest, p. 146 (1971).
(6) Simler, S., Randrianarisoa, H., Lehmann, A. and Mandel, P., in "Int. Symposium Pathogeny of Epilepsy", in press.
(7) Kuriyama, K., Roberts, E. and Rubinstein, M.K., Biochem. Pharmacol. 15: 221 (1966).
(8) Ciesielski, L. and Maitre, M., personal communication.
(9) Sidman, R.L., Dickie, M.M. and Appel, S.H., Science 144: 309 (1964).
(10) Meier, H. and McPike, A.D., Exp. Brain Res. 10: 512 (1970).
(11) Nussbaum, J.L., Neskovic, N. and Mandel, P., J. Neurochem. 16: 927 (1969).
(12) Nussbaum, J.L., Neskovic, N. and Kostic, D., in "Les mutants pathologiques chez l'animal, leur intérêt dans la recherche bio-médicale", Ed. M. Sabourdy, Editions du CNRS, Paris, n° 924, p. 34 (1970).
(13) Kostic, D., Nussbaum, J.L. and Neskovic, N., Acta Med. Jug. 24: 20 (1970).
(14) Suzuki, K., Poduslo, S.E. and Norton, W.T., Biochim. Biophys. Acta 144: 375 (1967).
(15) Korey, S.R. and Gonatas, J., Life Sci. 2: 296 (1963).
(16) Svennerholm, L., in Handbook of Neurochemistry, Ed. A. Lajtha, Plenum Press, New York, 3: 425 (1970).

(17) Neskovic, N.M., Nussbaum, J.L. and Mandel, P., Brain Res. 21: 39 (1970).
(18) Neskovic, N.M., Nussbaum, J.L. and Mandel, P., FEBS Letters 8: 213 (1970).
(19) Nussbaum, J.L. and Mandel, P., unpublished data.
(20) Sarlieve, L.L., Neskovic, N.M., Nussbaum, J.L. and Rebel, G., 3rd Int. Meeting Neurochem., Budapest, p. 346 (1971).
(21) Sarlieve, L.L., Neskovic, N.M. and Mandel, P., FEBS Letters, in press.
(22) Kandutsch, A.A. and Saucier, S.E., Arch. Biochem. Biophys. 135: 201 (1969).
(23) Nussbaum, J.L., Neskovic, N.M. and Mandel, P., J. Neurochem. 18: 1529 (1971).
(24) Baumann, N.A., Jacque, C.M., Pollet, S.A. and Harpin, M.L., Europ. J. Biochem. 4: 340 (1968).
(25) Drummond, G.I., Iyer, N.T. and Keith, J., J. Biol. Chem. 237: 3535 (1962).
(26) Kurihara, T. and Tsukada, Y., J. Neurochem. 14: 1167 (1967).
(27) Kurihara, T., Nussbaum, J.L. and Mandel, P., J. Neurochem. 17: 993 (1970).
(28) Zanetta, J.P., Benda, P., Gombos, G. and Morgan, I.G., J. Neurochem., in press.
(29) Nussbaum, J.L. and Neskovic, N.M., 3rd Int. Meeting Neurochem., Budapest, p. 27 (1971).

SECTION I

CHEMISTRY AND METABOLISM OF GLYCOPROTEINS AND MUCOPOLYSACCHARIDES OF THE NERVOUS SYSTEM

CHEMISTRY AND METABOLISM OF GLYCOPEPTIDES DERIVED FROM BRAIN GLYCOPROTEINS

E. G. BRUNNGRABER

Illinois State Psychiatric Institute

1601 West Taylor Street - Chicago, Illinois 60612

Research advances in the field of neural glycoproteins have been reviewed (1, 2, 3). The present article will focus attention on the chemistry and metabolism of the heteropolysaccharide chains that are associated with the brain glycoproteins. Another article in this volume will focus attention on changes induced in these substances as a consequence of neural pathology. A glycoprotein is a protein that has attached to it one or more carbohydrate groups. Such carbohydrate groups may be quite simple, consisting of a single carbohydrate residue or simple di-, tri- or oligosaccharides. At the other extreme, glycoproteins may contain one or more complex, highly branched, heteropolysaccharide chains. As a rule, the predominant sugars in such heteropolysaccharide chains are N-acetylneuraminic acid (NANA), mannose, galactose, N-acetylglucosamine, and fucose. N-acetylgalactosamine is often present in such chains. Glucose has been reported to be present in several glycoproteins, but this sugar is only rarely present in glycoprotein material.

The partial purification of several soluble glycoproteins from brain tissue has been reported (4-15). Several of these glycoproteins are brain-specific antigens. Little or nothing is known regarding the nature of their heteropolysaccharide chains. The soluble glycoproteins account for only a small proportion of the total brain glycoprotein material; approximately 80% of the glycoprotein carbohydrate of brain tissue is bound to membranous material and cannot be solubilized by the action of aqueous buffers and severe homogenization. Partial purification of insoluble glycoproteins has been attempted (16-19) but such efforts have met with limited success. With the exception of one report (16), such studies have for the most part involved analytical electrophoresis of glycoproteins that had been solubilized by the use of detergents. Ultimately, it will be necessary to devise preparative methods for the separation of glycoproteins so that sufficiently large amounts of material become available for structural work. This will be especially necessary if structural changes as a consequences of biological activity appear to be associated with a particular glycoprotein.

Neural tissue consists of nerve cell populations which differ from each other morphologically, functionally, and in their mode of development. Furthermore, individual neurones have specialized structures composed of membranous elements which are involved in distinct biological processes. Consequently, one must be concerned with glycoproteins that originate from the synaptic membrane, the synaptic cleft, the axon, the plasma cell membrane, the axon hillock, the dendritic structures, as well as various subcellular organelles such as mitochondria, nuclei, lysosomes, synaptic vesicles, the Golgi apparatus, the endoplasmic reticulum, and the cell sap. Neural tissue also contains various types of glial cells each of which probably also contains specialized subcellular regions and various types of glycoprotein material (1, 2, 3). A priori, one must accept the hypothesis that neural tissue contains a multitude of glycoproteins the isolation of each of which will prove to be a formidable task.

Several hypotheses (1, 2, 20-23) have been proposed suggesting that glycoproteins play important roles in brain development and function and in neuropathology (24). If such hypotheses are to be tested at this time, it appears to be justified to develop experimental procedures to isolate the heteropolysaccharide chains of all of the glycoprotein material of brain as glycopeptides which are produced by the proteolytic digestion of defatted brain tissue samples. In this way, it is possible to recover all of the glycopeptides quantitatively since no irreversible losses can occur during the procedure. One may be confronted with a formidable array of glycopeptides the complete separation of which may prove to be exceedingly difficult. Nevertheless, it may be possible to show whether certain types of glycopeptide material are altered as a consequence of functional or pathological changes and efforts can subsequently be directed toward the development of isolation procedures for the glycoproteins which contain the altered heteropolysaccharide chain of the isolated glycopeptide(s). Furthermore, it may be possible to discover at an early stage in this area of research, whether simple structural relationships exist between the isolated heteropolysaccharide chains. For example, it is not unlikely that although brain may contain a multitude of glycoproteins, many of these glycoproteins may contain similar or identical heteropolysaccharide chains. Differences between glycoproteins may therefore reside largely in the protein portion of the molecule, or in the number and type of chains attached to the polypeptide backbone. If we may use an analogy: man has developed many types of machines and structures. However, the number of the types of screws used in their manufacture is considerably smaller than the types of machines and structures which they hold together. It is reasonable to suppose that the biosynthetic machanisms in nerve tissue may be limited in the number of the type of heteropolysaccharide chains they are capable of synthesizing. That is to say, there may be many more glycoprotein types than there are heteropolysaccharide types. A study of the glycopeptides obtained from whole brain tissue may yield evidence supporting or voiding this hypothesis.

ISOLATION OF GLYCOPEPTIDES FROM BRAIN TISSUE

The procedure for the isolation of the glycopeptides has been described in detail elsewhere (25-32). The defatted tissue sample is subjected to the proteolytic action of papain. The insoluble papain-resistant material is retrieved by centrifugation. The supernatant is dialyzed under standardized conditions. The nondialyzable glycopeptide material is treated with cetylpyridinium chloride to remove by precipitation the glycosaminoglycans and nucleic acids. Excess cetylpyridinium chloride is removed by extraction with amyl alcohol. The nondialyzable glycopeptides are separated from impurities which absorb ultraviolet light by subjecting the preparation to gel filtration on Sephadex G-50, using water as eluant (32). The dialyzable glycopeptides are recovered from the dialyzate by concentration and gel filtration on Sephadex G-15 to remove most of the amino acids, peptides, salts and other impurities (31).

The distribution of various NANA- and hexosamine-containing materials is shown in Table I. Hexosamine in the glycopeptides ac-

Table I - NANA- and Hexosamine-containing Constituents of Rat Whole Brain.

Constituent	NANA		Hexosamine	
	μM/g	%	μM/g	%
Gangliosides	2.22	67.5	0.96	23
Nondialyzable Glycopeptides	0.64	19.4	1.43	34
Dialyzable Glycopeptides	0.23	7.1	0.81	19
Papain-resistant Fraction	0.08	2.4	0.32	8
Free NANA	0.12	3.6	-	-
Free Hexosamine	-	-	0.14	3
Glycosaminoglycans	-	-	0.55	13
Totals	3.29	-	4.21	-

counts for 53% of the total content of hexosamine in rat brain. Some NANA and hexosamine invariably remains associated with the papain-resistant fraction from rat brain. This fraction from bovine brain contains little NANA or hexosamine (30) and that from human brain contains no NANA (25).

Chloroform-methanol extracts prepared from rat brain do not contain glycoprotein material of the type known to be present in the defatted tissue residue (25). Approximately 95% of the NANA in the extracts can be accounted for as gangliosidic NANA (33). Such extracts do not contain glucosamine, mannose, or fucose. These sugars are prominent constituents of the rat brain glycoproteins.

THE NONDIALYZABLE GLYCOPEPTIDES

The nondialyzable glycopeptide preparation is 68% in carbohydrate content. The predominant amino acids are aspartic acid,

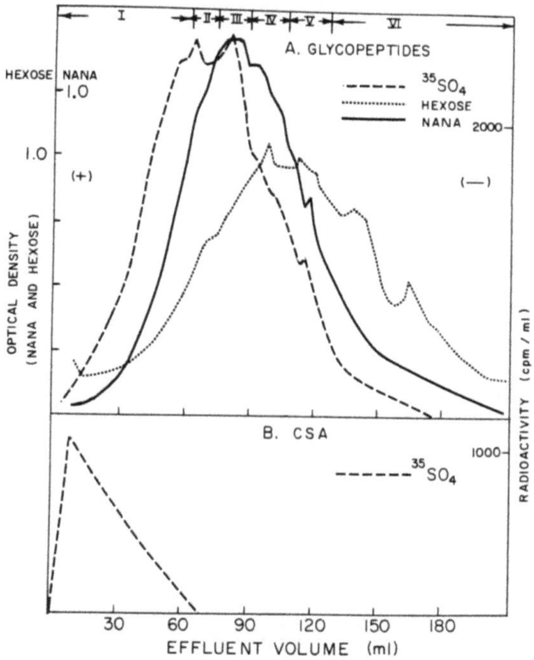

Fig. 1. A. Rats were injected intraperitoneally with $Na_2{}^{35}SO_4$ 24 hours prior to sacrifice. The nondialyzable glycopeptides were isolated and fractionated by column electrophoresis utilizing the LKB column apparatus No. 3340. Electrophoresis was conducted at 3° in 0.1 M glycine-NaOH buffer, pH 10.3, at 200 V and 25-27 mA for 55-60 hours. The column was eluted with the same buffer and 3 ml fractions were collected. The fractions were analyzed for NANA, hexose, and radioactivity. Contents of test tubes (see top of graph) containing materials corresponding to fractions I through VI were combined, dialyzed, and concentrated. The fractions were analyzed for carbohydrate composition and molecular weight (see Table II). B. Chondroitinsulfuric acid obtained from rat brains 24 hours after intraperitoneal injection with carrier-free $Na_2{}^{35}SO_4$ was subjected to column electrophoresis under conditions that were identical to those used in experiment depicted in A above.

threonine, serine, glutamic acid, and proline, which respectively account for 19, 12, 8, 12, and 10 residues per 100 amino acid residues recovered from the preparation. Aromatic amino acid content is low; tyrosine, phenylalanine, and histidine account for only

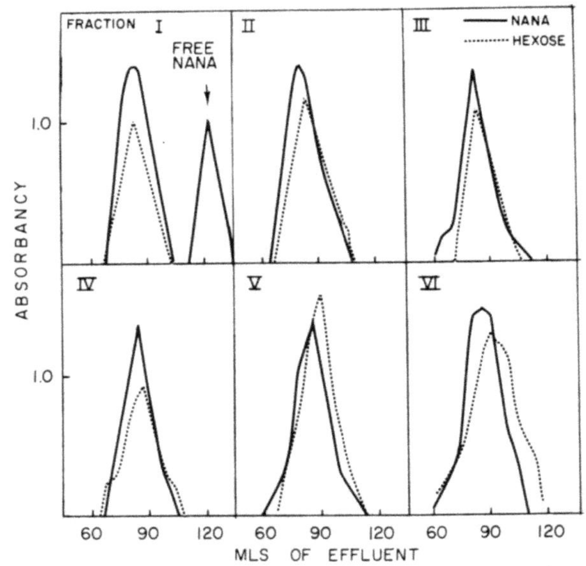

Fig. 2. Chromatogram obtained when fractions I through VI (see Fig. 1) were subjected to gel filtration on standardized columns of Sephadex G-50 (1 x 90 cm), in 0.05 M Tris-chloride buffer, pH 7.5, containing 0.1 M NaCl. Columns were previously standardized with glycopeptides of known molecular weight (Glycopeptide B from thyroglobulin and glycopeptide from transferrin). The position of free NANA is shown in figure at top, left hand corner.

1, 2, and 1 residue per 100 amino acid residues. Moderate amounts of glycine, alanine, valine, isoleucine, and leucine were also present (5, 6, 5, 3, and 4 residues per 100 amino acid residues respectively). The basic amino acids lysine and arginine accounted for 3 and 1-2 residues per 100 residues. Moderate amounts of cystine/cysteine and methionine (5 and 6 residues respectively) were recovered.

The nondialyzable glycopeptide preparation was subjected to column electrophoresis (Fig. 1). An indication of the absence of glycosaminoglycans (acid mucopolysaccharides) in the glycopeptide preparation is provided by a comparison of the electropherogram obtained when $^{35}SO_4$-labelled glycopeptides (Fig. 1, upper graph) is compared to that of $^{35}SO_4$-labelled chondroitinsulfuric acid prepared from rat brain (Fig. 1, lower graph). The glycosaminoglycan has a considerably greater mobility than that of the glycopeptides.

Table II - Carbohydrate Composition and Average Molecular Weight of Nondialyzable Glycopeptides obtained by Column Electrophoresis.
Contents of Test Tubes corresponding to Fractions I through VI (See Fig. 1) were combined and analyzed for Carbohydrate Components. Each fraction was subjected to gel filtration on standardized columns of Sephadex G-50 in order to estimate its average Molecular Weight (See Fig. 2). Each fraction was subjected to mild acid hydrolysis (0.1 N sulfuric acid, 80° C, 1 hour) for removal of NANA, and the Molecular Weight of the NANA-free glycopeptides was estimated.

Fraction	Molar Ratios					Average M. W.	Average M. W. after mild hydrolysis	Loss in M. W. after mild acid hydrolysis
	Mannose	Galactose	N-Acetyl-glucosamine	NANA	Fucose			
I	1	2.3	2.8	1.7	0.43	5200	3650	1550
II	1	2.3	3.4	2.1	0.43	4950	3500	1450
III	1	1.4	1.7	1.1	0.36	4850	3650	1200
IV	1	1.1	2.1	0.9	0.46	4800	3750	1050
V	1	1.0	1.8	0.6	0.50	4550	3450	1100
VI	1	0.8	1.5	0.3	0.52	4300	3650	650
Totals	1	1.2	1.9	0.8	0.47			

The absence of glucuronic acid in the nondialyzable glycopeptide preparation provides additional proof that glycosaminoglycans are absent.

Contents of test tubes containing material corresponding to fractions I through VI of the nondialyzable glycopeptides obtained by column electrophoresis (Fig. 1, upper graph) were combined and analyzed for sugar components (Table II). Each fraction was subjected to gel filtration on Sephadex G-50 before (Fig. 2) and after mild acid hydrolysis in order to estimate the average molecular weight of the fractions before and after the removal of NANA. In order to estimate their molecular weight, glycopeptides of known molecular weight were used to standardize the Sephadex G-50 column according to the method of Bhatti and Clamp (34). The elution volume of dextran Blue (V_o), glucose (V_{100}), and that of unknowns or standards (V_e) was determined. The value of (V_e-V_o/V_{100}-V_o) 100 was calculated. The relation between this value and molecular weight (Fig. 3) found for branched glycopeptides differed considerably from these found for linear polymers such as raffinose, stachyose, and oligosaccharides recovered from hyaluronidase digested oligosaccharides. The lower curve, using raffinose and stachyose as standards, was linear and extrapolated to a molecular weight of 4200, as was reported by Bhatti and Clamp (34). The upper curve was determined using glycopeptides obtained from thy-

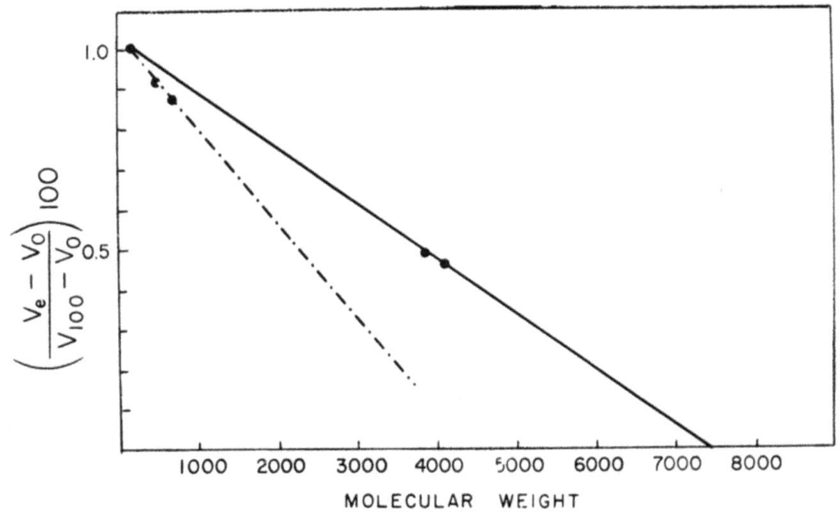

Fig. 3. Standardization of the Sephadex G-50 column used to estimate the molecular weight of the nondialyzable glycopeptides. Conditions for gel filtration were described under Fig. 2. The elution volumes of dextran Blue (V_0), glucose (V_{100}) and standards (V_e) were determined. In order of increasing molecular weight, upper curve is line drawn through points obtained for glucose, glycopeptide from transferrin, and glycopeptide B of thyroglobulin. Lower curve is line drawn through points obtained for glucose, raffinose, and stachyose.

roglobulin (35) and transferrin (36). This curve extrapolated to a value of approximately 7400. This is in fairly good agreement with the finding of Barnes and Partridge (37) who reported that a glycopeptide of molecular weight 7800 and containing NANA, galactose, mannose, fucose, and hexosamine isolated from human thoracic aorta was excluded from the gel matrix of similar Sephadex G-50 columns. Peptides cannot be used for molecular weight determinations of glycopeptides since they gives erroneously high values (31). The difficulty in using proteins as standards for molecular weight determinations of glycoproteins has been pointed out by Andrews (38).

The molecular weight of the nondialyzable glycopeptides (Table II) decreases as the charge (electrophoretic mobility) and NANA content decreases. The glycopeptide fractions appear to differ in molecular weight largely because of differences in their content of terminal NANA groups. If one assumes that fractions I and II contain two molecules of mannose, then each glycopeptide fraction would contain 4 molecules of NANA. The loss of these NANA residues would decrease the molecular weight of fractions I and II by 1240. If one assumes that fractions III through VI contain respectively 3, 3, 2, and 1 molecules of NANA, a loss in molecular weight of 930, 930, 620, and 310 would be expected. The decrease in molecular weight observed experimentally is somewhat greater than these values. This can be explained by the finding that the mild acid hydrolysis procedure known to cleave all of the NANA residues also cleaves approximately 8 and 4% of the hexose and fucose residues respectively. Approximately 5% of the hexosamine is also liberated (39).

It should be recognized that each of the six fractions most probably consists of a mixture of glycopeptides which are very similar in molecular weight and electrophoretic mobility and that the values for their sugar composition and molecular weight are average values for the mixture.

The finding that the glycopeptides of fractions I and II contain only two mannose residues per molecule is of interest. The structures of several glycopeptides, which are similar to the ones studied here inasmuch as they appear to be highly branched and contain the same sugar residues, have been studied. These include glycopeptides from fetuin (40), orosomucoid (41) and α_2-macroglobulin (42). In these three cases, the glycopeptides recovered from the parent glycoproteins have all been reported to contain three mannose residues. It has been suggested (29) that all of the glycopeptides of the nondialyzable preparation may be related, differing only in the number of terminal NANA, fucose, and sulfate groups attached to a common structure. This speculation assumes that the structure of the hexose-hexosamine portion of all of the glycopeptides are identical. Consequently, one would have anticipated that fractions I through VI would all have contained three mannose residues per glycopeptide molecule. If the glycopeptides of fractions I and II contain three mannose residues, the data of Table II indicate that these glycopeptides would contain six NANA groups. This is unlikely, since removal of all of the NANA causes a decrease in molecular weight of only 1550 and 1450 for fractions I and II respectively. If we assume that the glycopeptides of fraction I and II contain 4 NANA groups and that their internal structure is identical to that of fraction III, the glycopeptides of fractions I and II should contain 3 mannose, 4 galactose, 5 glucosamine, 4 NANA and 0.4 fucose residues. The actual analytical data for fractions I and II reveal that the amount of galactose, glucosamine, and NANA in these fractions is too large to accomodate this hypothesis. A possible explanation for these findings is that fractions I and II may contain another class of glycopeptides which differ markedly in structure from the predominant glycopeptides which contain three mannose residues per molecule. The amount of excess galactose, glucosamine, and NANA in fractions I and II suggests that this new class of glycopeptide material would contain NANA, galactose, and hexosamine in a ratio of approximately 1:1:2. From the available quantitative data (2) it can be calculated that this "impurity" would account for 6% of the total hexose and 13% of the total NANA of the nondialyzable glycopeptide preparation. An alkali-labile glycopeptide(s) containing NANA, galactose, glucosamine, galactosamine and sulfate has been reported by Margolis and Margolis (43). The isolation of this NANA-rich glycopeptide, characterized by a high ratio of NANA/hexose, from the nondialyzable glycopeptide preparation by these workers provides some experimental support for the above considerations.

We have obtained some experimental evidence which supports the above interpretation of our analytical data of Table II. The ratio of aspartic acid/threonine did not change after incubation of the nondialyzable glycopeptide preparation for 17 hours at 36° in 0.1 N NaOH, indicating that the linkage of the heteropolysaccharide chains to the protein is predominantly the alkali-stable β-aspartylglycosaminylamine linkage. Similar results were reported by Arima et al. (44). However, the glycopeptide mixture does contain a minor fraction that is susceptible to cleavage by the alkaline treatment. This was noted by subjecting the alkali-treated nondialyzable glycopeptide preparation to gel filtration on standardized columns of Sephadex G-50 (Fig. 4). The molecular weight of most of the NANA and hexose associated with the nondialyzable glycopeptides was not affected. However, approximately 11-13% of the total NANA appeared in a polysaccharide of decreased molecular weight. Presumably, this material formed as a consequence of the cleavage of a galactosamine-threonine (serine) linkage. This material may account for the small amount of galactosamine in the nondialyzable glycopeptide preparation. Approximately 90-95% of the hexosamine consists of glucosamine; the remainder is galactosamine. This procedure of

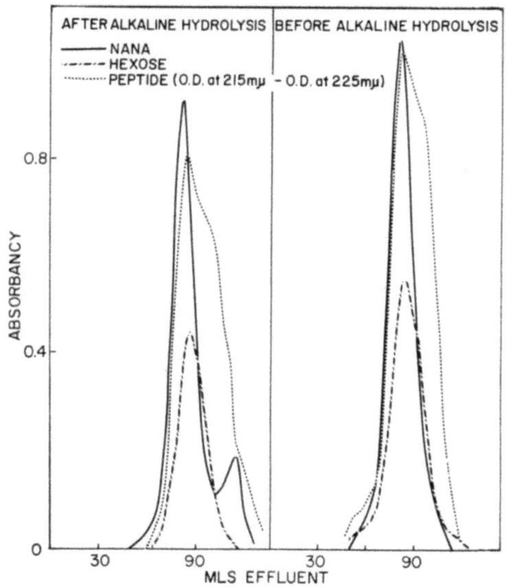

Fig. 4. A nondialyzable glycopeptide preparation from rat brain was divided into two portions. One portion was subjected to alkaline degradation (0.1 N NaOH, 17 hours, at 36°). Both samples were adjusted to contain 0.05 M Tris--chloride buffer containing 0.1 M NaCl and subjected to gel filtration on standardized columns of Sephadex G-50 (1 x 90 cm) as described under Fig. 2.

alkaline hydrolysis followed by gel filtration may be a satisfactory method of removing the minor contaminant.

The presence of a discrete group of glycopeptides differing from the bulk of the brain glycopeptides was also indicated by the following series of experiments. The nondialyzable glycopeptides were subjected to preparative gel filtration utilizing columns of Sephadex G-50. The glycopeptides were labelled with $^{35}SO_4$ by injecting rats 24 hours prior to sacrifice. If the experiment is conducted using water as the eluting solution (Fig. 5, upper graph), two radioactive peaks are obtained. These are indicated as fractions A and B respectively. This separation cannot be achieved if the eluting solution contains 0.05 M Tris-chloride buffer in 0.1 M NaCl (Fig. 5, lower graph). Fractions A and B were subjected to column electrophoresis (Fig. 6). The heterogeneity of both fractions is demonstrated by an examination of the electropherograms. A major difference between A and B is the finding that the amount of NANA relative to hexose is considerably greater in fraction A

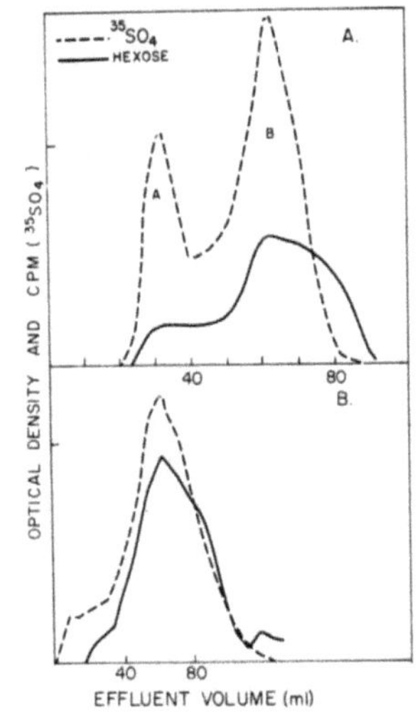

Fig. 5. Gel filtration of nondialyzable glycopeptide preparations that had been labelled with $^{35}SO_4$ on columns of Sephadex G-50. A. Water was used as eluting solution. Two radioactive peaks (fractions A and B) were obtained. B. Tris-chloride buffer (0.05 M, pH 7.5) containing 0.1 M NaCl was used as eluting agent.

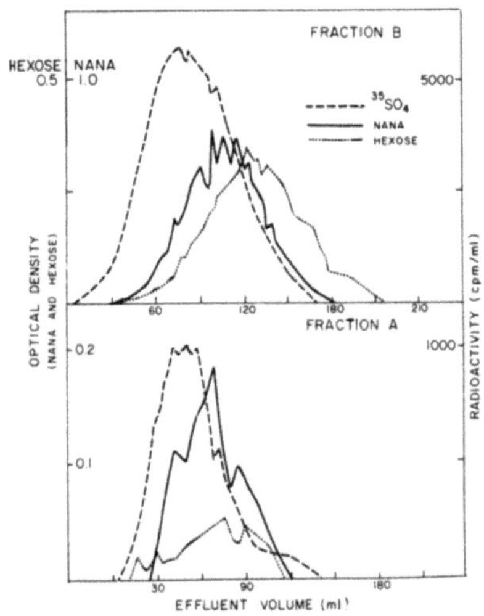

Fig. 6. Fractions A and B of the nondialyzable glycopeptides obtained by gel filtration using water as eluting agent (see Fig. 5) were subjected to column electrophoresis under conditions described in the legend for Fig. 1.

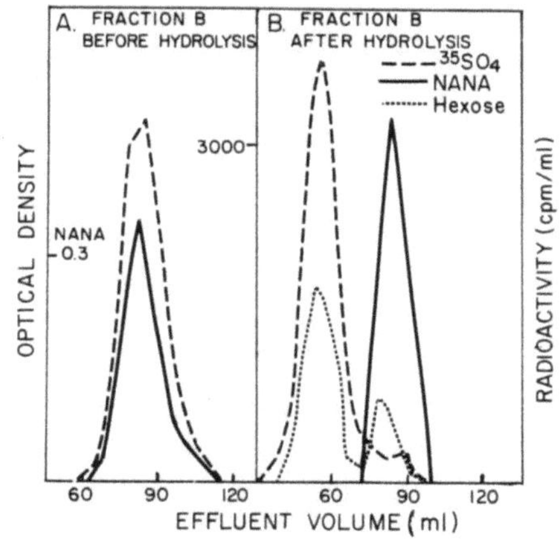

Fig. 7. Fraction B of the nondialyzable glycopeptides (see Fig. 5 A) was subjected to gel filtration on Sephadex G-50 before (A) and after (B) acid hydrolysis in 0.1 N sulfuric acid at 80°C for 1 hour. Experiment was conducted on standardized columns of Sephadex G-50 utilizing Tris-chloride (0.05 M, pH 7.5) containing 0.1 M NaCl as eluting solution. Before hydrolysis, incorporated sulfate is eluted in the same position as NANA and hexose. After hydrolysis, label remains with the hexose portion of the nondialyzable glycopeptides.

than in B. Subjections of fraction A to gel filtration on standardized columns of Sephadex G-50 have indicated that the average molecular weight of the NANA-containing glycopeptides of this fraction is approximately 5600, which corresponds closely to the average molecular weight of the glycopeptides of fraction I obtained by column electrophoresis of the total nondialyzable glycopeptide preparation (Fig. 1, Table II). The NANA-containing glycopeptides of fraction A therefore appear to be similar to the glycopeptides which appear in fractions I and II obtained by column electrophoresis of the total nondialyzable glycopeptide preparation in regard to molecular size, high value for the ratio of NANA/hexose and electrophoretic mobility. Fraction B contains respectively approximately 80 and 90% of the total NANA and hexose of the nondialyzable glycopeptide preparation. This material was subjected to gel filtration on Sephadex G-50 using Tris-chloride buffer as eluting agent (Fig. 7). After mild acid hydrolysis, the de-sialidized glycopeptides were separated from the free NANA released by the mild acid hydrolysis. The radioactivity is associated with the de-sialidized hexose-containing

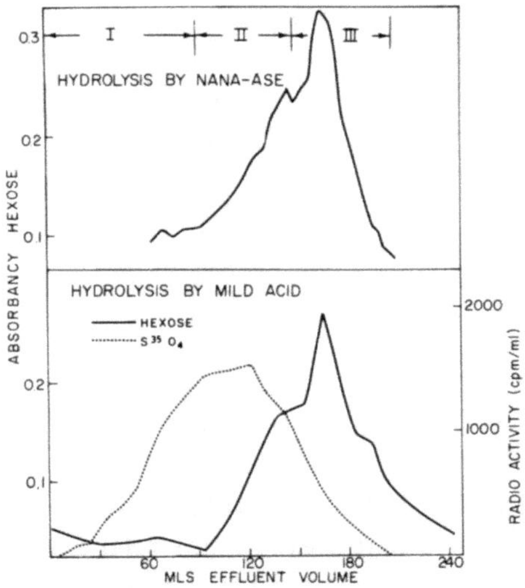

Fig. 8. Nondialyzable glycopeptide preparations from which all of the NANA had been removed were subjected to column electrophoresis under conditions described under Fig. 1. Electropherogram in upper graph was obtained by subjecting nondialyzable glycopeptides that had been treated with neuraminidase to column electrophoresis. Lower graph represents an electropherogram in which NANA had been liberated from the nondialyzable glycopeptides by mild acid hydrolysis. In both cases, the liberated NANA was removed from the glycopeptide preparation by means of gel filtration prior to column electrophoresis.

glycopeptides. Although considerable hexose (in the form of mono- and oligosaccharides) is liberated by means of mild acid hydrolysis, only small amounts of radioactive sulfate appears in these low molecular weight materials.

In the experiments shown in Figures 1, 5, 6, and 7, it appears that labelled sulfate appears in all of the glycopeptide material of the nondialyzable glycopeptide preparation, since radioactivity appears to co-exist with hexose in all of the fractions eluted from columns after gel filtration or column electrophoresis. To determine how much of the glycopeptide material is in fact sulfated, the nondialyzable glycopeptide preparation was treated by mild acid hydrolysis and the liberated NANA was removed by gel filtration. The de-sialidized glycopeptides were then subjected to column e-

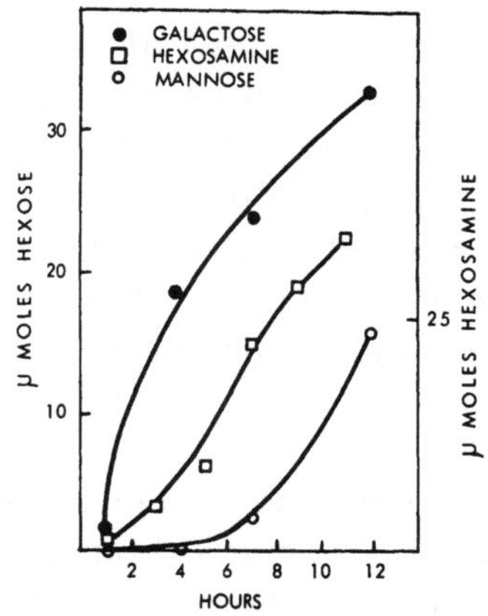

Fig. 9. Nondialyzable glycopeptides were subjected to mild acid hydrolysis (0.05 N HCl, 100°) for various periods of time. The liberated monosaccharides were isolated by gel filtration on Sephadex G-15 and analyzed for glucosamine, galactose, and mannose. Paper chromatography (49) was used to distinguish between mannose and galactose. Glucosamine was estimated by the method of Boas (50) after acid hydrolysis to remove the N-acetyl groups.

lectrophoresis (Fig. 8). Approximately 59% of the glycopeptide--hexose lacked labelled sulfate and these glycopeptides were represented by the peak which was eluted at 170 ml. The small peak eluted at about 190 ml probably corresponds to mono- and oligosaccharides that had been released by mild acid hydrolysis. This peak is not present if the NANA had been removed by the action of neuraminidase (Fig. 8, upper graph). With this exception, the electropherogram obtained by subjecting nondialyzable glycopeptides that had been de-sialidized by mild acid hydrolysis to column electrophoresis resembled that obtained by column electrophoresis of neuraminidase-treated glycopeptides.

Treatment of gangliosides with neuraminidase (Vibrio cholerae) for extended digestion periods removed about 70% of the NANA associated with the glycolipid material. On the other hand, all of the NANA associated with both the nondialyzable and dialyzable (see below) glycopeptides was removed after 24 hours of digestion.

The nondialyzable glycopeptide preparation was subjected to mild acid hydrolysis in 0.1 N HCl at 100° and the monosaccharides that were released at various time intervals were isolated by means of gel filtration on Sephadex G-15. All of the NANA is released in one hour. 75% of the fucose is liberated by 4 hours. An immediate release of galactose occurs (Fig. 9) suggesting that galactose is located in an external position. Glucosamine release shows an S-shaped curve suggesting that glucosamine is located both in an external and in an internal position. The long lag period for mannose release is suggestive for an internal position for this sugar.

The structure of several glycopeptides obtained from glycoproteins such as fetuin and orosomucoid have been studied (40, 41). In general, it has been found that terminal NANA groups are linked to galactose, which is in turn linked to a glucosamine residue. The external portion of the heteropolysaccharide unit thus consists of a NANA-galactose-N-acetylglucosamine residue. NANA may be replaced by fucose in some of these external positions. These chains are linked to the internally located portion of the heteropolysaccharide chain which consists of mannose and N-acetylglucosamine residues. The order of release of monosaccharides from the brain nondialyzable glycopeptide preparation is therefore consistant with the hypothesis that the bulk of the glycopeptide material is made up of heteropolysaccharide chains that are quite similar to those isolated from other sources.

Glycopeptide from Orosomucoid (41):

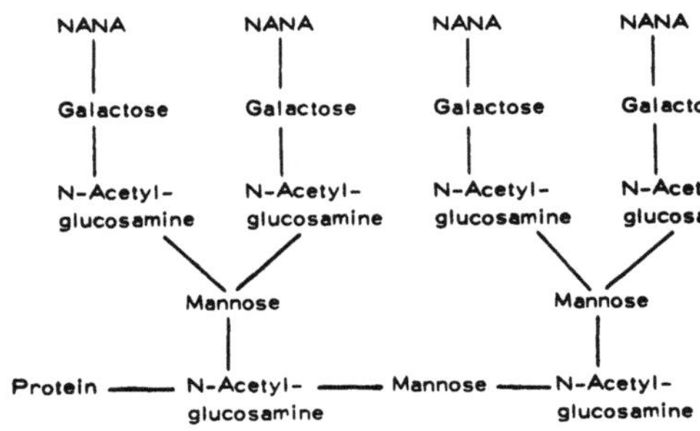

Glycopeptide from Fetuin (40):

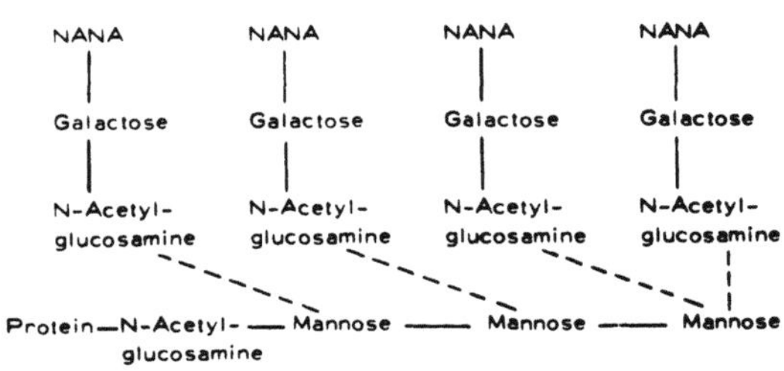

The major difference between the proposed glycopeptide structures for glycopeptides derived from fetuin and orosomucoid is that in the latter structure no two mannose residues are linked to each other. Since Öckermann (45) has reported the accumulation of oligosaccharides containing three mannose residues linked to a terminal glucosamine residue in a case of mannosidosis, one is inclined to suggest that the structure of the brain nondialyzable glycopeptides resembles that proposed for the glycopeptides isolated from fetuin. A structure similar to that proposed for the glycopeptide from orosomucoid appears to be less likely.

The presence of sulfate ester groups clearly distinguishes the brain nondialyzable glycopeptides from glycopeptides derived from glycoproteins such as fetuin, orosomucoid, and α_2-macroglobulin. Sulfate ester groups appear to be located in external positions (linked to galactose) as well as internal positions (linked to N-acetylglucosamine) (46). It has been shown (46) that both sulfated and nonsulfated nondialyzable glycopeptides contain mannose, galactose, glucosamine, and fucose as constituent sugars and that the molar ratios of these sugars are approximately the same in sulfated and nonsulfated glycopeptide preparations. Sulfated glycopeptides which contain large amounts of mannose, fucose, and NANA are clearly more closely related in composition to the heteropolysaccharide chains of glycoproteins than to the classical keratin sulfate which is a linear polymer of glucosamine and galactose (47, 48). Consequently, the term "keratosulfate" to describe these glycopeptides should be discontinued.

Clearly, the nondialyzable glycopeptide preparation is a complex mixture of heteropolysaccharide groups. The bulk of the material appears to consist of heteropolysaccharide chains the molecular weights of which differ largely due to the number of NANA groups associated with the molecule. All of the fractions recovered by column electrophoresis after removal of NANA appear to have the same average molecular weight (Table II). Preliminary experiments on the effects of graded acid hydrolysis suggest that these have a fundamental structure that resembles that of known glycopeptides which also contain NANA, fucose, glucosamine, galactose, and mannose as constituent sugars. A complicating factor is that all of the glycopeptide fractions recovered by column electrophoresis contain radioactive sulfate ester groups, although only about 40% of the glycopeptide material appears to be sulfated. It seems reasonable to suppose that sulfated and nonsulfated glycopeptides have the same fundamental structure, with the exception that sulfate groups are attached to some of the glycopeptide material. The ratio of mannose:galactose: hexosamine:fucose is approximately the same in both sulfated and nonsulfated glycopeptides. The nondialyzable glycopeptide preparation also appears to contain a glycopeptide or glycopeptides that contain NANA, galactose, and hexosamine (galactosamine and glucosamine) which are characterized by a high ratio of NANA/hexose, absence of fucose and mannose, presence of sulfate, and which may contain alkali-labile protein-carbohydrate linkages. They also appear to have a high electrophoretic mobility. These glycopeptides may represent 6% of the total hexose and 13% of the total NANA of the nondialyzable glycopeptide preparation.

The present investigation has revealed that glycopeptide fractions separable by charge (electrophoretic mobility) consist of materials which have similar molecular sizes. If in fact each fraction consists of a mixture of glycopeptides which differ considerably in fundamental structure, it may prove to be exceedingly difficult to separate these from each other in order to perform structural studies. If this is so, it may prove to be necessary to purify individual glycoproteins prior to elucidation of the structural features of the heteropolysaccharide chains. If our assumption that the nondialyzable glycopeptide preparation consists predominantly of a glycopeptide population of similar fundamental structure proves to be correct, one may anticipate that structural studies utilizing nondialyzable glycopeptides derived from whole brain tissue as starting material may prove to be fruitful. Evidence suggesting that brain tissue may contain a limited number of heteropolysaccharide chain types in spite of the probability of the existence of a multitude of glycoproteins is provided by the finding that the sugar ratios (NANA: fucose:hexose:hexosamine) are not significantly different between different brain areas (Table III) (30). One might have expected a diversity of values if the heteropolysaccharide chains of glycoproteins in different functional areas were markedly different. The ratios of the sugars for the glycopeptides derived from soluble and insoluble glycoproteins are also similar, although soluble glycoproteins contain less fucose and probably more radioactive sulfate ester groups per hexose and hexosamine molecule. However, fucose and sulfate are terminal groups which may be attached to internal hexose- and hexosamine-containing structures which are quite similar in the soluble and insoluble glycoproteins. An analysis of

Table III - Carbohydrate Composition of Nondialyzable Glycopeptides from Different Bovine Brain Areas (30).

	Molar Ratios					
	NANA/ Fucose	Hexosamine/ Fucose	Hexose/ Fucose	Hexose/ Hexosamine	Hexosamine/ NANA	Hexose/ NANA
1. Cerebellum	2.24	4.53	5.47	1.21	2.02	2.45
2. Medulla	2.00	4.36	5.64	1.30	2.17	2.82
3. Pons	2.30	4.80	6.00	1.25	2.09	2.61
4. Caudate nucleus, Globus pallidus, Putamen	2.30	3.96	4.78	1.21	1.72	2.08
5. Corpus Callosum	2.13	4.00	5.70	1.42	1.88	2.68
6. Thalamus	2.16	4.72	5.11	1.08	2.18	2.36
7. Cerebral Gray	2.00	4.26	5.23	1.23	2.13	2.61
8. Cerebral White	2.06	5.19	5.94	1.14	2.51	2.88
Average Value:	2.15 ± 0.01	4.48 ± 0.33	5.48 ± 0.34	1.23 ± 0.07	2.09 ± 0.16	2.56 ± 0.19

glycopeptides obtained from soluble and insoluble glycoproteins, glycopeptides from white and gray matter, and glycopeptides obtained from synaptosome-enriched and microsomal fractions is underway. The data obtained may provide some index of the heterogeneity of the heteropolysaccharide chains derived from whole rat brain tissue.

DIALYZABLE GLYCOPEPTIDES

The dialyzable glycopeptides were subjected to column electrophoresis (Fig. 10). Contents of test tubes containing fractions I through VII were combined and subjected to gel filtration on standardized columns of Sephadex G-50. Unlike the chromatographic profiles obtained for the nondialyzable glycopeptides (Fig. 2), the various fractions exhibited considerable heterogeneity upon gel filtration (Fig. 11). Several fractions (IA, IIA, IIIA, IVA, and V) had properties that were identical to glycopeptide fractions which appear in the nondialyzable glycopeptide preparation. For example, fractions IA, IIA, IIIA, IVA, and V (Fig. 11) have molecular weights and carbohydrate composition that are similar to those of fractions I, II-III, III-IV, V, and VI obtained by column electrophoresis of the nondialyzable glycopeptide preparation (Fig. 1, Table II). These do not contain glucose. Glucose appears to be a constituent sugar only of the dialyzable glycopeptides (IB, IIB, IIC, IIIC, and possibly VI and VII). The presence of glucose-containing glycopeptides is of interest in view of the report by Riddle and Leonard (51) who provided

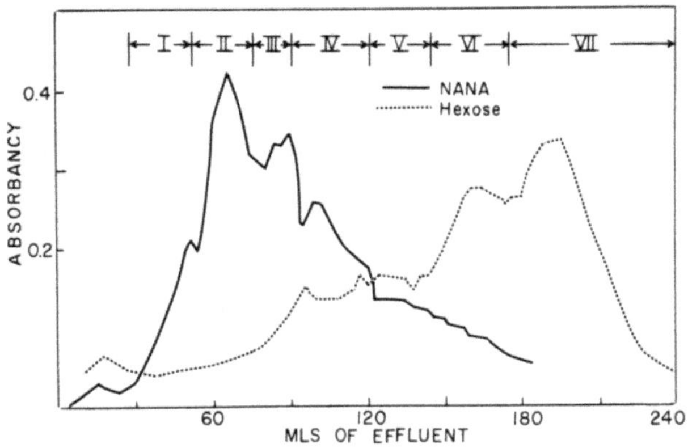

Fig. 10. Electropherogram obtained when rat brain dialyzable glycopeptides were subjected to column electrophoresis. Conditions of electrophoresis were identical to those described under Fig. 1. Contents of test tubes containing material corresponding to fractions I through VII (see top of graph) were combined, concentrated, desalted by gel filtration, and subjected to gel filtration on standardized columns of Sephadex G-50 (see Fig. 11).

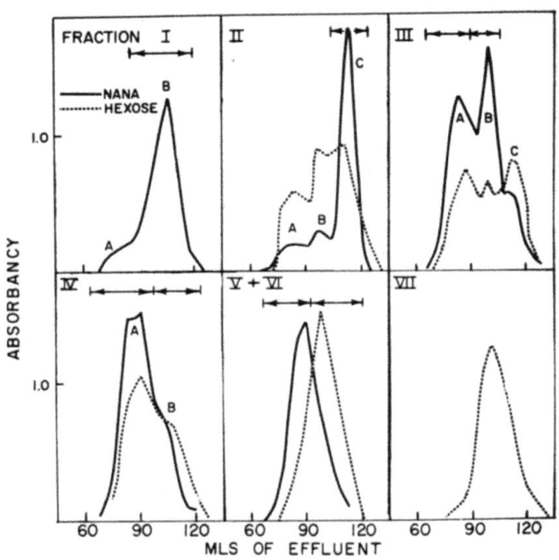

Fig. 11. Chromatogram obtained when fractions I through VII (see Fig. 10) were subjected to gel filtration on standardized columns of Sephadex G-50. Conditions are the same as those described under Fig. 2.

evidence that glucose-containing glycopeptides derived from brain glycoproteins cause coma upon injection into experimental animals. The hexosamine content of fractions IB, IIC, IIIB, and IIIC is partly or predominantly galactosamine. Glucosamine is the sole hexosamine in fractions IIIA, IVA, IVB, V, VI, and VII. Some of the dialyzable glycopeptides (IIB, IIIB, and IVB) may contain sulfate-ester groups.

Dialyzable glycopeptides of fractions VI and VII differ markedly from other glycopeptides examined because of their high content of mannose, absence of NANA, and very low levels (if any) of fucose. If one assumes that the glycopeptides of fractions VI and VII are 70% in carbohydrate content, the heteropolysaccharide chain would have a molecular weight of 2129. This would accomodate 8 mannose and 4 N-acetylglucosamine residues. One fucose and one galactose residue may also be present. The behavior of the hexosamine in these glycopeptides in certain colorimetric methods used to distinguish between glucosamine and galactosamine led to the suggestion that the hexosamine may be mannosamine (52). The use of the Gardell column in combination with spectrophotometric methods (52) have indicated that the hexosamine is exclusively glucosamine (25).

METABOLISM OF THE NONDIALYZABLE GLYCOPEPTIDES

All of the glycopeptides contain hexosamine, and the preparations are known to be free from other hexosamine-containing constituents such as glycosaminoglycans and gangliosides. Rats were injected with D-glucosamine-1-^{14}C and the nondialyzable glycopeptides were isolated (39). These were subjected to mild acid hydrolysis and the liberated NANA and mono- and disaccharides were recovered by gel filtration on Sephadex G-15. NANA was separated from the mono- and disaccharides by the use of Dowex 1-X8 (formate) anion exchange columns. The desialidized glycopeptides, which were not retarded by gel filtration on Sephadex G-15, were isolated and hydrolyzed to liberate amino acids, neutral sugars, and hexosamines. The neutral sugars (hexose and fucose) were recovered by passage of the hydrolyzate through coupled columns of Dowex 50-X4 (hydrogen) and Dowex 1-X8 (formate). Another aliquot of the hydrolyzate was applied to Dowex 50-X4 (hydrogen) columns. Acidic and neutral amino acids appeared in the effluent. The amino sugars and basic amino acids were adsorbed and were eluted with 2 N HCl. The latter fraction was acetylated and again passed through the Do-

Table IV - Distribution of Radioactivity in Monosaccharides and Amino Acids Recovered from the Nondialyzable Glycopeptide Preparation.
Distribution of total counts in various substances recovered from the nondialyzable glycopeptide preparation, as percentage of total counts in the original nondialyzable glycopeptide preparation.

Substance	Time after injection (hours)			
	1	4	24	72
NANA	17	19	23	16
Hexosamine	54	58	61	58
Neutral sugars	1	1	2	2
Neutral and acidic amino acids	3	2	4	4
Mono- and disaccharides released by mild acid hydrolysis	11	6	4	5
Substances adsorbed to Dowex 50-X4, hydrogen form, before and after acetylation	6	11	13	12

wex 50-X4 (hydrogen) columns. The effluent in this case contained the acetylated amino sugars. With the exception of a small amount of label associated with materials the nature of which is unknown and which were adsorbed to Dowex 50-X4 (hydrogen) both before and after acetylation, all of the radioactivity recovered could be accounted for by its presence in hexosamine and NANA (Table IV). Very little appears in the amino acids and neutral sugars. Fucose has often been used as a precursor for glycoproteins, since this sugar does not appear in any other brain constituent. However, as we have seen, not all glycopeptides contain fucose. Glucosamine is a sugar that is common to all of them, and all glycopeptide classes would be labelled if this precursor is used.

After intraperitoneal injection, maximum activity is attained 4 hours after injection in the 15-day old rat (Fig. 12). The rate of decline in radioactivity in the glycopeptides after maximal incorporation suggests the presence of at least two components,the ap-

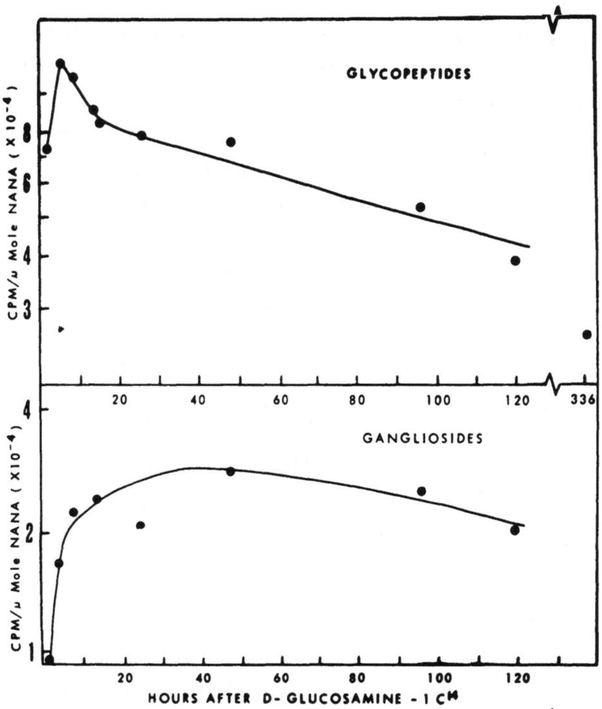

Fig. 12. Incorporation of D-glucosamine-1-^{14}C into non-diffusible glycopeptides and gangliosides recovered from brain tissue of 15-day old rats as a function of time after injection. D-glucosamine-1-^{14}C was administered intraperitoneally (0.4 μCi/g body wt). Radioactivity and NANA content were determined and reported as cpm/μmole glycopeptide NANA or gangliosidic NANA. Similar kinetic curves were obtained in four and two experiments for glycopeptides and gangliosides, respectively, and the average values for each time point were used to plot the data.

parent half-lives of which were approximately 18 hours and 4 days. In contrast, maximal activity in the gangliosides was not attained until 24 hours after injection, and loss of label after maximal incorporation proceeded at a slower rate. The incorporation data provided evidence that the carbohydrate portion of the glycoproteins turns over at a higher rate than that of the gangliosides. In the adult rat, the incorporation of D-glucosamine-1-14C reaches nearly maximal values by 8 hours after injection. The decline in radioactivity from its peak value at 24 hours suggests the presence of two components. The fast and slow components appear to possess half-lives of approximately 40 hours and 12.5 days respectively (39).

Table V - Incorporation of label from D-glucosamine-1-^{14}C into nondialyzable glycopeptide fractions obtained by column electrophoresis.

Fraction	CPM in fraction/μmoles hexosamine + NANA in fraction
I	22,800 ± 4400
II + III	20,200 ± 1800
IV	26,500 ± 3200
V	31,100 ± 600
VI	29,350 ± 1400

The results are the average of two experiments.

Fifteen-day old rats were injected intraperitoneally with glucosamine-1-^{14}C. The animals were sacrificed 4 hours later and the nondialyzable glycopeptides were isolated. The material was subjected to column electrophoresis and divided into fractions as shown in Fig. 1. Earlier, we had postulated that the nondialyzable glycopeptides contain two classes of glycopeptides. It was proposed that the nondialyzable glycopeptide preparation contained a minor component of high electrophoretic mobility and characterized by a high ratio of NANA/hexose and absence of mannose. It was therefore of interest to note (Table V) that the fractions containing this material showed significantly lower radioactivity 4 hours after the administration of the isotope than did the fractions of lower electrophoretic mobility. Since only one point was studied, it cannot be determined from these experiments whether the turnover rate of the minor component is greater or less than that of the major component. The presence of two glycopeptide populations may account for the biphasic nature of the radioactive decay curves depicted in Fig. 12.

It has been reported (23) that the amount and metabolism of glycoproteins in the pigeon brain are affected by training. These experiments have been criticized on biochemical (2) and behavioral grounds (53). We have attempted (54) to determine whether the metabolism and levels of the nondialyzable glycopeptides from rat brain are affected by stress and during the process of memory consolidation. Rats were trained for 3-5 minutes on a simple one--trial passive avoidance of foot shock task (55). A three-minute retention test was run 24 hours after original learning. Glucosamine-1-^{14}C was injected intraperitoneally 15 minutes before original learning (experiments 1 and 2), before retention (experiment 3), or before 10 minutes of inescapable foot shock (experiment 4). Control animals were handled in the same manner as experimental subjects except that the former never received footshock. Sacrifice occurred one hour (experiments 1, 3, and 4) and 24 hours (experiment 2) after injection. There were no significant differences between controls and experimental animals in the incorporation of isotope into the heteropolysaccharide chains isolated from brain glycoproteins. The amounts of NANA, hexose, hexosamine, and fucose in the heteropolysaccharide chains were also unaffected.

Irwin and Sampson (56) reported changes in the content of di- and trisialogangliosides as well as changes in ganglioside turnover in rat brain following behavioral stimulation. A possible role for glycoproteins was suggested by a lower specific activity of the residue obtained after extraction of the gangliosides 9 and 24 hours after injection of labelled glucosamine in rats that had been taught a swim-escape task.

RELATION BETWEEN NONDIALYZABLE AND DIALYZABLE GLYCOPEPTIDES

Analysis of the dialyzable glycopeptide preparation indicated that approximately 70% of the hexosamine associated with these materials is bound in mannose-rich glycopeptides which contain little or no fucose, NANA and galactose (Fig. 10 and 11, fractions VI and VII). Much of the remaining hexosamine in this preparation is associated with nondialyzable glycopeptides which appear in the dialyzable glycopeptide preparation under the used conditions of dialysis. The mannose-rich glycopeptides appear to possess a molecular weight of approximately 3100, based on gel filtration experiments using glycopeptides from purified glycoproteins (transferrin and thyroglobulin) as standards. Whether these standards are the appropriate ones to use in order to estimate the molecular weight of glycopeptides of smaller size is not known. If we assume that 3100 is a maximum molecular weight, and knowing that the ratio of hexose to hexosamine is 2, we have estimated (see above) that these molecules contain 8 mannose and 4 N-acetylglucosamine residues. If the molecular weight is in fact lower, possible values for the ratio of mannose to N-acetylglucosamine would be 6:3, 4:2, or 2:1. Gel filtration experiments (Fig. 11) suggest that these glycopeptides consist of a mixture of glycopeptides of approximately similar molecular size. The actual ratio of mannose to N-acetylglucosamine must therefore be only one of the four ratios suggested. It is also possible that the experimental value for the ratio of mannose:N-acetylglucosamine may be an average value for a population of glycopeptides all of similar size but which vary in the number of mannose and N-acetylglucosamine residues which they contain. This variation would be such that the sum of mannose and N-acetylglucosamine would be the same for each of the glycopeptide components in the fraction.

It is not possible at this time to decide whether these heteropolysaccharide chains may serve as metabolic precursors for the larger NANA- and galactose-containing glycopeptides of the nondialyzable glycopeptide preparation. The analytical data available at this time would suggest that this is unlikely. In any case, it is of interest to know whether they form a part of the same glycoprotein molecule as the larger nondialyzable NANA- and galactose-containing heteropolysaccharides, or whether they are the prosthetic groups of a different glycoprotein or glycoproteins. In earlier work (27) it was found that the percentage of the total particle-bound nondialyzable glycopeptide hexosamine in the nuclear, mitochondrial, and microsomal fractions was 16, 46, and 38%, respectively. On the other hand, the percentage of the total particle-bound dialyzable glycopeptide-hexosamine in the nuclear, mitochondrial, and microsomal fractions was 28, 42, and 30%, respectively. The data suggest a slight enrichment of the mannose-rich glycopeptide material in the crude nuclear fraction which contains blood elements such as red blood cells and capillary fragments, as well as large cell fragments, possibly including glial fragments. The crude nuclear fraction also contains large amounts of myelin with enclosed axons. In later work (57) in which the crude mitochondrial fraction was subfractionated according to the method of Whittaker (58), it was shown that the ratio of nondialyzable glycopeptide-hexosamine/dialyzable glycopeptide-hexosamine was approximately the same in P2A (myelin-enriched fraction), P2B (synaptosome-enriched fraction) and P2C (mitochondria).

Triton X-100 is capable of solubilizing approximately 70 to 80% of the insoluble glycoprotein material of the combined mitochondrial and microsomal fractions. Such extracts were subjected to column chromatography (16) on calcium hydroxylapatite columns. This resulted in a 2 to 3 fold purification of the glycoprotein material in the extracts. 100% of the glycoprotein applied to the column was recovered. The degree of purification of glycoproteins that yielded nondialyzable glycopeptide-hexosamine upon proteolytic digestion was identical to the degree of purification of glycoproteins that yielded dialyzable glycopeptide-hexosamine. Since the degree of purification of both nondialyzable and dialyzable glycopeptides coincided in all of the fractions recovered from the column, the tentative conclusion derived from these experiments was that the same glycoprotein or glycoproteins probably contain both types of heteropolysaccharide chains.

Analysis of the soluble glycoproteins extracted by severe homogenization of tissue indicated that 73% of the nondialyzable NANA- and galactose-containing glycopeptides remain in the precipitate while 80% of the dialyzable mannose-rich glycopeptides remain insoluble. Thus the soluble glycoproteins contain somewhat less of the dialyzable glycopeptide-hexosamine in relation to nondialyzable glycopeptide-hexosamine than the insoluble membrane-bound glycoproteins.

In bovine brain (30), the ratio of nondialyzable glycopeptide--hexosamine to dialyzable glycopeptide-hexosamine increases in the following order: corpus callosum (0.83), pons (0.86), thalamus (1.23), medulla (1.36), caudate nucleus, globus pallidus, and putamen (1.54), cerebellum (1.57), cerebral gray (2.40) and cerebral white (2.44). Although cerebral white matter falls out of sequence, it appears that as a rule, white matter contains a greater amount of dialyzable hexosamine relative to nondialyzable hexosamine than does gray matter. Human gray matter and white matter contain respectively 1.03 and 1.08 μmoles of dialyzable glycopeptide-hexosamine per gram tissue. The values for nondialyzable glycopeptide--hexosamine were 1.93 and 1.10 for gray and white matter respectively. The ratio of nondialyzable glycopeptide-hexosamine/dialyzable glycopeptide-hexosamine is therefore 1.9 in gray matter and 1.0 in white (59).

Conceivably, a larger part of the glycopeptides of low molecular weight are derived from glycoproteins present in glial and/or axonal elements. Breckenridge et al. (60) have provided evidence that glycoproteins containing the large NANA- and galactose-containing glycopeptides can be separated from glycoproteins containing heteropolysaccharide chains rich in mannose and N-acetylglucosamine. Consequently, the two types of heteropolysaccharide chains may be attached to different glycoproteins. Triton X-100 was found to be incapable of extracting glycoproteins containing the mannose-rich glycopeptide from purified synaptosomal membrane fractions. It had been shown earlier (16) that this detergent will extract most of the glycoproteins (60%) containing these chains from the particulates consisting of the combined mitochondrial and microsomal fractions. This need not be considered as a conflicting datum. Rather, it may indicate that the mannose-rich dialyzable glycopeptides represent a population of glycopeptides of similar molecular size and carbohydrate composition, most of which are associated with glial and/or axonal elements and some of which may be associated with synaptosomal membranes. It is equally likely, of course, that the synaptosomal membrane preparation of Breckenridge et al. (60) contains axonal and/or glial membranes as well as synaptosomal membranes.

SUBCELLULAR AND ANATOMICAL DISTRIBUTION OF THE GLYCOPROTEINS

Published investigations on the subcellular distribution of glycoproteins (27, 57, 61, 62) have been extensively reviewed (1, 2). The synaptosomal and microsomal preparations from rat brain account for most of the glycoprotein material of the tissue. However, it has been shown that the usual synaptosomal preparation is contaminated with structures that are morphologically similar to axonal and possibly glial fragments (61, 62, 63) and these may contain glycoproteins. The mitochondria contain very little glycoprotein material, part of which is probably associated with the lysosomes (13) which sediment in this fraction. Lysosomes contain several acid hydrolases which are glycoproteins, although the lysosomal membrane does not appear to contain NANA (64). Purified brain mitochondrial preparations contain very little, if any, glycoprotein material as judged by their NANA content (Dr. Neville Marks, personal communication). The crude nuclear preparation accounts for only 20% or less of the total brain glycoproteins. Some of these glycoproteins are associated with red blood cells and blood vessels which sediment in this fraction. The crude nuclear fraction also contains considerable amounts of myelin with enclosed axons as well as synaptosomal material, mitochondria, and possibly glial fragments. As we have noted, approximately 20% of the brain glycoproteins appear in the soluble fraction after severe homogenization in dilute aqueous buffers. The normally prepared cell sap (mild homogenization in 0.32 M sucrose) contains about half of this amount of glycoprotein material. Purified myelin contains little, if any, glycoprotein material.

Brain tissue samples that consist largely of cell bodies contain a higher concentration of glycoproteins and gangliosides than areas that are enriched in fiber tracts (30, 65). The concentration of glycoproteins relative to gangliosides is somewhat more elevated in white matter than in gray matter. Since the concentration of glycoproteins relative to gangliosides is greater in the synaptosomal fraction than in axonal fragments, it was suggested (2, 30) that oligodendroglial cells in white matter contain glycoproteins and that fragments originating from these cells contaminate the used synaptosomal fraction. This was in agreement with morphological evidence (61, 62, 63).

A large part of the nondialyzable glycopeptide content, per gram rat brain, was synthesized prior to the acceleration of ganglioside deposition which occurs between 6 and 29 days of age (39). There is a slow and steady increase in glycopeptide deposition between 6 and 29 days of age. Quarles and Brady (66) studied total glycoprotein-NANA content and obtained similar results. Nearly adult levels are reached by 29 days of age. The composition of the nondialyzable glycopeptides derived from the 15-day old rat showed qualitative differences compared to those isolated from the adult rat brain (39).

FUNCTION OF BRAIN GLYCOPROTEINS

There has been considerable speculation on the function of brain glycoproteins, particularly in regard to information storage (1, 2, 3, 20, 21, 23) and development (20, 21, 22). This speculation is solely based on the carbohydrate portions of the glycoprotein molecule. We feel that short term memory is probably due to rapid conformational changes in the macromolecular structure of the neuronal membrane (21). On the other hand, long term memory may be a consequence of an alteration of the neuronal membrane due to the deposition of glycoprotein material the heteropolysaccharide chains of which had been altered in structure as a consequence of neural activity. Since the carbohydrate portion of the glycoproteins is synthesized within the Golgi apparatus (see review, 3), changes in the structure of the heteropolysaccharide chains of the glycoproteins due to stimulation would occur during the passage of the glycoprotein material through the biosynthetic channels of the smooth endoplasmic reticulum. The chemical composition of the heteropolysaccharide chains may be affected by small changes in pH, ion composition of the medium, or even in spatial arrangements within the cisternae as a consequence of neural activity. The report of Whetsell and Bunge (67) that exposure of the sensory ganglia to ouabain caused the Golgi apparatus to swell, may be of interest in this connection. The effect was reversible.

REFERENCES

(1) Brunngraber, E. G., in Handbook of Neurochemistry, Ed. A. Lajtha, Plenum Press, New York, vol. 1, p. 223 (1969).

(2) Brunngraber, E. G., in Protein Metabolism of the Nervous System, Ed. A. Lajtha, Plenum Press, New York, p. 383 (1970).

(3) Brunngraber, E. G., in The Symposium on Proteins of the Nervous System, 3d Meeting of the International Society of Neurochemistry, Budapest, Pergamon Press, in press (1971).

(4) Di Benedetta, C., Chang, I. and Brunngraber, E. G., Italian J. Biochem., in press.

(5) Swanborg, R. H. and Shulman, S., Immunology 19: 31 (1970).

(6) Liakopoulou, A. and MacPherson, C. F. C., J. Immunology 105: 512 (1970).

(7) Hatcher, V. B. and MacPherson, C. F. C., J. Immunology 104: 633 (1970).

(8) Rajam, P. C., Gaudreau, C. J., Grady, A. and Rundlett, St., Immunology 17: 813 (1969).

(9) Kashkin, K. P., Sharetskii, A. N., Surimov, B. P. and Bochkova, D. N., Vop. Med. Khim. 15: 235 (1969).

(10) Shulman, S., Milgrom, F. and Swanborg, R. H., Immunology 16: 25 (1969).

(11) Warecka, K., J. Neurochem. 17: 829 (1970).

(12) Saraswathi, S. and Bacchawat, B. K., Biochim. Biophys. Acta 212: 170 (1970).

(13) Koenig, H., in The Lysosomes in Biology and Pathology, Ed. J. T. Dingle and H. B. Fell, North-Holland, Amsterdam, vol. 2, p. 110 (1969).

(14) Shelanski, M.L., Berry, R., Feit, H., Albert, S., Daniels, M., Slusarek, L. and Slusarek, W., Abstracts, 3d International Meeting of the International Society for Neurochemistry, Budapest, p. 431 (1971).
(15) Falxa, M.L. and Gills, T.J., Arch. Biochem. Biophys. 135: 194 (1969).
(16) Brunngraber, E.G., Aguilar, V. and Aro, A., Arch. Biochem. Biophys. 129: 131 (1969).
(17) Bosmann, H.B., Case, K.R. and Shea, M.B., FEBS Lett. 11: 261 (1970).
(18) Susz, J., Fed. Proc. 28: 830 (1969).
(19) Dutton, G.R. and Barondes, S.H., J. Neurochem. 17: 913 (1970).
(20) Brunngraber, E.G., Perspect. Biol. and Med. 12: 467 (1969).
(21) Brunngraber, E.G., in Biogenesis, Evolution, and Homeostasis, Ed. A. Locker, Springer-Verlag, Heidelberg, in press.
(22) Dische, Z., in Protides of the Biological Fluids, Ed. H. Peeters, Elsevier Publishing Co., Amsterdam, vol. 13, p. 1 (1966).
(23) Bogoch, S., The Biochemistry of Memory, Oxford University Press, New York (1968).
(24) Brunngraber, E.G., J. Pediatrics 77: 166 (1970).
(25) Brunngraber, E.G., Brown, B.D. and Hof, H., Clin. Chim. Acta 32: 159 (1971).
(26) Brunngraber, E.G. and Brown, B.D., Biochim. Biophys. Acta 69: 581 (1963).
(27) Brunngraber, E.G. and Brown, B.D., J. Neurochem. 11: 449 (1964).
(28) Brunngraber, E.G. and Brown, B.D., Biochim. Biophys. Acta 83: 357 (1964).
(29) Brunngraber, E.G. and Brown, B.D., Biochem. J. 103: 65 (1967).
(30) Brunngraber, E.G., Brown, B.D. and Aguilar, V., J. Neurochem. 16: 1059 (1969).
(31) Di Benedetta, C., Brunngraber, E.G., Whitney, G., Brown, B.D. and Aro, A., Arch. Biochem. Biophys. 131: 404 (1969).
(32) Brunngraber, E.G. and Whitney, G., J. Chromatography 32: 749 (1968).
(33) Suzuki, K., J. Neurochem. 12: 629 (1965).
(34) Bhatti, T. and Clamp, J.R., Biochim. Biophys. Acta 170: 206 (1968).
(35) Spiro, R.G., J. Biol. Chem. 240: 1603 (1965).

(36) Jamieson, G. A., J. Biol. Chem. 240: 2914 (1965).
(37) Barnes, M. J. and Partridge, S. M., Biochem. J. 109: 883 (1968).
(38) Andrews, P., Biochem. J. 96: 595 (1965).
(39) Holian, O., Dill, D. and Brunngraber, E. G., Arch. Biochem. Biophys. 142: 111 (1970).
(40) Spiro, R. G., J. Biol. Chem. 239: 567 (1964).
(41) Wagh, P. V., Bornstein, I. and Winzler, R. J., J. Biol. Chem. 244: 658 (1969).
(42) Dunn, J. T. and Spiro, R. G., J. Biol. Chem. 242: 5556 (1967).
(43) Margolis, R. U. and Margolis, R. K., Abstracts, 3d International Meeting of the Society for Neurochemistry, Budapest, p. 276 (1971).
(44) Arima, T., Muramatsu, T., Saigo, K. and Egami, F., Jap. J. Exptl. Med. 39: 301 (1969).
(45) Öckermann, P. A., J. Pediatrics 77: 168 (1970).
(46) Margolis, R. K. and Margolis, R. U., Biochemistry 9: 4389 (1970).
(47) Matthews, M. B. and Cifonelli, J. A., J. Biol. Chem. 240: 4140 (1965).
(48) Hirano, S., Hoffman, P. and Meyer, K., J. Org. Chem. 26: 5064 (1961).
(49) Pridham, J. B., Anal. Chem. 28: 1967 (1956).
(50) Boas, N. F., J. Biol. Chem. 204: 553 (1953).
(51) Riddle, D. and Leonard, B. E., Neuropharmacology 9: 283 (1970).
(52) Brunngraber, E. G., Aro, A. and Brown, B. D., Clin. Chim. Acta 29: 233 (1970).
(53) Glassman, E., Ann. Rev. Biochem. 38: 605 (1969).
(54) Holian, O., Routtenberg, A. and Brunngraber, E. G., Life Sci. 10: 1029 (1971).
(55) Routtenberg, A., Zechmeister, E. B. and Benton, O., Life Sci. 9: 909 (1970).
(56) Irwin, L. N. and Samson, F. E., J. Neurochem. 18: 203 (1971).
(57) Brunngraber, E. G., Dekirmenjian, H. and Brown, B. D., Biochem. J. 103: 73 (1967).
(58) Whittaker, V. P., Biochem. J. 72: 694 (1959).
(59) Brunngraber, E. G., Brown, B. D. and Chang, I., J. Neuropath. and Exptl. Neurology 30: 525 (1971).
(60) Breckenridge, W. C., Breckenridge, J. E. and Morgan, I., Abstracts, 3d International Meeting of the International Society for Neurochemistry, Budapest, p. 431 (1971).

(61) Dekirmenjian, H. and Brunngraber, E. G., Biochim. Biophys. Acta 177: 1 (1969).
(62) Dekirmenjian, H., Brunngraber, E. G., Johnston, N. L. and Larramendi, L. M. H., Exptl. Brain Res. 8: 97 (1969).
(63) Johnston, N. L. and Dekirmenjian, H., Exptl. Brain Res. 11: 392 (1970).
(64) Sellinger, O. Z. and Nordrum, L. M., J. Neurochem. 16: 1219 (1969).
(65) Roukema, P. A. and Heijlman, J., J. Neurochem. 17: 773 (1970).
(66) Quarles, R. H. and Brady, R. C., J. Neurochem. 17: 801 (1970).
(67) Whetsell, Jr. W. O. and Bunge, R. P., J. Cell Biol. 42: 490 (1969).

CHEMISTRY AND METABOLISM OF GLYCOSAMINOGLYCANS OF THE NERVOUS SYSTEM

B. K. BACHHAWAT, K. A. BALASUBRAMANIAN,
A. S. BALASUBRAMANIAN, M. SINGH, E. GEORGE
and E. V. CHANDRASEKARAN

Neurochemistry Laboratory, Department of Neurological Sciences, Christian Medical College Hospital, Vellore, S. India

Glycosaminoglycans are hexosamine containing polymers widely distributed in connective tissue. The structure of glycosaminoglycans as known today are shown in Fig. 1. All the glycosaminoglycans except hyaluronic acid are sulphated. Kerato sulphate does not have uronic acid in the molecule. Heparin is unique in that it has both sulphamino as well as sulphate ester groups. Heparan sulphate (not shown in the figure) is similar to heparin except for the presence of both N-acetyl and N-sulphated group in the molecule. Although the presence of glycosaminoglycans in brain has been shown by various workers, studies on their characterisation have been incomplete. From biochemical analysis, the presence of acidic glycosaminoglycans in brain was demonstrated by Meyer (1) and his associates and by Brante (2). Sulphate incorporation studies by Ringertz (3) suggested the presence of these compounds in mouse brain. In later years, Szabò and Roboz-Einstein (4) isolated glycosaminoglycans from bovine brain and identified them as chondroitin-4-sulphate and hyaluronic acid whereas Clausen and Hansen (5) isolated glycosaminoglycans from human brain and characterized them from electrophoretic and infra-red analysis as hyaluronic acid, chondroitin-6-sulphate and dermatan sulphate. Recently, Margolis (6) has isolated from bovine brain, chondroitin-4-sulphate,

Abbreviation used: GAG = glycosaminoglycan.

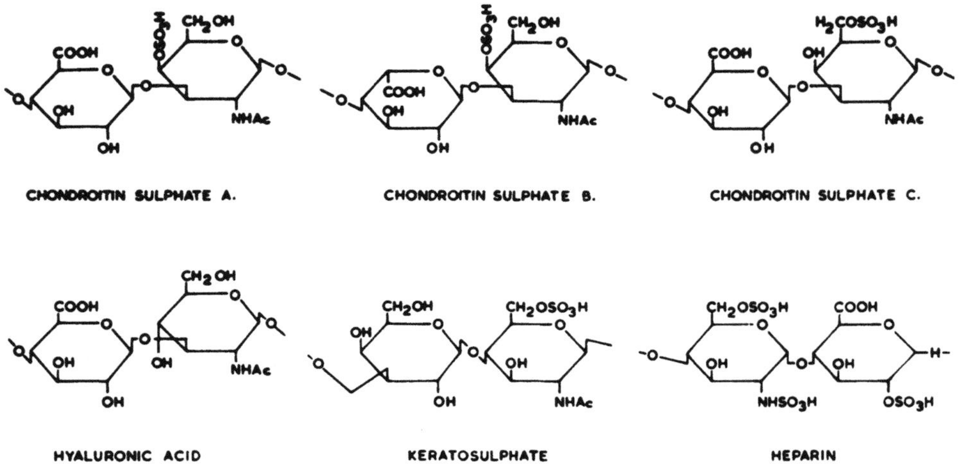

Fig. 1. Structure of glycosaminoglycans.

chondroitin-6-sulphate and hyaluronic acid. These results suggest that either are species specific glycosaminoglycans or that the methods of isolation procedure employed by these workers are different. We have been interested in the metabolism of glycosaminoglycans in brain for a number of years and we have isolated and characterized these compounds (7) by the procedure as summarized in the following table. Table I will show the summary of the procedure by which the extraction and fractionation of glycosaminoglycans were carried out. Briefly, the acetone powder of the tissue was prepared and then it was extracted with chloroform methanol to remove lipids and the dry defatted tissue was employed for subsequent methods of extraction. The defatted material was then subjected to proteolytic digestion with papain and pronase, and alkali digestion to remove polypeptide by β-elimination. Glycosaminoglycans were precipitated by cetyl pyridinium bromide (CPB). CPB-glycosaminoglycan complex was extracted with various concentrations of NaCl according to the method of Dorfman (8) and his associates. These fractions were precipitated with alcohol and further fractionated on DEAE-Sephadex A-25 column according to the method of Schmidt (9) and various fractions were collected using 0.5 M, 0.9 M, 1.5 M and 2.0 M sodium chloride. We obtained four subfractions out of fractions I and II. Constituent analysis was carried out by subjecting the various fractions to hyaluronidase digestion (10, 11) and determining the relative concentration of various amino sugars (12), total amino sugar (13) and

Table I - Dry defatted tissue was subjected to papain and pronase digestion. After alkali digestion and deproteinization, glycosaminoglycans (GAG) were precipitated with CPB. CPB GAG complex was extracted successively with 0.4 M NaCl, 1.2 M NaCl and 2.1 M NaCl according to the method of Dorfman and his associates (8).
After precipitating GAG with alcohol each of the NaCl fraction was chromatographed on DEAE-Sephadex column according to the method of Schmidt (9).

Fraction I (0.4M NaCl extract)				Fraction II (1.2M NaCl extract)			
0.5M NaCl	0.9M NaCl	1.5M NaCl 0.01 N HCl	2M NaCl 0.01 N HCl	0.5M NaCl	0.9M NaCl	1.5M NaCl 0.01 N HCl	2M NaCl 0.01 N HCl
↓	↓	↓	↓	↓	↓	↓	↓
I-SA	I-SB	Absent	Absent	Absent	II-SB	II-SC	Absent

Table II - Constituent Analysis.

1. Uronic acid was estimated by carbazole method of Dische (14) as modified by Bitter and Muir (15).
2. Total hexosamines (13), glucosamine and galactosamine (12) were determined by the method of Ludowieg and Benmaman after hydrolysis of glycosaminoglycans. Amino sugars were further identified by paper chromatography.
3. Total sulphate was estimated by the method of Dodgson and Price (16). N-sulphate was determined by estimating the sulphate released after hydrolysing the glycosaminoglycan for 90 min. at 100°C with 0.04N HCl.
4. G-25 Sephadex gel filtration was carried out after hyaluronidase digestion to estimate the hyaluronidase resistant material. Chondroitin-4-sulphate and chondroitin-6-sulphate were estimated by the method of Mathew and Inouye (10, 11).
5. Sulfamino group was estimated by nitrous acid - indole method of Lagunoff and Warren (17).
6. For the determination of dermatan sulphate uronic acid estimation was carried out with and without borate.

Table III - The Nature and Level of the Various Glycosaminoglycans in the DEAE-Sephadex Fractions.

Fractions	μg uronic acid/g dry lipid-free brain			
	Monkey	Man	Chicken	Sheep
I-SA				
Hyaluronic acid	1176	528	236	208
I-SB				
Heparan sulphate	73	66	33	25
Galactosamine-GAG	42	18	16	50
II-SB				
Heparan sulphate	253	157	77	56
Chondroitin-6-sulphate	95	62	20	12
Chondroitin-4-sulphate	180	77	12	42
II-SC				
Hyaluronidase-resistant galactosamine-GAG	292	6	91	20
Chondroitin-6-sulphate	136	66	60	9
Chondroitin-4-sulphate	364	430	31	81

uronic acid (14, 15) as shown in Table II. On the basis of these analyses the concentrations of the different glycosaminoglycans in the various fractions are shown in Table III. It is of interest to note that Fraction I as obtained by the method of Dorfman and his associates contains not only hyaluronic acid but also some sulphated glycosaminoglycans.

Recently, Onodora et al. (18) reported that rabbit brain does not contain any hyaluronic acid. Considering our observation that hyaluronic acid is one of the principal glycosaminoglycans in nervous tissue, this observation was rather surprising and this led us to investigate the nature and level of glycosaminoglycans in

Table IV - The Nature and Level of the Various Glycosaminoglycans in Brain of Different Species.

Glycosaminoglycans	Concentration in brain as the percentage of total GAG					
	Rat	Monkey	Man	Chicken	Sheep	Rabbit
Hyaluronic acid	35.8	45.0	37.4	41.0	41.3	32.7
Heparan sulphate	12.6	12.5	15.8	19.1	16.1	13.6
Low sulphated chondroitin sulphate	2.0	1.6	1.2	2.7	10.0	1.6
Chondroitin-6-sulphate	9.7	8.8	9.0	13.9	4.1	16.3
Chondroitin-4-sulphate	25.4	20.8	36.0	7.4	24.1	35.8
Hyaluronidase resistant galactosamine-GAG	14.6	11.3	0.6	15.9	4.1	-
Total GAG/mg uronic acid/gm LFDT	3415	2611	1485	593	565	477

various species (19). In Table IV these data are summarized and it is apparent from this that rabbit brain does contain considerable amount of hyaluronic acid although the total glycosaminoglycans is very low. The lowest concentration of glycosaminoglycans was found in rabbit brain. It is of interest to note that besides heparan sulphate there was very little hyaluronidase resistant material present in human brain although this was not true for other species. However, the nature of this hyaluronidase resistant material was not established from our work. Some of the sulphated glucosamine containing glycosaminoglycans in human brain was found to be degraded by hyaluronidase. The nature of this glycosaminoglycan was not further characterized. The presence of hyaluronic acid, chondroitin-4-sulphate, chondroitin-6--sulphate, dermatan sulphate and heparan sulphate has also been reported by Cunningham and Goldberg (20) in various species. But this is in contradiction to the work of Margolis (6) who was unable

Table V - The Nature and Level of the Glycosaminoglycans in the Peripheral Nerve, Spinal Cord and Brain of Monkey.

Glycosaminoglycans	Concentration of individual GAG as the percentage of total		
	Peripheral nerve	Spinal cord	Brain
Hyaluronic acid	68.0	64.3	45.0
Heparan sulphate	5.8	7.6	12.5
Chondroitin-6-sulphate including low sulphated chondroitin-4-sulphate	1.7	3.6	5.2
Chondroitin-6-sulphate	2.7	4.4	5.2
Chondroitin-4-sulphate	16.5	18.1	20.8
Hyaluronidase resistant galactosamine GAG	5.3	2.0	11.3
Total GAG/mg uronic acid/gm LFDT	1456	709	2611

to detect any heparan sulphate in bovine brain. The presence of heparan sulphate has been shown by the work of Elam et al. (21) on rapid axonal flow as well as by Dorfman (22) and his associates in their studies with glial cell tissue culture. Egami (23) and his associates and recently Margolis (24) have also shown the presence of this glycosaminoglycan.

We also studied the nature and level of various glycosaminoglycans in spinal cord and peripheral nerve. This is shown in Table V. It can be seen that the total concentration of glycosaminoglycans was low in spinal cord and moreover the concentration of hyaluronic acid was very high. Peripheral nerve had a comparatively higher concentration of glycosaminoglycans than spinal cord. Dermatan sulphate was definitely present in this tissue.

The concentration of glycosaminoglycans seems to change considerably with the development of brain as shown in Fig. 2 (26).

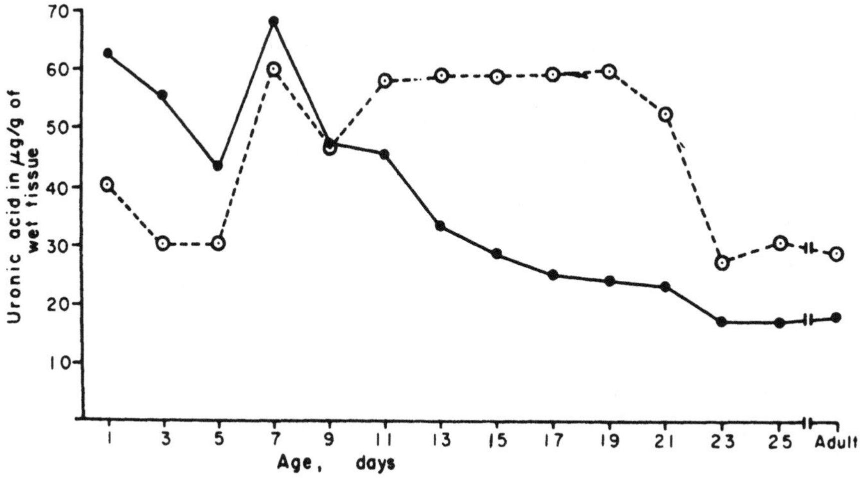

Fig. 2. Variation of hyaluronic acid (●—●—●) and chondroitin sulphate (o--o--o) with age in the rat brain.

Table VI - (^{35}S)Sulphate Incorporation into Glycosaminoglycans in Various Subcellular Fractions of Developing Rat Brain.

Fraction	Counts/min/g brain			II/I
	F-I	F-II	Total	
Seven days				
H	8660	27590	36250	3.2
P_2B	1546	2046	3592	1.3
P_3	410	2786	3196	6.7
SN	1518	10214	11732	6.7
Twenty days				
H	5878	18000	23878	3.1
P_2B	1991	3570	5561	1.8
P_3	548	1486	2034	2.8
SN	1003	8021	9024	8.0
Sixty days				
H	617	3860	4477	6.3
P_2B	185	438	623	2.4
P_3	ND	1520	1520	-
SN	140	1918	2058	13.6

H = Homogenate; P_2B = Synaptosomes; P3 = Microsomes; SN = Supernatant; ND = not done.

Fraction I which contains predominantly hyaluronic acid was very high in young rat brain compared to that of the 60 day old rat brain. The sulphated glycosaminoglycans are high at 7 days after birth. In Table VI data are presented in relation to the in vivo sulphate incorporation into various subcellular fractions and also the fractionation of these glycosaminoglycans with sodium chloride and CPB (27). (^{35}S)Sulphate incorporation into glycosaminoglycan was maximum at 7 day brain homogenate and decreased gradually at 20 and 60 days. The decrease in total (^{35}S)Sulphate incorporation into glycosaminoglycans was noticed both in microsomes and supernatant with increasing age. Among the various fractions studied, supernatant had the highest amount (32-45%) of (^{35}S)Sulphate incorporation into glycosaminoglycans. Comparing the radioactivity incorporated into glycosaminoglycans of the different subcellular fractions in any one age group, it is observed that highest amount of radioactivity incorporated in Fraction I was in synaptosomes followed by supernatant, thereby suggesting that synaptosomes are the best organelles for (^{35}S)Sulphate incorporation into Fraction I, particularly for the 20 and 60-day old rat brain. Only negligible amount of (^{35}S)Sulphate was incorporated to microsomal glycosaminoglycans. However, 17-22% of total radioactivity incorporated into glycosaminoglycans was in the supernatant fraction.

These results lead one to speculate that sulphation of glycosaminoglycan occurs in more than one site in brain. Synaptosomes were found to be the location for the sulphation of low sulphated chondroitin sulphate and heparan sulphate (Fraction I). Fraction I is of further interest for the fact that synaptosomes is known to contain considerable amount of glucosamine containing glycosaminoglycan as reported by Vos et al. (28). They had used cetyl pyridinium bromide procedure for the fractionation of individual glycosaminoglycan at subcellular level. On the basis of the amino sugar analysis they concluded that high concentration of glycosaminoglycan in synaptic vesicles is due to hyaluronic acid. The present observation that (^{35}S)Sulphate is incorporated into Fraction I of glycosaminoglycans in synaptosomes indicates that glucosamine containing glycosaminoglycan, probably a heparan sulphate, is present in this fraction.

Considering our earlier work (29) on the high level of incorporation of radioactive sulphate into brain glycosaminoglycans at

Table VII - Activity of the Enzyme from Young and Adult Rat Brains Towards Various Glycosaminoglycan Acceptors.

Glycosaminoglycan acceptor	Activity of the enzyme (')		
	1	2	3
Heparan sulphate	6867	4856	763
Dermatan sulphate	4248	3582	1530
Chondroitin-4-sulphate	2152	2012	781
Chondroitin sulphate (Bovine nasal septa)	2408	2179	885
Keratosulphate	1308	844	204
Chondroitin-6-sulphate	398	370	98
Hyaluronic acid	400	378	94
No acceptor	372	260	89

(') Activity of the enzyme expressed as counts/100 sec of (^{35}S) glycosaminoglycan formed per mg of enzyme protein. 1 = Enzyme prepared from 5-day-old rat brain. 2 = Enzyme prepared from 16-day-old rat brain. 3 = Enzyme prepared from 2-month-old rat brain. (^{35}S)PAPS (233,000 counts/100 sec) was used.

birth and the concomitant decrease in the rate of incorporation as the brain matures led us to investigate the nature of sulphotransferase activities at various stages of brain development. This investigation was carried out using (^{35}S)Sulphate labelled 3'-phosphoadenosine 5'-phosphosulphate (PAPS) and enzyme preparation from brain of various age groups of rats. For the purpose of convenience, we chose to study brain from rats at the premyelination stage (7 days after birth), myelination stage (15 to 20-day old), and adult stage (60 days after birth). The enzyme assay was carried out by the method of Balasubramanian and Bachhawat (30). As can be seen from Table VII with the young rat brain, i.e. 5-7-day old, the enzymic activity towards heparan sulphate as an acceptor was very high compared to the adult. Hyaluronic acid was not a good acceptor for this enzyme. This study was extended further with respect to the distribution

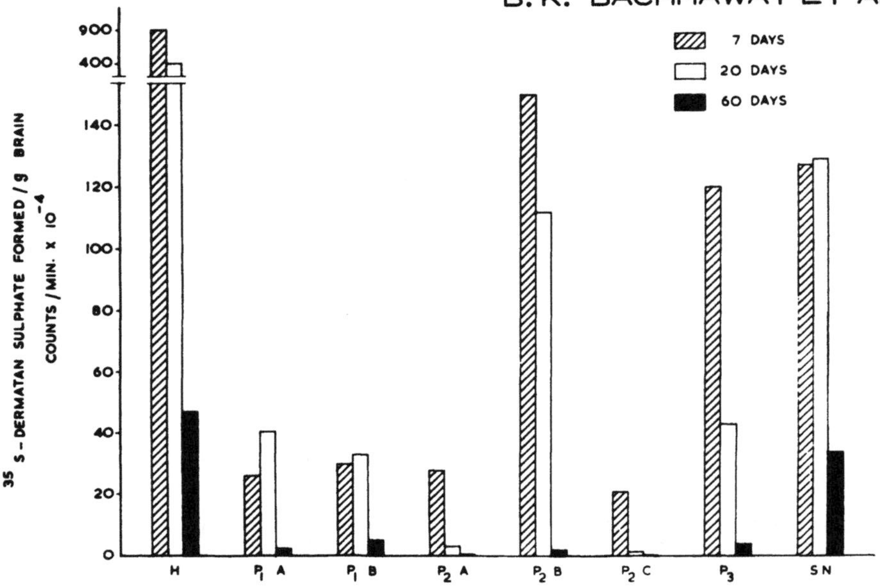

Fig. 3. The distribution of glycosaminoglycan sulphotransferase activity/g brain in various subcellular fractions of developing rat brain. H = Homogenate; P_1A = Large myelin; P_1B = Nuclei; P_2A = Small myelin; P_2B = Synaptosomes; P_2C = Mitochondria; P_3 = Microsomes; SN = Supernatant.

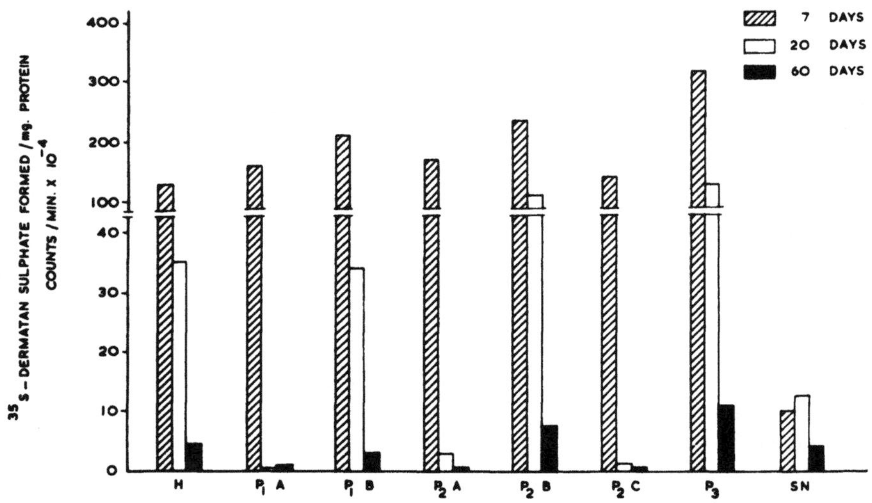

Fig. 4. Specific activity of glycosaminoglycan sulphotransferase in various subcellular fractions of developing rat brain. (Abbreviations are as in Fig. 3).

Table VIII - HCl (0.04 N) Hydrolysis of Heparan (^{35}S)Sulphate Obtained after In Vitro Sulphation Using Brain Sulphotransferase of Different Ages.

Brain	Percentage (^{35}S)Sulphate liberated on hydrolysis	N-sulphate/ O-sulphate ratio
Human		
Premature baby	91	10.0
Neonatal brain	86	6.0
7 Year old child	76	3.0
13 Year old child	66	1.9
Rat		
1 day old	86	6.0
10 day old	65	1.9
Adult (2 month old)	41	0.7

of the glycosaminoglycan sulphotransferase in various subcellular fractions at various ages using dermatan sulphate as the sulphate acceptor (27). This is shown in Figures 3 and 4. This study shows that although the specific activity of the enzyme was very high in a number of fractions, the concentration of the sulphotransferase activity was highest in synaptosomes on the basis of the total activity followed by microsomes. Supernatant also had considerable enzyme activity. This is consistent with the observation which was shown in the earlier figure indicating that the sulphate incorporation was high in synaptosomal fraction. In Table VIII the results are presented to indicate that there are variations in enzymic N--sulphation and O-sulphation during the brain development (31). This study was carried out using heparan sulphate as the sulphate acceptor and protamine treated enzyme fraction from various age groups of human and rat brain tissue. One of the interesting observations was that N-sulphation was very high at early stages of brain development and the ratio of N-sulphation to O-sulphation continued to decrease till it reached adult stage, indicating significant differences in the sulphation of the amino group and the hydroxyl group, when the brain matures.

Recently we have been directing our attention to the various changes in glycosaminoglycans associated with the diseases such as

Table IX - Glycosaminoglycans in Urine.

Patient	Age (years)	Total uronic acid mg/24 hours	% of total		
			Dermatan sulphate	Heparan sulphate	Chondroitin sulphate
Normal	2 ½	2 to 5	2	5	80
San Filippo					
1	7	54.0	7	87	6
2	8	31.0	-	84	16
3	9	51.0	-	47	-
Hurler	4	20.2	80	20	-

Table X - Concentration of Glycosaminoglycans in Various DEAE -Sephadex Fractions from a Hurler Brain.

Specimen	mg uronic acid/gm LFDT				
	I-SA	I-SB	II-SB	II-SC	Total
Hurler (3 year old)	0.365	0.1	1.0	1.4	2.865
Normal (1 year old)	1.495	0.165	0.203	0.615	2.478

mucopolysaccharidoses as well as protein calorie malnutrition. Some of the mucopolysaccharidoses were identified by the analysis of urinary glycosaminoglycans. The nature and the level of the glycosaminoglycans excreted in urine in some of the mucopolysaccharidoses are shown in Table IX. It is clear from the data presented that heparan sulphate is the major glycosaminoglycan which is excreted in high amounts in San Filippo syndrome (32). In a case of urine from Hurler patient there was considerable increase in dermatan and heparan sulphate. We had the opportunity to examine one Hurler brain biopsy sample for the characterization of glycosaminoglycans (33). The concentration of glycosaminoglycans in the various fractions in this disease are shown in Table X, and it has been compared with a brain from a one year

Table XI - Glycosaminoglycans in Brain.

Glycosaminoglycans	San Filippo brain (1)			Normal (1)
	I 7 Yrs	II 8 Yrs	III 9 Yrs	Average of 10 & 11 years
Heparan sulphate	927 (57.3%)	870.0 (54%)	827.0 (50%)	155.0 (10.6%)
Hyaluronic acid	ND	242.0	ND	614.0 (Hyaluronic acid + Chondroitin-6-sulphate)
Chondroitin-6-sulphate	ND	87.0	ND	
Chondroitin-4-sulphate	ND	392.0	ND	628.0
Dermatan sulphate	205	0	ND	65.0
Total	1603	1611.0	1655.0	1462.0

(1) μg uronic acid/gm dry defatted tissue; ND = not done.

Table XII - Analysis of Glycosaminoglycans in Cerebrospinal Fluid.

Specimen	uronic acid μg/100 ml	Constituent analysis (expressed as molar ratio of uronic acid)		
		Gal-NH_2	Glu-NH_2	N-SO_4
San Filippo	400.0	0.1	0.9	0.6
Normal (1)	8-10			

(1) From Constantopoulos, G. and Dekaban, A. S., J. Neurochem. 17: 111 (1970).

old control child. Since the biopsy material available was quite small, a comparative study has only been made. It can be said that Fraction II-SC contains galactosamine and the carbazole method of analysis with and without borate suggested that it was rich in dermatan sulphate. It is of interest to note in this brain that although the concentration of total glycosaminoglycans did not change, there was a marked variation in the ratio of the concentration of sulphated to that of the non-sulphate glycosaminoglycans. There was a significant increase in the concentration of sulphated glycosaminoglycans in this disease, which is consistent with the earlier report of Meyer (34) and his associates who had also made similar observations. We also investigated samples from three different cases of San Filippo syndrome (35). The analysis of glycosaminoglycans from this brain is presented in Table XI and compared with control brain. It is clear from the data that heparan sulphate is the major glycosaminoglycan which is accumulated in brain in this disease with concomitant decrease in chondroitin sulphate as well as hyaluronic acid. As observed earlier in the case of Hurler brain the total amount of glycosaminoglycan was not changed markedly compared to that of the control. We had enough cerebrospinal fluid (CSF) (36) available from one patient with San Filippo syndrome and the analysis of glycosaminoglycans in the CSF is shown in Table XII. These data indicate that there was a 40 fold increase in the total concentration of glycosaminoglycans. The CSF analyses give support to the predominance of N-sulphated glucosamine containing glycosaminoglycan in San Filippo patient. Constantopoulos and Dekaban (37) reported the concentration of glycosaminoglycans in normal CSF as 8 to 10 $\mu g/100$ ml although in earlier work Friman (38) was unable to show the presence of glycosaminoglycan in normal CSF. The former workers found excessive accumulation of glycosaminoglycans in CSF of patients with Hurler and Hunter syndromes and the major glycosaminoglycans were dermatan sulphate and heparan sulphate. The intriguing questions of the sources of glycosaminoglycans and mechanism responsible for accumulation of glycosaminoglycans in CSF of San Filippo syndrome remain unanswered.

In our next series of experiments we have analyzed sulphotransferase activity with chondroitin-4-sulphate, dermatan sulphate and heparan sulphate as substrates in the San Filippo brain and compared this with the control brain (35). We had three biopsy samples for this purpose. The results are shown in Table XIII. In all these three cases the sulphotransferase activity towards chondroitin-4-sulphate was markedly decreased whereas the sulphotransferase activity towards heparan sulphate was same as control indicating that in this diseased condition there is a substantial change in the sulphotransferase activity in the brain. The analysis of the relative sulphate transfer to the amino group as well as the hydroxy group of heparan sulphate by sulphotransferase obtained from normal and San Filippo brain indicated that although the sulphotransferase activity was not increased, the N-sulphation was

Table XIII - Glycosaminoglycan Sulphotransferase in Brain.

Biopsy	Age (years)	Specific activity (units)		
		Chondroitin-4-sulphate	Dermatan sulphate	Heparan sulphate
San Filippo	8	0	25.0	254.0
Normal	10	8.3	30.0	292.0
San Filippo	7	0	12.0	260.0
Normal	11	8.0	20.0	393.0
San Filippo	9	1.72	18.4	219.8

1 Unit is 1,000 counts of (^{35}S)GAG formed/mg enzyme protein.

Table XIV - Sulphate Analysis of the Heparan Sulphate from San Filippo Brain and Urine.

Fractions	Percentage of total GAG	Expressed as the molar ratio of uronic acid	
		Total sulphate	N-sulphate
<u>Urine</u>			
Fraction I-SB	50	0.99	0.53
Fraction II-SB	21	1.0	0.35
<u>Brain</u>			
Fraction I-SB	6	0.97	0.9
Fraction II-SB	38	1.23	0.29

Standard chondroitin sulphate (Sigma) gave 15.6% release of sulphate in 0.04N HCl hydrolysis.

more predominant than the O-sulphation in the case of San Filippo brain. This observation that N-sulphation was preferred in San Filippo syndrome was further confirmed by analysis of various heparan sulphate fractions from an autopsy San Filippo brain. As shown in the Table XIV one of the fractions which constitute 15% of the total heparan sulphate contained high amount of sulphamino group compared to O-sulphate group. Urinary fraction also gave two types of heparan sulphate - one had a comparatively higher sulphamino group but not as high as the fraction obtained from the brain (unpublished data). Recently a number of reports have become available regarding the lack of β-galactosidase in some of the mucopolysaccharidoses including the San Filippo syndrome (39, 40). This had led to the suggestion that β-galactosidase deficiency might be responsible for the excessive accumulation of various glycosaminoglycan- protein complexes in these conditions. Since the protein glycosaminoglycan linkage region is the same for all glycosaminoglycans except hyaluronic acid and keratan sulphate both chondroitin sulphate and heparan sulphate may be expected to accumulate in San Filippo syndrome. On the other hand, only heparan sulphate is found to be accumulated. Thus, the β-galactosidase deficiency alone cannot account for the preferential accumulation of heparan sulphate in San Filippo syndrome. There must exist some other mechanism by which this glycosaminoglycan is accumulated at the expense of other glycosaminoglycans. The following observations namely the complete absence or very low activity of chondroitin-4-sulphate sulphotransferase, normal values for sulphotransferase activity towards heparan sulphate and the increased sulphotransferase activity to transfer sulphate to the amino group as compared to the hydroxy group of heparan sulphate are of interest. Perhaps the availability of specific glycosaminoglycan sulphotransferases may determine which glycosaminoglycan may be preferentially accumulated.

In view of the intimate association of protein in the biosynthesis of glycosaminoglycans (41) we have recently diverted our attention towards the nature and level of glycosaminoglycans in nutritional deficiency of early childhood (kwashiorkor) (42). We have so far analysed four autopsy samples from various age groups of 1, 3, 4 and 9 years old kwashiorkor children. Table XV shows the concentration of various glycosaminoglycans in kwashiorkor brain as compared to control. It is clear from the data that there is a marked decrease in the concentration of glycosaminoglycans in this condition. One of the most remarkable observations was the low concentration of total chondroitin sulphate and the considerable decrease in total glycosaminoglycans in the case of the nine-year-old child who had a history of repeated kwashiorkor. On a percentage basis, hyaluronic acid fraction was considerably increased whereas the decrease in the chondroitin sulphates was very marked. In the case of the 4-year-old brain the protein calorie malnutrition

Table XV - Concentration of Various Glycosaminoglycans in Normal and Kwashiorkor Brain.

Glycosaminoglycans	Concentration of various glycosaminoglycans in brain as the percentage of total					
	Kwashiorkor				Normal	
	1 year	3 years	4 years	9 years	New born	10 years
Hyaluronic acid	73.5	74	27.5	78.5	60	42
Heparan sulphate	8.8	12.5	10.2	5.2	4.4	10.6
Hyaluronidase resistant galactosamine-GAG	2.3	2	1.7	-	-	4.4
Low sulphated chondroitin sulphate	5.0	2.5	31.2	9.1	10	-
Chondroitin sulphate	9.7	9	28.7	6.2	24.8	43

was noticed only at a later stage of the disease and as such the results seem to be at variance with the other three brains. Nevertheless total concentration of glycosaminoglycans is decreased very markedly in all the cases. The implication of this finding is yet to be ascertained. However, it may be mentioned here that we have also analysed the glycolipids in this fraction and in the case of one year old and four year old children the decrease in glycolipid fraction was not marked compared to the control of the same age group indicating that the decrease in glycosaminoglycan concentration in protein calorie malnutrition is one of the early symptoms compared to the decrease in myelin lipids (43).

In recent reports a number of observations on the physiological function of glycosaminoglycans in brain have been made. Elam et al. (21) observed that heparan sulphate and chondroitin sulphate are important constituents in the rapid axonal flow and may be important in carrying the nutrients. Marked changes in the brain activity in cat by repeated hyaluronidase injection in the brain have been observed by Custod and Young (44). They had also shown that

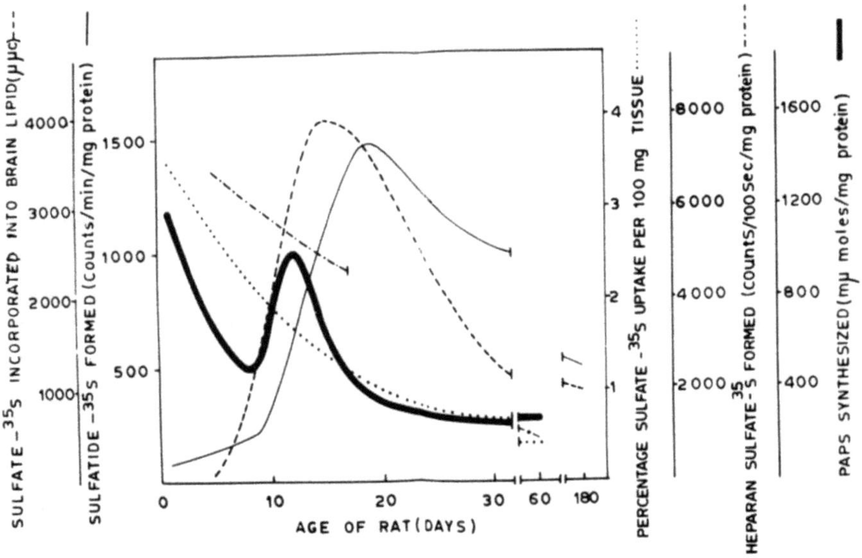

Fig. 5. Synthesis of PAPS, sulphated glycosaminoglycans and sulphatide in developing rat brain. (▬▬) Enzymatic synthesis of PAPS. (............) Incorporation of (35S)Sulphate into glycosaminoglycans. (-·-·-·-) Enzymatic transfer of (35S)Sulphate from (35S)PAPS to heparan sulphate. (-----) Incorporation of (35S) Sulphate into sulphatide. (———) Enzymatic synthesis of (35S) sulphatide from (35S)PAPS.

primarily there is a decrease in the chondroitin sulphates and hyaluronic acid fraction in these cases, indicating again that the reduction in the glycosaminoglycans in brain leads to a considerable change in the brain function. Adey (45) and his associates have shown that electrical impedence was markedly affected by hyaluronidase treatment. The present observation of an increased concentration of heparan sulphate in synaptosomal fraction and an increased heparan sulphate concentration in brain in some of the mucopolysaccharidoses which are associated with mental dysfunction indicate that heparan sulphate may have an important role in the neuronal activity. Whether the chondroitin sulphates and hyaluronic acid have different roles to play in relation to brain function is as yet to be ascertained. But it may be mentioned here that our preliminary work indicates that in experimental allergic encephalomyelitis (46) there is a marked decrease in the total glycosaminoglycans. The increased sulphate incorporation into glycosaminoglycans at the acute stage of experimental allergic encephalomyelitis (EAE) indicates that there is a rapid turnover of glycosaminoglycans especially sulphated glycosaminoglycans in this disease. The various aspects of the synthesis and concentration of glycosaminoglycans in comparison to developmental pattern are summarized in Fig. 5 (47). This figure indicates that sulphate incorporation into glycosaminoglycans as well as sulphate activating enzyme and the glycosaminoglycan sulphotransferase activity are most active prior to myelination whereas sulphate incorporation into sulphatides and enzymic sulphation of cerebroside sulphate are most active during myelination. This study thus indicates that at premyelination periods, sulphation process of glycosaminoglycan is rapid. What role, if any, these glycosaminoglycans have in relation to the process of myelination is yet to be ascertained. However, it may be of interest to note that recently it has been reported that heparin has an important role in the control of differentiation in sea urchin (48). Considering this aspect of the role of glycosaminoglycans in the process of differentiation and the observation that prior to myelination there is a rapid increase in the concentration of glycosaminoglycans one may speculate that these acidic macromolecules may have an important role in the regulation of myelinogenesis by controlling the differentiation of oligodendroglial cells.

ACKNOWLEDGEMENTS

The work was supported by grants from the National Multiple Sclerosis Society (U. S. A.), National Institutes of Health (U. S. A.) and Council of Scientific and Industrial Research (India).

REFERENCES

(1) Meyer, K., Grumbach, M.M., Linker, A. and Hoffman, P., Proc. Soc. Exptl. Biol. Med. 97: 275 (1958).

(2) Brante, G., in Metabolism of Central Nervous System, Ed. D. Richter, Pergamon Press, New York, p. 112 (1957).

(3) Ringertz, N.R., Exp. Cell Res. 10: 230 (1956).

(4) Szabò, M.M. and Roboz-Einstein, D., Arch. Biochem. Biophys. 98: 406 (1962).

(5) Clausen, J. and Hansen, A., J. Neurochem. 10: 165 (1963).

(6) Margolis, R.U., Biochim. Biophys. Acta 141: 91 (1967).

(7) Singh, M. and Bachhawat, B.K., J. Neurochem. 15: 249 (1968).

(8) Schiller, S., Slover, G. A. and Dorfman, A., J. Biol. Chem. 236: 983 (1961).
(9) Schmidt, M., Biochim. Biophys. Acta 63: 346 (1962).
(10) Mathews, M. B. and Inouye, M., Biochim. Biophys. Acta 53: 509 (1961).
(11) Schmidt, M. and Dmochowski, A., Biochim. Biophys. Acta 83: 137 (1964).
(12) Ludowieg, J. J. and Benmaman, J. B., Carbohyd. Res. 8: 185 (1968).
(13) Gatt, S. and Berman, E. R., Analyt. Biochem. 15: 167 (1966).
(14) Dische, Z., J. Biol. Chem. 167: 189 (1947).
(15) Bitter, T. and Muir, H., Analyt. Biochem. 4: 330 (1962).
(16) Dodgson, K. S. and Price, R. G., Biochem. J. 84: 106 (1962).
(17) Lagunoff, D. and Warren, G., Arch. Biochem. 99: 396 (1962).
(18) Onodera, K., Hirano, S., Hiriuchi, F. and Nashimura, N., Carbohyd. Res. 3: 234 (1966).
(19) Singh, M., Chandrasekaran, E. V., Cherian, R. and Bachhawat, B. K., J. Neurochem. 16: 1157 (1969).
(20) Cunningham, W. L. and Goldberg, J. M., Biochem. J. 110: 35 (1968).
(21) Elam, J. S., Goldberg, J. M., Radin, N. S. and Agranoff, B. W., Science 170: 458 (1970).
(22) Dorfman, A. and Pei Le Ho, Proc. Nat. Acad. Sci. 66: 495 (1970).
(23) Saigo, K. and Egami, F., J. Neurochem. 17: 633 (1970).
(24) Margolis, R. E. and Margolis, R. U., Biochemistry 9: 4389 (1970).
(25) Chandrasekaran, E. V. and Bachhawat, B. K., J. Neurochem. 16: 1529 (1969).
(26) Singh, M. and Bachhawat, B. K., J. Neurochem. 12: 519 (1965).
(27) Saxena, S., George, E., Kokrady, S. and Bachhawat, B. K., Ind. J. Biochem. Biophys. 8: 1 (1971).
(28) Vos, J., Kuriyama, K. and Roberts, E., Brain Res. 12: 172 (1969).
(29) Guha, A., Northover, B. J. and Bachhawat, B. K., J. Sci. Ind. Res. 19C: 289 (1960).
(30) Balasubramanian, A. S. and Bachhawat, B. K., J. Neurochem. 11: 877 (1964).
(31) George, E., Singh, M. and Bachhawat, B. K., J. Neurochem. 17: 189 (1970).
(32) Abraham, J., Shetty, G., Bhakthaviziam, A., Chakrapani, B., Kokrady, S. and Bachhawat, B. K., Ind. J. Med. Res. 58: 1073 (1970).

(33) Abraham, J., Singh, M., Chakrapani, B., Kokrady, S. and Bachhawat, B.K., Ind. J. Med. Res. 57: 1761 (1969).

(34) Meyer, K., Hoffman, P., Linker, A., Melvin, M. and Sampson, P., Proc. Soc. Exp. Biol. & Med. 102: 587 (1959).

(35) George, E. and Bachhawat, B.K., Clin. Chim. Acta 30: 317 (1970).

(36) George, E. and Bachhawat, B.K., Neurology (India), in press.

(37) Constantopoulos, G. and Dekaben, A.S., J. Neurochem. 17: 117 (1970).

(38) Friman, Clin. Chim. Acta 5: 378 (1967).

(39) Van Hoff, F. and Hers, H.G., Europ. J. Biochem. 7: 34 (1968).

(40) Öckerman, P.A., Hultberg, B. and Erikson, O., Clin. Chim. Acta 25: 97 (1969).

(41) Telser, A., Robinson, H.C. and Dorfman, A., J. Biol. Chem. 239: 3623 (1964).

(42) Chandrasekaran, E.V., Mukherji, K.L. and Bachhawat, B.K., J. Neurochem. (1971), in press.

(43) Kokrady, S., Balasubramanian, K.A., Gopalakrishnamurthy and Bachhawat, B.K., submitted for publication.

(44) Custod, J.T. and Young, I.J., J. Neurochem. 15: 809 (1968).

(45) Wary, H.H. and Adey, W.R., Exptl. Neurol. 25: 70 (1969).

(46) Vasan, N.S., Abraham, J. and Bachhawat, B.K., J. Neurochem. 18: 59 (1971).

(47) Balasubramanian, A.S. and Bachhawat, B.K., Brain Res. 20: 341 (1970).

(48) Kinoshita, S., Exp. Cell Res. 64: 403 (1971).

BIOSYNTHESIS OF BRAIN GLYCOPROTEINS

P. LOUISOT

University of Lyon, UER Medicine Lyon-Sud

69 - Oullins, France

INTRODUCTION

The biosynthesis of brain glycoproteins is presently of great interest from a double standpoint. First of all, these macromolecules are numerous and, secondly, there is a great variety of membranous structures which are normally transglycosylation sites in cerebral cells.

According to Brunngraber (1), one gram of cerebral tissue contains 1.2 mg of protein-bound carbohydrates and, approximately, 100 mg of protein. Since the most well-known glycoproteins extracted from various tissues contain between 15 per cent to 30 per cent carbohydrates, it is easy to calculate that 5 to 12 per cent of all cerebral proteins are glycoproteins.

Most of the experimental results, obtained using a wide variety of cellular fractionating techniques in sucrose gradient, led us to believe that synaptosomes, axons and a great number of poorly identified cell membrane fragments contain glycoproteins as structural elements. Yet the highest glycoprotein content has been found in the membranous material associated with the subcellular fraction which is rich in synaptosomes. High amounts of these substances have been also found in the non-purified microsome fraction. These particulate elements result from the degradation of many organized structures, such as synaptosomes, dendrites, axons and plasma membranes.

All the carbohydrate segments present in these cerebral glycoproteins contain N-acetyl-neuraminic acid and glucosamine. There is an interesting distinction between these glycoproteins and cerebral gangliosides (which also contain N-acetyl-neuraminic acid) where the hexosamine is galactosamine.

Since 1965, the predominant role of endoplasmic membranes in the mechanism of the fixation of the sugar to glycoproteins has been confirmed. Undoubtedly, the liver is the most studied organ. First of all, we will briefly summarize what is known about this organ, in order to facilitate the understanding of the mechanisms occurring in brain.

I. GENERAL MECHANISMS OF GLYCOPROTEINS BIOSYNTHESIS

The basic ideas were formulated in 1964, when Sarcione et al. (2) and then Helgeland (3) demonstrated, with the aid of ^{14}C-glucosamine injected in vivo, that the principal incorporation site of the radioactive carbohydrate precursor is located in the soluble deoxycholate subcellular fraction (Fig. 1). These concise initial data had already attracted attention towards the role played by endoplasmic membranes, which had been dissolved by deoxycholate. Further information were given, a year later, by Molnar, Robinson and Winzler (4) who used a subcellular fraction which was more elaborate and better controlled. In addition, the same authors performed a kinetic study of glucosamine incorporation. Their results (Fig. 2) showed that:

- ribosomes, "rough membranes" and "smooth membranes" are hexosamines incorporation sites
- collectively, membranes are quantitatively the largest incorporation site.

At this point the idea arose of the existence of many subcellular sites for the incorporation of carbohydrate units into glycoproteins. This was a function of the place occupied by these carbohydrate units in the ramified chains attached to the polypeptide. In 1966, Lawford and Schachter (5), and later (1967) Molnar and Sy (6) demonstrated that glucosamine was incorporated by enzymatic systems present in all endoplasmic membranes, while mannosamine, constitutive of sialic acid in terminal position on the chains, was mainly incorporated only by enzymes located in the "smooth membranes".

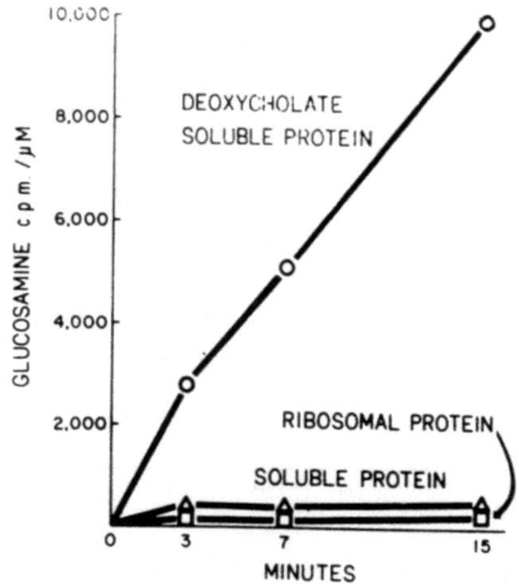

Fig. 1. Incorporation of ^{14}C-glucosamine into perchloric acid-insoluble protein fractions of subcellular components of sequential lobes of rat liver after addition of a tracer dose of D-^{14}C-glucosamine (50 μC) to blood perfusing the isolated rat liver _in vitro_. ^{14}C-glucosamine was added at zero time. The microsomes were fractionated with sodium deoxycholate.
Data from E. J. Sarcione, M. Bohne and M. Leahy (2).

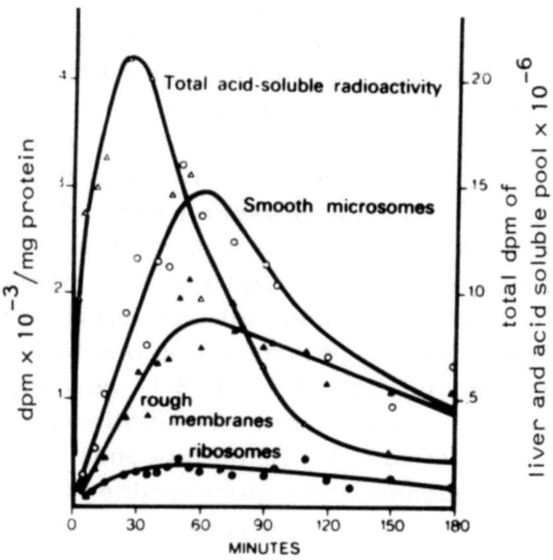

Fig. 2. Incorporation of 1-^{14}C-D-glucosamine in endoplasmic membranes of rat liver cells.
Data from J. Molnar, G. B. Robinson and R. J. Winzler (4).

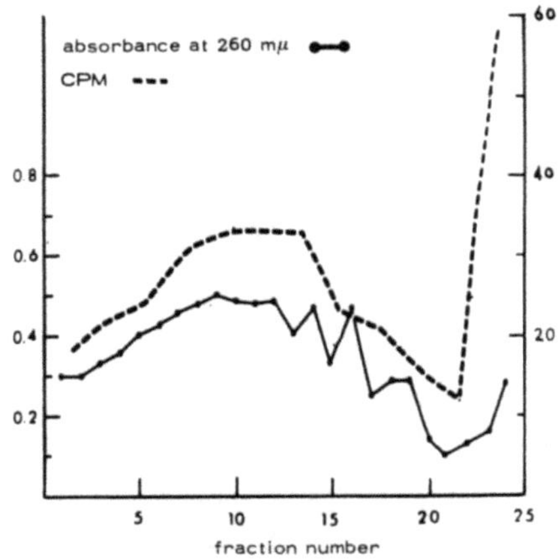

Fig. 3. Zone centrifugation at 25,000 rpm for 2 hours of polyribosomes prepared from the liver of a rat given 25 μC of uniformly labelled ^{14}C-glucosamine in the femoral vein 30 min. prior to killing.
Data from G.R. Lawford and H. Schachter (5).

The possible role played by ribosomes in glycoprotein biosynthesis is more difficult to determine.

It is known that ^{14}C-glucosamine can be detached from ribosomes by using puromycin, when the cells are treated with the antibiotic after a given time of biosynthesis. This proves that at least one fraction of glucosamine is put into place while the polypeptide is still fixed to the polysome. In addition, when labelled glucosamine incorporating polysomes is sedimented in a density gradient, one notes a certain quantity of radioactivity in these polysomes (Fig. 3). Obviously, this type of experiment had often been criticized. The most banal criticism considered that ribosomes and polysomes were contaminated by residual membrane fragments during cellular fractionation. Other cellular systems favour the idea that this phenomenon exists. However, no one has ever isolated the specific glucosaminyl-aminoacyl-t-RNA responsible for the fixation of the hexosamine-amino acid ramification complex onto the polypeptide. It is believed that, if the glucosamine fixation really takes place in the ribosome, this glucosamine is the "initial glucide" responsible for the glucide-polypeptide association.

The work performed in our laboratory on monolayer cell cultures and on cell suspensions has given the same results (7-15). Yet, we were able to provide evidence for the occurrence of three

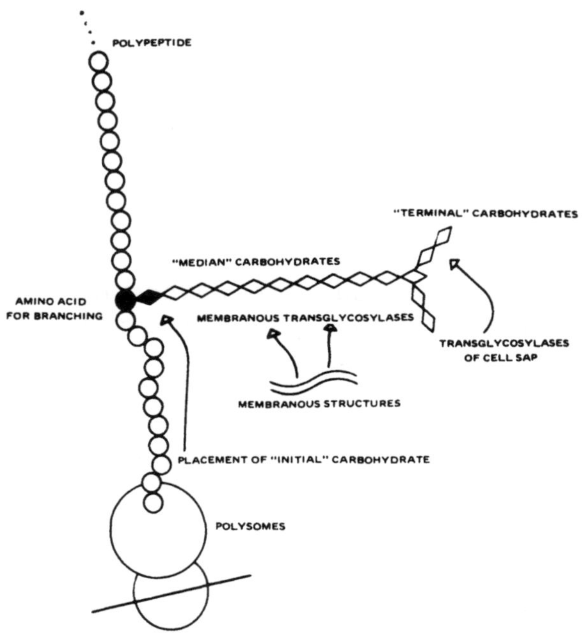

Fig. 4. Schematic representation for the "three sites of transglycosylation" hypothesis in subcellular structures.

distinct subcellular sites of incorporation, which led us to broadly interpret the biosynthesis of glycoprotein glucidic groups in the following manner (Fig. 4.):

- an <u>internal site</u>, more-than-likely represented by polysomes, insures the placement of the <u>initial hexosamine</u>. This hexosamine is responsible for the attachment of the carbohydrate chain onto the polypeptidic sequence;
- an extensive <u>median site</u>, located in the <u>endoplasmic membranes</u>, allows the placement of the carbohydrates which constitute the <u>median</u> part of the chains;
- a <u>peripheral site</u>, probably located in the soluble <u>cytoplasmic phase</u> or in "smooth membranes", is responsible for the placement of the terminal carbohydrates of the chains for example sialic acid.

Whatever subcellular site is responsible for the placement of a given sugar, the involved enzymatic system is always a glycosyltransferase or transglycosylase. These systems have a certain number of similar basic properties for practically all the cellular systems studied. In details (16-18):

- They are particulate systems, closely associated with subcellular structures. Their proper functioning depends on the integrity of these structures.
- The sugar, or its incorporated derivative, can only be taken from its active coenzymatic form: GDP-mannose, UDP-galactose, UDP-N-acetyl-glucosamine, UDP-N-acetyl-galactosamine, etc. These coenzymatic forms have sufficient molecular activation to permit transglycosylation without need of additional energy.
- The affinity of the individual transglycosylases for their respective substrates is very large, as shown by the small Michaelis constants: 0.2 μM as in the case of splenic-cell mannosyl-transferase.
- The quasi-totality of the glycosyl-transferases function only in the presence of sufficient bivalent cation concentrations, particularly manganese.
- Glycosyl-transferases are generally inhibited by nucleoside-diphosphates (UDP, CDP, etc), while are activated by nucleoside-triphosphates (ATP, GTP, etc.) and nucleotides. These last two classes do not act as energy suppliers, since their non-metabolizable structural analogues (for example β-γ-methylene-GTP or β-γ-methylene ATP) produce the same effect. These nucleoside-triphosphates act on the glycosyl-transferases by their conformational character.
- Many transglycosylases in *in vitro* acellular systems are activated by detergents. Triton X-100, for instance, has a two-fold activating action. In low concentrations, it modifies the enzyme-substrate ratio by apparently allowing a better access of the enzyme to an exogenous polypeptidic acceptor. On the contrary, in high concentrations, it strongly modifies the architectural environment of the enzyme in the structure which constitutes the endoplasmic membranes, the transglycosylation activity being completely inhibited.

Moreover, glycosyl-transferase function is now well-known in its purely enzymatic aspects. But the problem of the sequential organization of the carbohydrate units in the chains is still to be solved. In the glycoproteins of all the tissues there is a double sequential specificity: that of the polypeptide and that of the carbohydrate units.(Fig. 5). The former, as a function of the genetic code, is based on the specific codon-anticodon pairing responsible for the attachment of the amino acid specific t-RNA. The sugar sequence is more difficult to be coded. First of all it is important to know why the carbohydrate chain is linked to a specific threonyl, seryl or

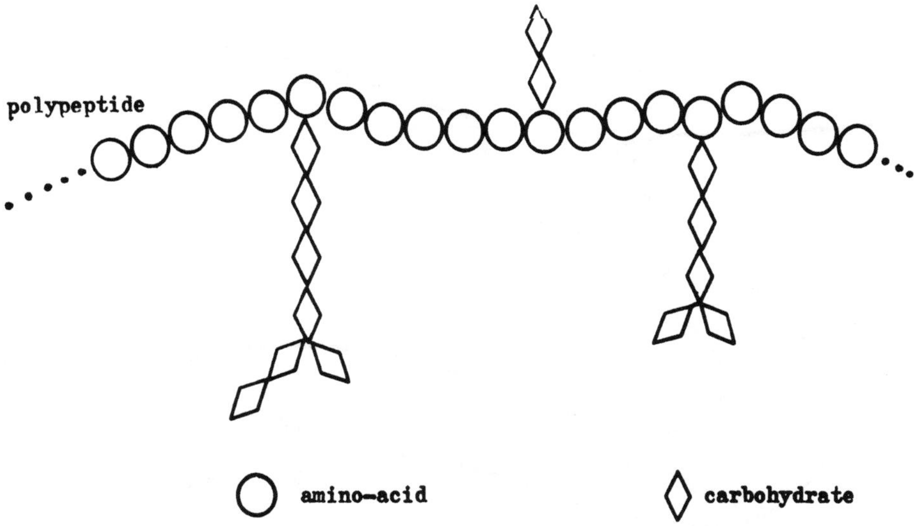

Fig. 5. Two sequential organizations (amino acids and carbohydrates) in glycoprotein structure.

asparaginyl residue of the polypeptide, and not to a neighbouring threonyl, seryl or asparaginyl residue. A t-RNA coding mechanism is to be excluded.

Very often the carbohydrate chain is attached to asparagine by a N-glycosydic linkage. We showed that asparagine was integrated into the following types of sequence:

asparagine - X - serine

or

asparagine - X - threonine

Where (X) is any amino acid. As a matter of fact all the four complete glycoprotein sequences presently known have the carbohydrate fragments N-glycosydic bound to the asparaginyl group of the above mentioned tripeptides. Thus the fixation site of the initial sugar on the polypeptides seems to depend on a well-defined local structural arrangement.

The second difficulty is to establish the conditions for the sequential addition of the individual carbohydrate units. From this point of view two concepts are debatable:

- The first concept is that the carbohydrate sequence is determined by the positional order of the enzymes in the endoplasmic

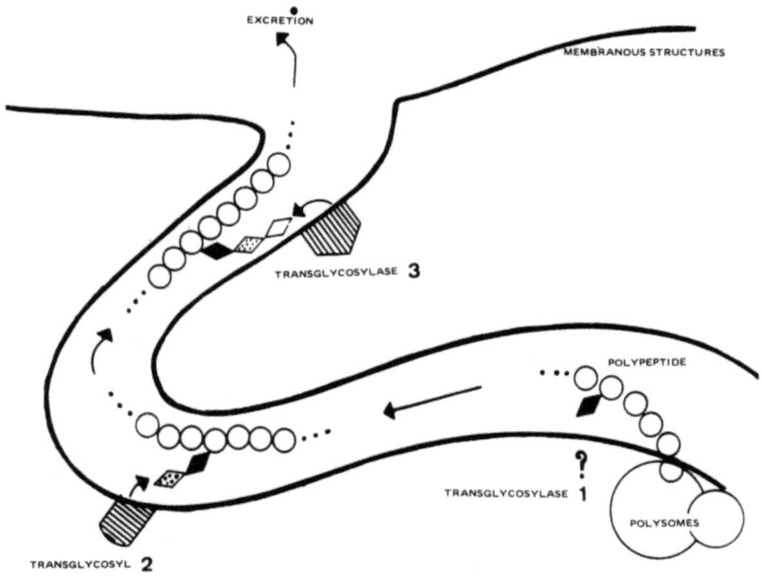

Fig. 6. Concept of Spiro (19) and Schachter (20) for transglycosylations.

reticular membranes (Fig. 6). The hypothesis was developed by Spiro (19) and taken up once again by Schachter (20).
The arguments supporting this hypothesis are not well-founded and lead us to doubt of the existence of this organization. Indeed, the architectural structure of the membranes should always be the same as far as the sequential arrangement of the enzymes is concerned, so that the cell always synthesizes the same glycoprotein types. This has never been demonstrated. The opposing argument could otherwise be tempered by the fact that the carbohydrate sequences present in the same glycoprotein could undergo modifications.

- The second concept, suggested by the author (Fig. 7) is that the primary structure of the polypeptide, genetically coded, plays the only basic role. The primary structure controls the tridimensional evolution of the protein into secondary and tertiary structures. As a result, not all of the possibly available sites for sugar fixation are accessible to transglycosylases. Moreover, the sequential addition of median sugars depends as well on the available sites for a given sugar in the locality of the polypeptide. It is no longer necessary to look for a coding mechanism since everything depends upon "available space" in the tertiary organization of the molecule. This is why I propose the term "theory of vital space" in order to

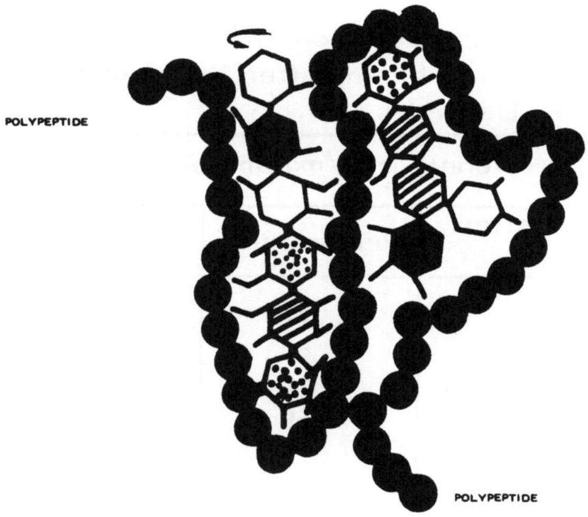

Fig. 7. Theory of vital space for transglycosylations. The sequential addition of carbohydrates depends on the available sites for a given sugar.

summarize this viewpoint. At the present time we don't have a direct argument in favour of this concept. However, it seems to me much more flexible than the preceeding one and is fundamentally based on the only known coding mechanism, that of amino acids.

II. BIOSYNTHESIS OF BRAIN GLYCOPROTEINS

The study on the biosynthesis of brain glycoproteins has widely profitted from previous acquisitions on other tissues. The first work was experimentally based on *in vivo* precursor incorporation, the pattern of which was followed at the subcellular level. In later investigations, *in vitro* acellular biosynthetic systems were used. I will now show the progress of the studies in this field.

A) *In vivo* Precursor Incorporation

The first results obtained by *in vivo* studies with radioactive sugar precursors were considerable. Brunngraber (1) concluded, after evaluation of the known enzyme activities that convert hexose, hexosamine and N-acetyl-neuraminic acid to the correspondent nucleoside-diphosphoglycosyl derivatives, that the metabolism of this nucleoside-diphosphoglycosyl derivatives proceeds at a rate suggesting that brain glycoprotein turnover is neither very rapid nor slow. The results of a few isotopic incorporation studies sup-

Table I - Kinetics of incorporation of ^{14}C-glucosamine into macromolecules of brain, after intracerebral injection. Data from Barondes (22).

Counts/min/mg protein		
Minutes after injection	Precipitate	Supernatant
0	39	91100
5	82	24500
15	226	13200
30	259	3300
60	575	2400
120	889	1100
240	1022	860

Table II - Incorporation of ^{14}C-glucosamine into macromolecules of subcellular fractions after intracerebral injection (3 h). Data from Barondes (22).

Fractions	Counts/min/mg protein
Soluble	330
Soluble of nerve endings	295
Microsomes	987
Mitochondria	242
Myelin	220
Pellet of nerve endings	547

Table III - Incorporation of ^{3}H-leucine into macromolecules of subcellular fractions of brain 3 hours after intracerebral injection. Data from Barondes (22).

Fractions	Counts/min/mg protein
Soluble	1560
Soluble of nerve endings	142
Microsomes	1441
Mitochondria	549
Myelin	540
Pellet of nerve endings	376

ported this view. Recently, D-glucosamine-1-^{14}C was injected intraperitoneally into rats (Holian, Dill and Brunngraber, 21), and brain glycoproteins were isolated. The time of maximal isotope incorporation was 4-8 hours. Thereafter, the amount of the incorporated isotope decreased at a rate consistent with a calculated half-life of approximately 6 days in the 15-day-old rats, and of 5 days in the 55-day-old rats. The in vivo experiments performed by Barondes (22) showed that glucosamine was rapidly and extensively incorporated into macromolecules which are associated with nerve endings. Table I shows the kinetics of ^{14}C-glucosamine incorporation into macromolecules of brain after intracerebral injection. Table II shows the incorporation of ^{14}C-glucosamine into macromolecules of various subcellular fractions 3 hours after intracerebral injection of the precursor, with a maximal specific radioactivity in microsomes and nerve endings (soluble + pellet). In contrast, there was only a slight incorporation of tritiated leucine into the soluble proteins from isolated nerve endings in the first few hours after administration, whereas the other subcellular fractions were maximally labelled at that time. Table III illustrates this phenomenon by the incorporation of tritiated leucine into macromolecules of subcellular fractions 3 hours after intracerebral injection. The data suggest that, unlike proteins which are largely transported to nerve endings in the axoplasm, there is an extensive incorporation of

Table IV - Effect of acetoxycycloheximide on incorporation of radioactive glucosamine into brain macromolecules. Data from Barondes and Dutton (23).

Counts/min/mg protein			
Fractions	Control	+ acetoxycyclo-heximide	% inhib.
Whole brain	1923	524	73 %
Soluble of whole brain	1279	376	71 %
Soluble of nerve endings	841	782	7 %
Particulate of nerve endings	1020	560	48 %
Microsomes	3220	641	80 %
Myelin	548	142	74 %
Mitochondria	1070	334	69 %

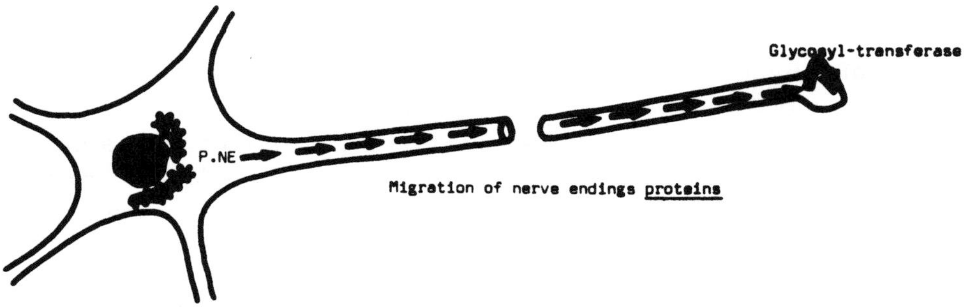

Fig. 8. Hypothesis for biosynthesis of nerve endings glycoproteins.

Whereas the protein component of a glycoprotein may be transported to the nerve ending from the perikaryon, the structure and function of this protein may be modified at the nerve ending by additional attachment of glucosamine, sialic acid and possibly other carbohydrates. Further studies of Barondes and Dutton (23) show (Table IV) the effect of acetoxycycloheximide on the *in vivo* incorporation of radioactive glucosamine into brain macromolecules. Acetoxycycloheximide inhibits radioactive carbohydrate incorporation into whole brain soluble glycoproteins but not in those of the soluble fraction of nerve endings. This suggests that the nerve ending protein into which the carbohydrates are incorporated has been synthesized before the glucosamine was incorporated. The studies suggest, as did the previous work, that some polypeptide chains are formed in the perikaryon, transported to the nerve endings and here modified by the addition of carbohydrates. This is shown in Fig. 8.

Although glucosamine is a useful precursor for the study of the biosynthesis of glycoproteins, it has several limitations when used *in vivo*. It may be metabolized by a number of pathways to other monosaccharides, including galactosamine, galactose, glucose or sialic acid, which may then be incorporated into gangliosides or other glycolipids. Thus, care must be taken particularly in brain tissue, to distinguish the synthesis of glycoproteins from the synthesis of glycolipids. In contrast to glucosamine, fucose is a precursor of far fewer products and is not readily metabolized into other sugars, so the radioactivity remains in its moiety throughout

Table V - Incorporation of ^3H-fucose into subcellular fractions of brain 3 hours after intracerebral injection. Data from Zatz and Barondes (24).

Fractions	Counts/min/ /mg protein
Whole homogenate	3519
Mitochondria	2490
Microsomes	7620
Soluble fraction of whole brain	1851
Soluble component of nerve ending fraction	1609
Pellet of nerve ending fraction	3794
Myelin	2820

Table VI - Relative specific activities of subcellular fractions of brain 3 hours after intracerebral injection of ^{14}C-glucosamine and ^3H-fucose. Data from Zatz and Barondes (24).

Fractions	Glucosamine	Fucose
Whole homogenate	1.00	1.00
Microsomes	1.77	1.88
Soluble fraction of whole brain	0.73	0.38
Soluble component of nerve ending	0.64	0.27
Pellet of nerve ending fraction	0.91	1.07

The table gives relative specific activities in the subcellular fractions, taking the counts/min/mg protein in whole homogenate as 1.00.

the experiment. Furthermore fucose is not a significant constituent of glycolipids and is absent in mucopolysaccharides. An additional advantage is that fucose occupies a terminal position on the carbohydrate chains of glycoproteins in which it is found, so that it might be particularly useful for studying reversible terminal modifications of glycoprotein structure. Accordingly, Zatz and Barondes (24) have investigated in vivo incorporation of radioactive fucose into brain glycoproteins. The results concerning the incorporation of tritiated fucose into brain subcellular fractions 3 hours after intracerebral injection are shown in Table V. The maximal specific radioactivity was obtained in microsomes and in

Table VII - Inhibition of brain glycoprotein biosynthesis by acetoxycycloheximide. Data from Zatz and Barondes (24).

Fractions	Percentage inhibition	
	Glucosamine	Fucose
Whole homogenate	63	84
Microsomes	67	79
Soluble fraction of whole brain	61	87
Soluble component of nerve ending fraction	29	83
Pellet of nerve ending fraction	56	89

nerve ending pellets. The data exposed in Table VI show the relative specific activities of brain subcellular fractions after intracerebral injection of ^{14}C-glucosamine and ^{3}H-fucose. The biosynthesis of brain glycoprotein is inhibited by acetoxycycloheximide (Table VII). Since fucose has been found exclusively in a terminal position on glycoprotein chains, it might have been anticipated that pretreatment with acetoxycycloheximide would inhibit the incorporation of fucose into glycoprotein less than that of glucosamine. The opposite was found to be the case by Zatz and Barondes (24): pretreatment with acetoxycycloheximide inhibited the incorporation of fucose into glycoprotein much more than that of glucosamine. Another major difference between these two precursors was the effect of pretreatment with acetoxycycloheximide on their incorporation into the soluble glycoproteins of nerve endings. Barondes and Dutton (23) have observed that incorporation of radioactivity from glucosamine into soluble glycoproteins of nerve endings was not markedly inhibited by the inhibition of cerebral protein synthesis for several hours. This observation was considered as indicating that polypeptide chains, which were in transit in the axon when the acetoxycycloheximide was administered, continued to arrive at the nerve ending and to provide acceptors for the incorporation of carbohydrates even while protein synthesis was inhibited. Zatz and Barondes (24) expected a similar effect with fucose.

The incorporation of fucose into the soluble glycoproteins of nerve endings appeared to be markedly inhibited by pretreatment with acetoxycycloheximide. The reason for this difference was not clear. One possibility was that acetoxycycloheximide directly interfered with the incorporation of fucose into glycoproteins. Another possibility was that fucose-containing glycoproteins, unlike glucosamine-containing glycoproteins, were made up in the perikaryon, but were transported very rapidly to the nerve endings. Evidence for the rapid transport of some proteins to nerve endings or the rapid appearance of proteins at nerve endings has been presented by Barondes (25) and by Droz and Barondes (27). Another possibility was the local synthesis of the polypeptide acceptors for fucose at nerve endings, and the inhibition of their biosynthesis by acetoxycycloheximide. Evidence for local biosynthesis of proteins in nerve ending fractions has been reported by Austin and Morgan (28) and by Autilio et al. (29). The remarkable inhibition by acetoxycycloheximide on the incorporation of N-acetyl-neuraminic acid into brain glycoproteins, observed by De Vries and Barondes (30), and the fact that incorporation into the soluble glycoproteins of the nerve ending fraction was inhibited least of all, lead to the same consideration.

B) Acellular in vitro Systems with Specific Transglycosylases

In vivo radioactive precursor incorporation studies have produced a large body of results, of which several examples have just been given. Additional results have been recently accomplished by a new approach: the biosynthesis of glycoproteins by specific transglycosylases using acellular in vitro systems.

In this domain we used two main techniques:

- In some cases, we prepared from cerebral tissue a certain number of specific transglycosylases (galactosyl-transferase, mannosyl--transferase, N-acetyl-neuraminyl-transferase, etc.) able to fix the correspondent carbohydrate on a protein acceptor or on an exogenous glycoprotein. This acceptor has often no relationship with a cerebral protein or glycoprotein but has the advantage of bein available in a large quantity.
- In other cases, we prepared brain subcellular fractions, which we carefully controlled at the electron microscope, and which showed enzymatic activity. We studied their transglycosylation ability on endogenous acceptors, proteins or locally-produced glycoproteins. This plan of action is closer to reality, but often limited by the availability of the acceptor substrate in the acellular system.

Table VIII - Glycosyl-transferases in embryonic chicken brain.

1 - Sialyl-transferase		
CMP-NAN + Gal-glycoprotein	→	Sialyl-Gal-Glycoprotein + CMP
CMP-NAN + Gal-Glu (lactose)	→	Sialyl-(2-3)-Gal-Glu + CMP
2 - Galactosyl-transferase		
UDP-Gal + Glu(NH_2)-N-ac. -Glycoprotein	→	Gal-Glu(NH_2)-N-ac. -Glycoprotein + UDP
UDP-Gal + Glu(NH_2)-N-ac.	→	Gal(1-4)-Glu(NH_2)-N-ac. + UDP
3 - Galactosyl-transferase		
UDP-Gal + Gal(NH_2)-N-ac. - mucin	→	Gal-Gal(NH_2)-N-ac. -mucin + UDP
4 - N-acetyl-glucosaminyl-transferase		
UDP-Glu(NH_2)-N-ac. + Man-Glycoprotein	→	Glu(NH_2)-N-ac. -Man-Glycoprotein + UDP

Data from Den, Kaufman and Roseman (31).

We will give examples of the above mentioned experimental approaches by citing the work of Den, Kaufman and Roseman (31), and our own work with P. Broquet (32).

The work of Den, Kaufman and Roseman (31) on embryonic chicken brain attempted to investigate the properties and subcellular localization of the transglycosylases which transfer sialic acid, galactose and N-acetylglucosamine to glycoproteins. The following enzyme activities present in the synaptosome-rich fraction were detected (Table VIII): sialyl-transferase, galactosyl-transferase, N-acetyl-glucosaminyl-transferase (exogenous glycoprotein acceptors prepared from α_1-glycoprotein pretreated with sialidase, β-glucosidase or N-acetyl-glucosaminidase were used). These enzymes synthesize the terminal trisaccharide sialyl→galactosyl→N-acetyl-glucosamine of serum-type glycoproteins by stepwise addition of the respective sugar moieties. Three other galactosyl-transferase activities were detected in the same synaptosome-rich fraction. These enzymes catalyze the transfer of galactosyl groups to the terminal N-acetyl-galactosaminyl residue of sialidase-treated mucins, to Tay-Sachs gangliosides and to xylosylserine.

A distinguishing characteristic of all these particulate enzymes is the strict necessity of the detergent Triton X-100 for their _in vitro_ activity. At early stages of embryonic brain development, ga-

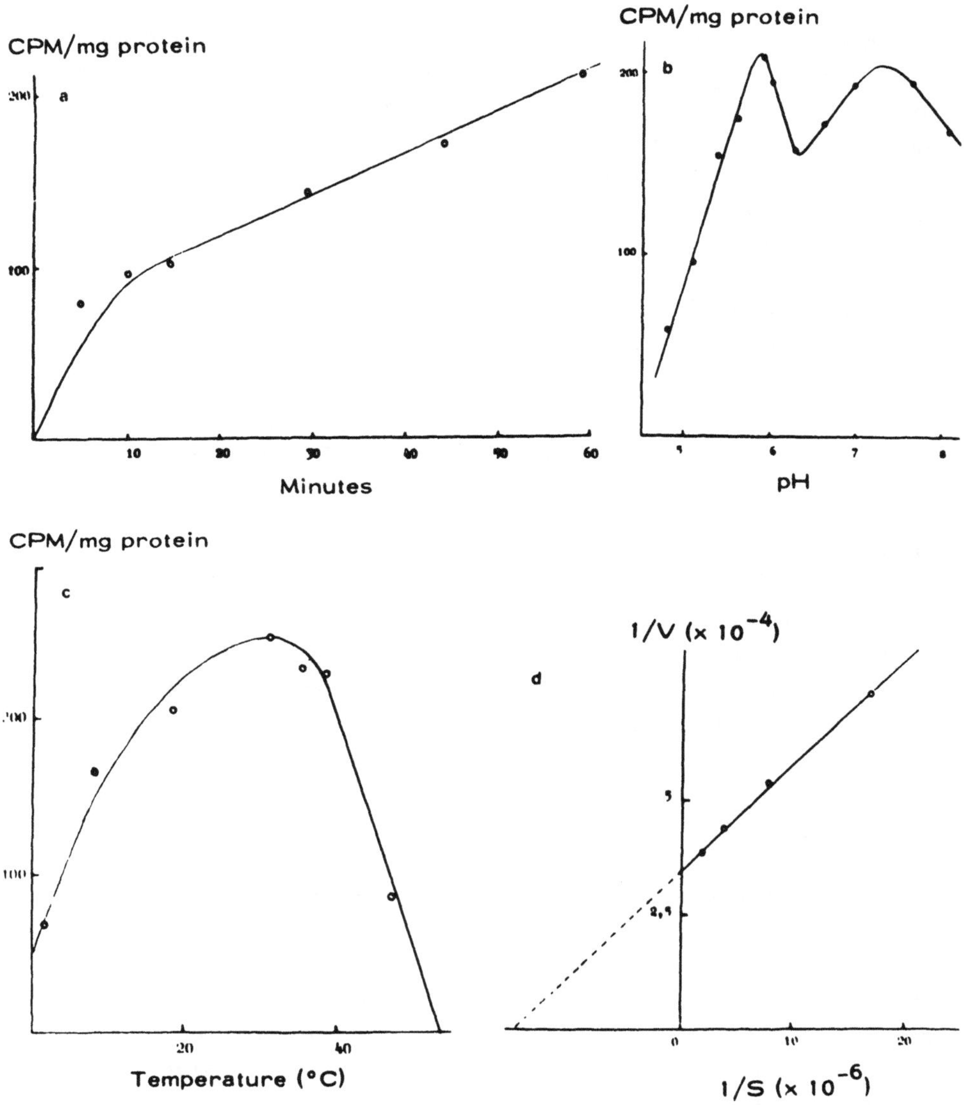

Fig. 9. Incorporation of ^{14}C-mannose, from GDP-^{14}C-mannose, to an endogenous proteinic acceptor, as a function of: a) incubation time; b) pH of the medium; c) temperature; d) GDP-mannose concentration (Lineweaver-Burk plot). All data from P. Broquet and P. Louisot (32).

lactosyl-transferase and N-acetyl-glucosaminyl-transferase activities on exogenous acceptors were found in soluble form in the fluid around the embryonic chicken brain.

We studied mannosyl-transferase activity on an endogenous acceptor in an in vitro acellular system. This activity, was located in the microsome-rich fraction and in the synaptosome fragments obtained from cerebral tissue. The mannose donor was GDP-mannose-^{14}C. Particulate mannosyl-transferase activity appears to be related to a complex enzymatic system. Fig. 9a shows the incorporation kinetics of ^{14}C-mannose as a function of incubation time. Fig. 9b shows the effect of pH of the reaction medium (between 4. 5 and 8) on the enzyme activity. A first peak of maximal activity at pH 5. 9 and a second at pH 7. 4 are present.

In Fig. 9c we see that the maximal in vitro incorporation was attained at 30°C, and that the enzymatic activity decreased quite rapidly when the supra-optimal incubation temperatures were reacted. The influence of GDP-mannose concentration was studied at pH 6 and represented according to the Lineweaver-Burk plot in Fig. 9d. The Michaelis constant was 0. 07 μM. The same study performed at pH 7. 2 resulted in the same Km value, but to a slightly higher maximal velocity. The Km determination before and after heat exposure for 60 minutes at supra-optimal temperatures of 35° or 45° gave the following results:

Km = 0. 04 μM after 60 minutes at 35°C

Km = 0. 15 μM after 60 minutes at 45°C.

The complexity of the system is even more apparent when one considers the thermal curve of the enzyme after a more or less prolonged exposure to supra-optimal temperatures (Fig. 10). At 35°, for example, the obtained curve seems to be the result of two opposing phenomena. The first is the classic thermal inactivation which increases with the time spent at supra-optimal temperatures. The second is the increased reaction rate due to the influence of higher temperature; the latter phenomenon can be interpreted in two ways: it could be due to the destruction of a specific inhibitor which is particularly sensitive to thermal denaturation, or it could be due to a progressively improved access of the particulate enzyme molecules to their substrate, as the consequence of the elimination of the inhibiting subcellular structures in the immediate vicinity of the enzyme.

The study of enzyme-substrate affinity constants after 60 minutes at 35°C or 45°C shows a decreased enzyme-substrate affinity as the temperature increases, while simultaneously the reaction rate increased. For this reason, we interpret the activation phenomenon due to prolonged exposure to supra-optimal temperatures as the result of the increased number of active sites which had access to the GDP-mannose substrate by environmental modifications, which also can well explain the initial thermal denaturation effect.

Fig. 11a shows how sensitive mannosyl-transferase is to the presence of manganese or magnesium ions. The maximal transglycosylation activity was obtained at 10^{-3}M manganese concentration. Under the influence of magnesium, the activity rises sharply as the concentration increases, representing a rare case of transglycosylation activity at an elevated ionic concentration.

The data exposed in Table IX show the inhibitory effect of nucleoside-diphosphates, guanosine-triphosphate and its structural analogue, β-γ-methylene-GTP. They all operate at a relatively high (millimolar) concentration. They are inhibitors of the non-competitive type, with the Km values being respectively 0.5 μM in the presence of GDP or of β-γ-methylene-GTP and 0.06 μM in the presence of GTP. The inhibition of GTP can be assumed to be due to the conformation of this nucleoside-triphosphate since its non-hydrolyzable structural analogue had the same effect.

Fig. 11b shows how the enzymatic activity is sensitive to Triton X-100. This detergent acts as an activator if the Triton/ /protein ratio is 0.15. It is known that this compound allows the solubilization of various microsomic transglycosylases of different origins while sparing their basic biological activity. On the other hand, spleen mannosyl-transferase, as observed by M. Richard in our laboratory, is inhibited by Triton X-100. In the case of cerebral mannosyl-transferase too, Triton X-100, at high concentrations, decreased the enzyme activity.

Thus at moderate concentrations of Triton X-100, the enzyme is more efficient because of the easier access to the substrate, while, at higher detergent concentrations, it is considerably altered and less active.

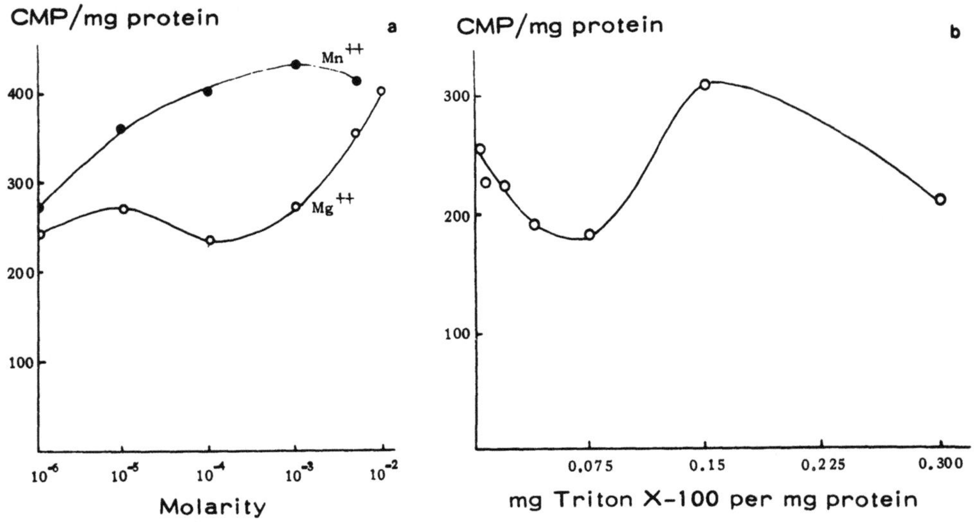

Fig. 10. Incorporation of ^{14}C-mannose at optimal temperature after enzyme exposure to supra-optimal temperatures. Data from P. Broquet and P. Louisot (32).

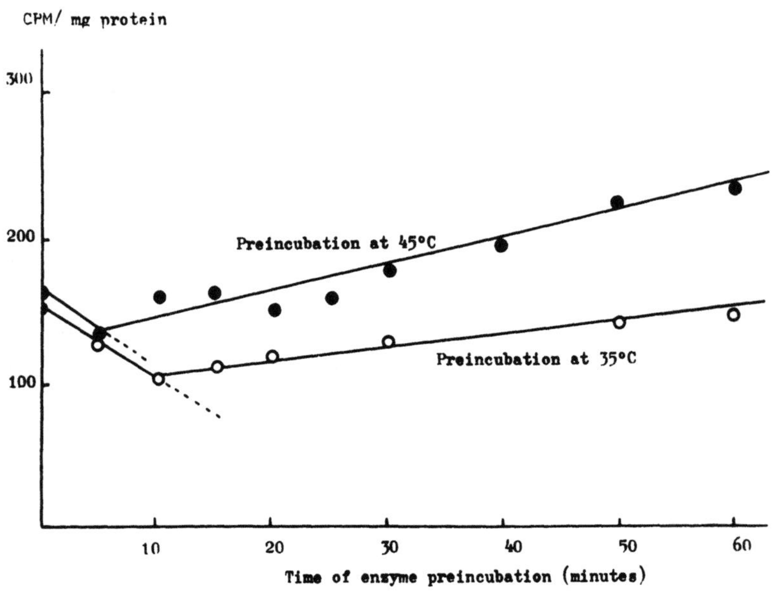

Fig. 11. Incorporation of ^{14}C-mannose as a function of:
a) concentration of Mn^{++} and Mg^{++};
b) concentration of Triton X-100 in the medium.
Data from P. Broquet and P. Louisot (32).

Table IX - Effect of nucleotides, nucleosides di- and tri-phosphates on the mannosyl-transferase activity. Tris--maleate buffer, 0.05 M, $MnCl_2$ 0.0001 M.

Effector / Concentration	AMP	3', 5'-cyclic AMP	ADP	GDP	ATP	CTP	UTP	GTP	β-γ-CH_2 ATP	β-γ-CH_2 GTP
10^{-3} M	-4%	0%	-72%	-82%	0%	+16%	+22%	-90%	-14%	-74%
10^{-4} M	-4%	0%	-14%	- 4%	0%	+12%	+ 2%	- 4%	-12%	-12%
10^{-5} M	+2%	0%	-14%	+ 8%	0%	+12%	- 8%	- 2%	-10%	-10%
10^{-6} M	-4%	0%	- 8%	0%	0%	0%	- 8%	0%	- 2%	- 2%

Data from P. Broquet and P. Louisot (32).

Table X - Effect of cycloheximide and puromycin on the cerebral mannosyl-transferase activity.

Concentrations / Antibiotics	10^{-3} M	10^{-4} M	10^{-5} M	10^{-6} M
Cycloheximide	0%	0%	0%	0%
Puromycin	-30%	-20%	-20%	-15%

Data from P. Broquet and P. Louisot (32).

Table XI - Distribution of three transglycosylase activities into brain subcellular fractions (Whittaker technique). Activities in CPM/mg of protein, after 60 minutes incubation at 30°C.

Fractions	Mannosyl-transf.	N-acetyl-glucosaminyl-transf.	Galactosyl-transferase
Whole homogenate	550	350	185
Post-nuclear supern.	840	310	320
Pellet mitochondria--synaptosomes	740	500	280
Pellet mitochondria--synaptosomes after osmotic lysis	700	625	300
Synaptosomal membran.	400	208	320
Synaptosomal membran. and non vesicular membranes	660	360	870
Synaptic vesicles	3100	2800	1400
Post-mitochondrial supernatant	940	476	320

Data from P. Broquet and P. Louisot (32).

Sodium deoxycholate completely inhibits the system at final concentrations between 0. 1 to 1 per cent.

The enzymatic system is sensitive to puromycin and not to cycloheximide (Table X). This is important because it is the direct proof that transglycosylase has maximal activity while the polypeptide is still fixed to the polysomes. When the puromycin detaches, the transglycosylase efficiency decreases significantly.

Particulate mannosyl-transferase activity of cerebral cells is the result of a complex enzymatic system, sedimenting at 100. 000 x g, possessing two maximal activities at two different pHs and

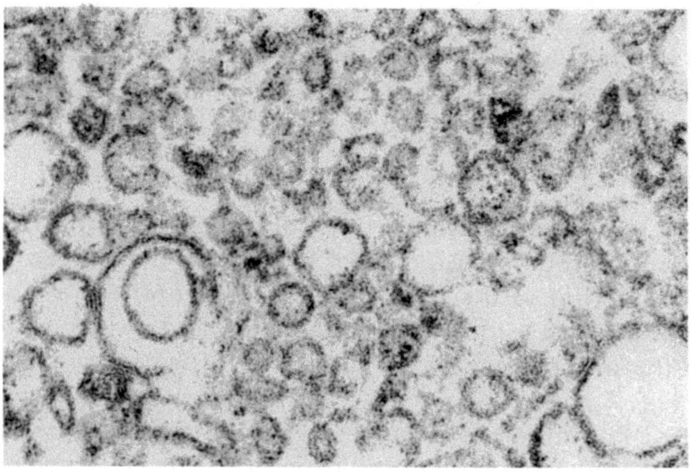

Fig. 12. Synaptic vesicles (fractions D and E from Whittaker) (X 100,000).

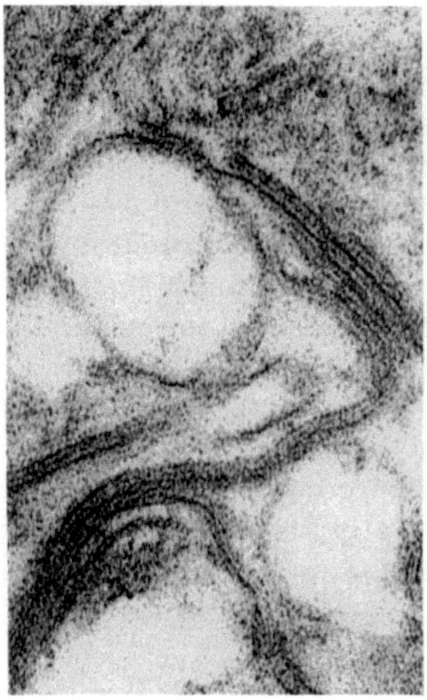

Fig. 13. Incompletely lysed synaptosomes and fragments of smooth membranes (fraction H from Whittaker) (X 100,000).

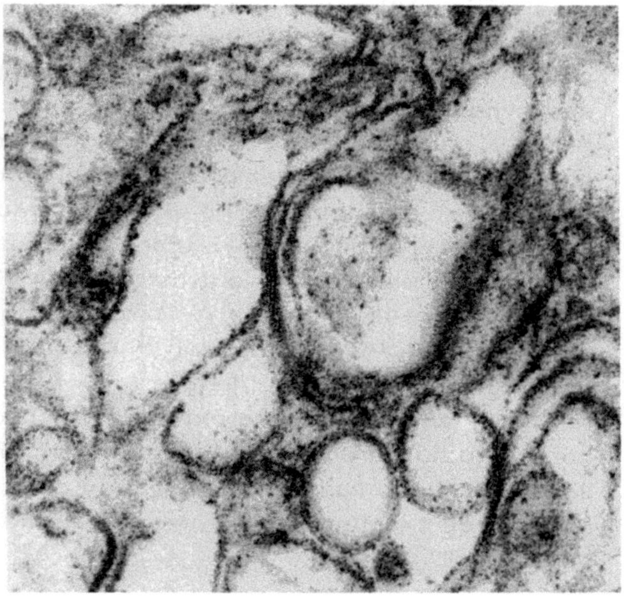

Fig. 14. Synaptosomal membranes (fraction G from Whittaker) (X 100,000).

All figures from P. Broquet and P. Louisot (32).

very sensitive to the architectural environment of the subcellular structures which support it, as shown by the effects of temperature and detergents in vitro. It is also sensitive to various effectors such as the cations Mg^{++} and Mn^{++}, and the nucleoside di- and tri-phosphates.

Subcellular fractionation in discontinuous gradient, performed (32) with a technique adapted from Whittaker confirmed the discontinuous repartition of the activity, with a preferential location of the enzymatic system in the synaptosomes and, even more specially, in the membranes of synaptic vesicles.

Table XI contains the results concerning the subcellular distribution of three transglycosylation activities. It shows clear-cut transglycosylase activity (concerning mannose, galactose and N--acetyl-glucosamine) in the synaptic vesicles. The electron microscope studies, carried out on the three most significant subcellular fractions obtained, showed that the fraction called "synaptic vesicles" (Fig. 12) actually contains a preponderance of double membrane vesicles, most of them of small diameter (400 to 500 Å) and with a clear content. On Fig. 13 we see that the uncompletely lysed synaptosomes fraction contains vesicles of size that vary greatly, surrounded by fragments of smooth membranes. Fig. 14 shows synaptosomal membranes, groups of membranes and synaptosome envelopes. The distribution of mannosyl-, galactosyl-, and N-acetyl-glucosaminyl-transferase activities in the different subcellular fractions is outlined in Fig. 15. It seems that mannosyl-transferase is specially located in the two fractions rich in synaptic vesicles (D and E of Whittaker). The galactosyl-transferase activity is also evident in one of the two parts (E) of the same fraction, and to a lesser extent, in a fraction containing membranous structures of undetermined origin. Finally, N-acetyl-glucosaminyl-transferase is more widely distributed in all subcellular structures examined, with no preferential location in the synaptic vesicles.

Thus, at the level of nerve cell function, it seems that the fixation of N-acetyl-glucosamine onto an endogenous acceptor (by a specific transglycosylase) is a property of all the subcellular structures. On the contrary, the fixation capacity for mannose and, perhaps, galactose is specifically located in the synaptic vesicles, while it is absent in the synaptosomal membranes (fraction

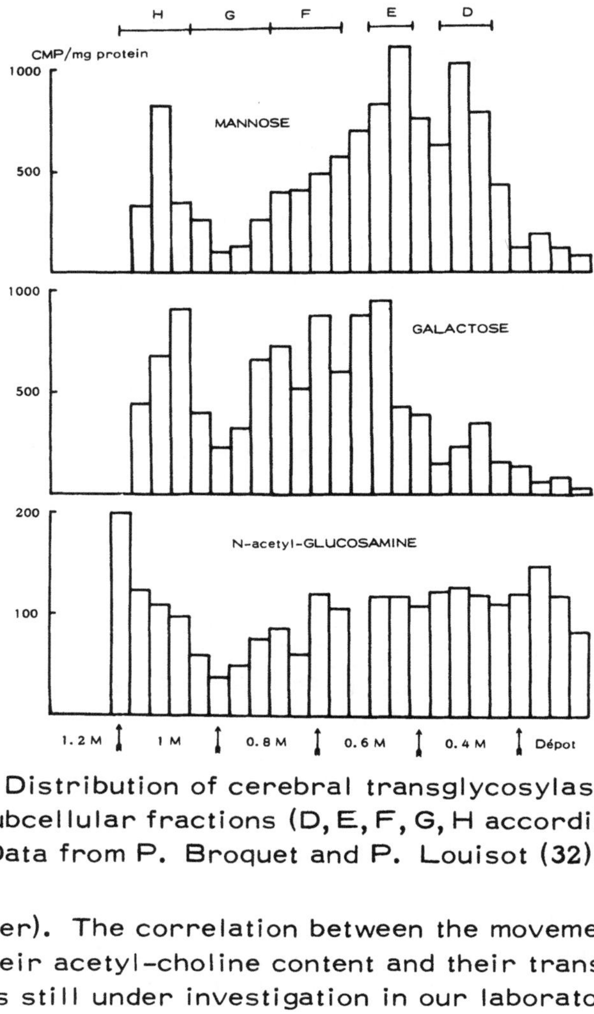

Fig. 15. Distribution of cerebral transglycosylase activities in subcellular fractions (D, E, F, G, H according to Whittaker). Data from P. Broquet and P. Louisot (32).

6 of Whittaker). The correlation between the movement of synaptic vesicles, their acetyl-choline content and their transglycosylation capacities is still under investigation in our laboratory.

CONCLUSION

The biosynthesis of cerebral glycoproteins seems today to be due to a series of complex reaction mechanisms.

Begun with the *in vivo* incorporation of labelled sugar precursors, these studies have greatly profitted from the availability of elaborate cellular fractionation techniques.

In vitro acellular systems and individual purified transglycosylases were also used.

These recent approaches, which are an example of how to study the structure-function relationships in cells, should lead, in the next years, to great progress in the knowledge of the functional brain mechanisms.

REFERENCES

(1) Brunngraber, E.G., in Protein Metabolism of the Nervous System, Plenum Press, New York, p. 383 (1970).

(2) Sarcione, E.J., Bohne, M. and Leahy, M., Biochemistry 3: 373 (1964).

(3) Helgeland, L., Biochim. Biophys. Acta 101: 106 (1965).

(4) Molnar, J., Robinson, G.B. and Winzler, R.J., J. Biol. Chem. 240: 1882 (1965).

(5) Lawford, G.R. and Schachter, H., J. Biol. Chem. 241: 5408 (1966).

(6) Molnar, J. and Sy, D., Biochemistry 6: 141 (1967).

(7) Got, R., Louisot, P., Frot-Coutaz, J. and Colobert, L., C.R. Acad. Sci. Paris 266: 286 (1968).

(8) Got, R., Frot-Coutaz, J., Colobert, L. and Louisot, P., Biochim. Biophys. Acta 157: 599 (1968).

(9) Frot-Coutaz, J., Louisot, P., Got, R. and Colobert, L., Experientia 24: 1206 (1968).

(10) Louisot, P., Frot-Coutaz, J. and Got, R., Bull. Soc. Chim. Biol. 50: 2533 (1968).

(11) Louisot, P., Lebre, D., Pradal, M.B. and Got, R., C.R. Acad. Sci. Paris 269: 1140 (1969).

(12) Pradal, M.B., Lebre, D., Louisot, P. and Got, R., Comp. Biochem. Physiol. 35: 31 (1970).

(13) Louisot, P., Lebre, D., Pradal, M.B., Gresle, J. and Got, R., Canad. J. Biochem. 48: 1082 (1970).

(14) Got, R. and Louisot, P., Exp. Ann. Biochim. Med. 30: 1 (1970).

(15) Louisot, P. and Got, R., Bull. Soc. Chim. Biol. 52: 455 (1970).

(16) Letoublon, R., Richard, M., Louisot, P. and Got, R., Europ. J. Biochem. 18: 194 (1971).

(17) Richard, M., Broquet, P., Got, R. and Louisot, P., Biochimie 53: 107 (1971).

(18) Richard, M., Letoublon, R., Louisot, P. and Got, R., Biochim. Biophys. Acta 230: 603 (1971).

(19) Spiro, R.G., New Engl. J. Med. 281: 991 (1969).

(20) Schachter, H. and Letts, P. J., 8th International Congress of Biochemistry, Symposium VIII, Montreux (1970).
(21) Holian, O., Dill, D. and Brunngraber, E. G., Arch. Biochem. Biophys. 142: 111 (1971).
(22) Barondes, S. H., J. Neurochem. 15: 699 (1968).
(23) Barondes, S. H. and Dutton, G. R., J. Neurobiol. 1: 99 (1969).
(24) Zatz, M. and Barondes, S. H., J. Neurochem. 17: 157 (1970).
(25) Barondes, S. H., Comm. Beh. Biology (part A) 1: 179 (1968).
(26) Barondes, S. H., J. Neurochem. 15: 343 (1968).
(27) Droz, B. and Barondes, S. H., Science 165: 1131 (1969).
(28) Austin, L. and Morgan, I. G., J. Neurochem. 15: 41 (1968).
(29) Autilio, L. A., Appel, S. H., Pettis, P. and Gambetti, P. L., Biochemistry 7: 2615 (1968).
(30) De Vries, G. H. and Barondes, S. H., J. Neurochem. 18: 101 (1971).
(31) Den, H., Kaufman, B. and Roseman, S., J. Biol. Chem. 245: 6607 (1970).
(32) Broquet, P. and Louisot, P., J. Neurochem., in press (1971).

GLYCOPROTEINS OF THE SYNAPTOSOMAL PLASMA MEMBRANE

G. GOMBOS,* I. G. MORGAN,* T. V. WAEHNELDT,◆
G. VINCENDON and W. C. BRECKENRIDGE ❖

Centre de Neurochimie du CNRS et Institut de Chimie
Biologique de la Faculté de Médecine
67 - Strasbourg, France

Plasma membranes contain glycoproteins (1-4) which are believed to account for part of the carbohydrate-containing material (heavily stained by the periodic acid-Schiff reaction), coating the outer surface of cells (5). Such coat material is present also in neurons (5), but, in the absence of suitable techniques for isolating highly purified neuronal plasma membranes, studies of their glycoproteins have been limited to the results of Brunngraber et al. (6-8), indicating that subcellular fractions enriched in synaptosomal plasma membrane are also enriched in protein-bound carbohydrate.

The preparation of large quantities of highly purified nerve cell body plasma membrane is a dubious and difficult task with present techniques, but the problem can be by-passed if we use a model for the neuronal plasma membrane, the plasma membrane of a specialized area of the neuron: the nerve-ending. Synaptosomal plasma membranes (SPM) were isolated from adult rat brain as described elsewhere (9) and shown to be approximately 80 %

* Attaché de Recherche au Centre National de la Recherche Scientifique.

* Boursier étranger de l'Institut National de la Santé et de la Recherche Médicale.

◆ Chargé de Recherche au Centre National de la Recherche Scientifique.

❖ Fellow of the Medical Research Council of Canada.

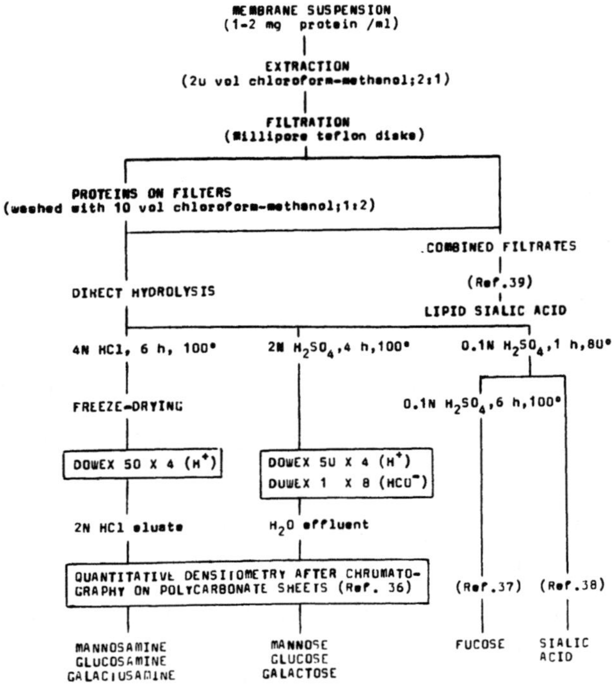

Fig. 1

neuronal plasma membrane. The analysis of the different subcellular fractions obtained from synaptosomes shows that SPM have the highest specific activity of protein-bound sialic acid (41.1 nmole/mg protein in the purest fraction: FL) of the synaptosomal subfractions (10).

The protein-bound carbohydrates in SPM were analyzed as described in Fig. 1. Fucose, sialic acid, mannose, galactose, glucosamine and galactosamine were detected (Table I). No mannosamine was found. Glucose was present, but it appeared to be derived from contamination with the Ficoll and sucrose used for the purification of the membranes. The carbohydrate content per mg protein of the purest SPM fraction (fraction FL) compared to that of total rat brain [1] was increased approximately ten-fold for ganglioside sialic acid and for protein-bound glucosamine, while protein-bound sialic acid or fucose were concentrated 4 times. Galactose and galactosamine were enriched 4 times while mannose was enriched only 1.5 times in the SPM.

---

(1) The values for whole rat brain were recalculated from Di Benedetta et al. (11).

Table I - Carbohydrate Composition of Rat Brain Synaptosomal Plasma Membranes (Fraction FL) Compared to that of Plasma Membranes of other Tissues.

Protein-bound	nmole carbohydrate per mg protein			molar ratios relative to glucosamine		
	SPM (')	LPM ('')	APM (°)	SPM (')	LPM ('')	APM (°)
Glucosamine	152.50	48.00	53.70	1.00	1.00	1.00
Galactosamine	40.80	traces	19.90	0.27		0.37
Mannose	40.60	155.00	57.20	0.27	3.25	1.07
Galactose	33.30		43.20	0.22		0.80
Fucose	21.40		16.10	0.15		0.29
Sialic acid	41.10	34.30	20.70	0.27	0.72	0.38
Lipid sialic acid	128.70		8.10			

(') Analyzed as described in Fig. 1. ('') Liver plasma membranes (2). (°) Ascites tumor plasma membranes (3).

SPM fractions prepared by the Whittaker procedure (12) contain less ganglioside and glycoprotein sialic acid (7, 8) per mg protein than our fractions. In addition, the ratios between protein-bound sialic acid and ganglioside sialic acid are higher than in our fractions. Although present in glial cell plasma membranes and in myelin (13, 14), gangliosides seem to be mainly concentrated in the neuronal plasma membrane (15, 16), whereas glycoproteins are also present in other intracellular membraneous structures. Thus the higher glycoprotein sialic acid to ganglioside sialic acid ratios found in the classical preparations could be due to a more marked contamination of such SPM by other subcellular structures.

Synaptosomal plasma membranes (table I), are slightly richer in protein-bound sialic acid, much richer in protein-bound hexosamines and poorer in hexoses than rat liver (2) or ascites cell plasma membranes (3). Furthermore SPM are slightly richer in protein-bound fucose and more than 26 times richer in glycolipid sialic acid than ascites tumor cell plasma membranes (3).

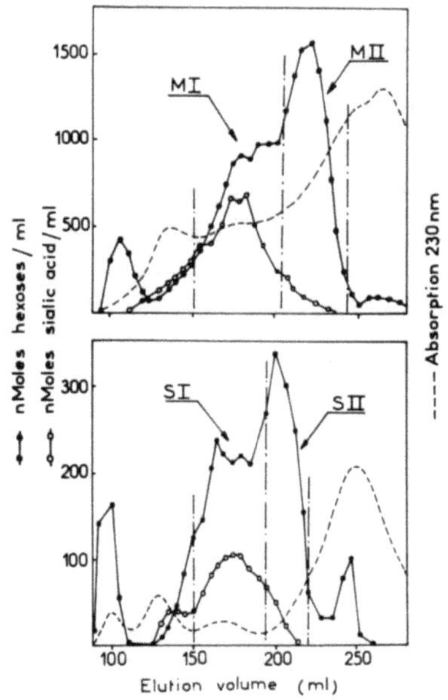

Fig. 2. Fractionation on Sephadex G-50 columns (2 x 85 cm) of synaptosomal plasma membrane glycopeptides (S-glycopeptides) and microsomal glycopeptides (M-glycopeptides). Glycopeptides were prepared from thoroughly delipidized fractions after prolonged pronase digestion as described in detail elsewhere (17). The first and last carbohydrate peaks are Ficoll or glycogen and sucrose respectively. The columns were equilibrated and eluted with 50 mM ammonium formate (pH 7.1).

When the synaptosomal plasma membranes were sequentially extracted with detergents, with a view to subsequent electrophoresis, the glycoproteins containing sialic acid were practically all solubilized by Triton X-100 (17). However more than one third of the protein-bound hexoses and hexosamines remained in the Triton-insoluble residue (17). This Triton-insoluble material could be partially, but not completely, solubilized with 0.1 % SDS (17). The electrophoretic profiles of the two extracts were shown to be different both staining with amido-black and the periodic acid-Schiff reaction (18).

These results suggest that two pools of synaptosomal plasma membrane glycoproteins can be defined; one rich in sialic acid with complex carbohydrate chains and another, with simple carbohydrate chains, which do not contain sialic acid.

The presence of glycoproteins with simple carbohydrate chains has been confirmed by the analysis of glycopeptides (S-glycopeptides) obtained by proteolytic digestion of SPM. When the glycopeptides were resolved on Sephadex G-50 (Fig. 2) two fractions were distinguished. The glycopeptides of high molecular weight, pooled into one fraction (fraction SI) (Fig. 1) contained fucose, galactose, and galactosamine, 36 % of the glucosamine and mannose, and about 90 % of the sialic acid. Glycopeptides of lower molecular weight were pooled into one fraction (fraction SII) (Fig. 2) which was poor in sialic acid and lacked fucose. This fraction consisted almost exclusively of mannose and glucosamine (Table II). Very little galactose and galactosamine (Table III) were found.

Fraction SII was separated into three major peaks (called SII_1, SII_2 and SII_3) by chromatography on DEAE-Sephadex (Fig. 3). Minor peaks containing some sialic acid, were not examined extensively. The three major fractions (Table III) contained large quantities of mannose and glucosamine, but the molar ratios changed from peak to peak. Low levels of galactose and galactosamine, but no sialic acid and fucose were detected.

Thus, from the glycopeptides, two classes of carbohydrate chains can be defined; one, of complex structure, rich in sialic

Table II - Carbohydrate Molar Ratios Relative to Glucosamine of Rat Brain Glycopeptides.

	SPM		Microsomal fraction	
	SI	SII	MI	MII
Glucosamine	1.00	1.00	1.00	1.00
Galactosamine	0.25	0.10	0.14	0.08
Mannose	0.36	0.71	0.25	3.05
Galactose	0.25	0.11	0.25	0.70
Fucose	0.25	-	0.11	-
Sialic acid	0.50	0.09	0.21	0.22

S- and M-glycopeptides were prepared and separated by gel filtration as shown in Fig. 2.

Table III - Carbohydrate Molar Ratios Relative to Glucosamine of the Subfractions of SII- and MII-Glycopeptides.

	SII-glycopeptides			MII-glycopeptides		
	SII1	SII2	SII3	MII1	MII2	MII3
Glucosamine	1.00	1.00	1.00	1.00	1.00	1.00
Galactosamine	0.10	0.07	0.05	0.04	0.16	0.04
Mannose	0.95	1.32	2.66	3.99	2.86	2.66
Galactose	0.06	0.15	0.17	0.09	0.25	0.49

Subfractions separated by chromatography on DEAE Sephadex A-25 as shown in Fig. 3.

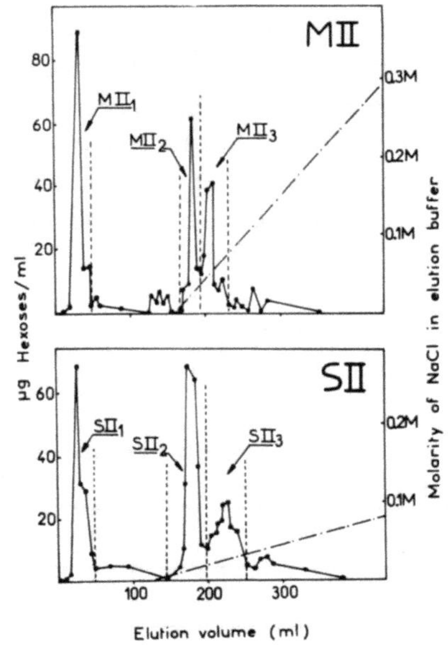

Fig. 3. Separation of MII and SII glycopeptide fractions (see Fig. 1) on DEAE Sephadex A-25 column (1 x 30 cm) equilibrated with 5mM Tris-HCl buffer pH 7.7, and eluted with a linear gradient of NaCl in the same buffer. The glycopeptide fractions eluted from Sephadex G-50 were digested with leucine aminopeptidase before chromatography on DEAE Sephadex to minimize the effect of the residual amino acids on chromatographic separation.

acid, fucose and galactose, and another which contains only mannose and glucosamine.

For comparative purposes, parallel studies were carried out on brain microsomal fractions. Rat brain microsomal glycopeptides (M-glycopeptides) were prepared and purified under the same conditions as those used for S-glycopeptides. The elution profile, from Sephadex G-50 of the M-glycopeptides was similar to that of synaptosomal plasma membrane glycopeptides (S-glycopeptides) (Fig. 2). The first and the last carbohydrate peaks were discarded since their analysis indicated that they were glycogen and sucrose respectively. The M-glycopeptides were pooled in two fractions, analogous to SI and SII (MI and MII).

Fraction MII contained more mannose (78 % of the total) less glucosamine (40 % of the total), and more sialic acid (20 % of the total) than fraction SII. Differences in carbohydrate ratios between SI and MI and between SII and MII were evident (Table II). Sialic acid and galactose were present in MII and practically absent from SII; furthermore, MII contained less galactosamine and more hexoses than SII.

It was possible to isolate from fraction MII (by ion exchange chromatography on DEAE-Sephadex) three major glycopeptide fractions (MII_1, MII_2 and MII_3) with chromatographic characteristics similar to the three fractions from SII (Fig. 3). The proportion of the three peaks of the M-glycopeptides were different from those of the S-glycopeptides.

The minor peaks eluted from the columns after peak MII_3 were not analyzed extensively, but they contained sialic acid. The analysis of carbohydrate composition of each of the major peaks (Table III) showed that sialic acid or fucose were absent and that only small amounts of galactose were present: they contained almost exclusively mannose and glucosamine. These two carbohydrates were present in different proportions in MII1 and SII1, MII2 and SII2 (Table III) while MII_3 and SII_3 had the same carbohydrate composition.

These results, with the exception of those for peaks MII_3 and SII_3, point to definite differences in the structure of these chromatographically similar classes of glycopeptides present in the microsomal and synaptosomal plasma membrane fractions. Brain microsomal fractions are very heterogeneous since they contain fragments of plasma membrane from the cell bodies, of neurons and glial cells, smooth and rough endoplasmic reticulum membranes, dendrite and axonal membranes, synaptosomes and probably Golgi apparatus membranes.

In view of the fact that neuronal plasma membrane fragments are present in brain microsomal fractions, we must assume that, unless the glycoproteins of the plasma membrane of the neuronal cell body are different from those of synaptosomal plasma membrane, each M-glycopeptide peak contains the S-glycopeptides and in addition glycopeptides derived from other subcellular structures. These additional glycopeptides would have markedly different carbohydrate compositions. Furthermore peaks SII_1 and SII_2 may also be heterogeneous. Separation of the various possible components of each peak will be necessary to elucidate this point.

Two classes of glycoproteins, one with complex carbohydrate chains and one with simple carbohydrate chains, were differentially extracted from a brain microsomal fraction, by using a procedure similar to that described for SPM. To determine if the two classes of glycopeptides corresponded to the two glycoprotein fractions defined by differential extraction, glycopeptides from the differential extracts of the microsomal fraction were separated on Sephadex G-50. The Triton insoluble fraction gave practically only fraction MII, while the Triton soluble fraction gave a profile similar to that of the glycopeptides of the whole membrane fraction, except that it was proportionally enriched in the sialic acid-containing glycopeptides (17).

The presence of glycoproteins, whose carbohydrate moieties are made only of glucosamine and mannose, in the microsomal fraction, is not surprising since this fraction contains fragments of the endoplasmic reticulum and of the Golgi apparatus. It is generally accepted (at least for soluble glycoproteins) that carbohydrates are added in a stepwise fashion to the polypeptide chains during their passage through the smooth endoplasmic reticulum and Golgi apparatus (for review see Ref. 19). Sometimes this process is continued in the soluble cytoplasm. Margolis (20) has shown that most brain glycoproteins contain the N-acetylglucosaminyl-asparagine linkage. In most glycoproteins the subsequent sugar is mannose. Thus the "core region" of most cerebral glycoproteins is probably composed of one glucosamine and several mannoses. Hexosamines and galactose are then attached to the "core", and the polysaccharide chain could be terminated by sialic acid or fucose. Taking into account these possibilities, it is reasonable to assume that at least some of the MII glycopeptides are incomplete "cores" of the more complex carbohydrate chains of glycoproteins is passage through the smooth endoplasmic reticulum and the Golgi apparatus.

The presence of glycoproteins containing only mannose and glucosamine in the SPM is of more immediate interest. Such glycoproteins have not been detected in liver cell plasma membranes (21) although the experiments which suggest that such glycoproteins are not present are by no means conclusive.

One of the possible interpretations is that these glycoproteins in the synaptosomal plasma membrane have incomplete carbohydrate chains. In fact, the phenomenon of microheterogeneity, the presence at one site on a given polypeptide chain of carbohydrate chains of different lengths and compositions, has been attributed to the presence of carbohydrate chains at different stages of completion (19). In other glycoproteins (i.e. thyroglobulin) however the composition of the smaller carbohydrate chain is such that it cannot be the "core" of the more complex chain (22). In this case it seems likely that the carbohydrate chains are found at two different sites.

If such is the case, the presence of incomplete carbohydrate chains in SPM glycoproteins might suggest that the synaptosome can complete the carbohydrate chains of its glycoproteins (23-27). On the contrary, we have evidence from the kinetics of incorporation of C^{14}-glucosamine into SPM glycoproteins (10) and from the determination of UDP-galactose: N-acetyl glucosamine galactosyl transferase activity (10) that the capacity of synaptosomes to synthesize or complete carbohydrate chains of glycoproteins is very limited, if it exists at all.

In addition, in view of the differential extraction of the carbohydrate chains, unless the solubility characteristics of the membrane glycoproteins depend primarily on the nature of the carbohydrate chain, it seems likely that at least part of the simple carbohydrate chains are on different polypeptide chains than are the more complex carbohydrate chains. In confirmation of this result, the protein profiles of the differential detergent extracts differ considerably.

These results, viewed as a whole, tend to suggest that some of the glycoproteins with simple carbohydrate chains (containing only mannose and glucosamine) in the synaptosomal plasma membrane are not the result of incomplete synthesis of more complex glycoproteins. Instead, it seems likely that these glycoproteins are of a type analogous to soluble glycoproteins such as ovalbumin (28).

Additional, although indirect, evidence for the presence in the SPM of glycoproteins with mannose, as terminal sugar of their carbohydrate moieties, is derived from experiments carried out with lectins (vegetal proteins that bind to carbohydrates). Concanavalin (Con-A), a lectin which binds preferentially to D-manno- and D-glucopyranosyl residues (29), agglutinates synaptosomes (unpublished results). This effect is inhibited by the addition of α-methyl mannoside, which inhibits the binding of Con-A to mannose (30). Since glucose is practically absent from SPM, the carbohydrate chain of the Con-A "receptor" glycoprotein probably has a terminal mannose. The glycopeptides containing only mannose and glucosamine, are good candidates as Con-A "receptors", although the more complex carbohydrate chains could have terminal mannose on side-branches. Experiments are in progress to ascertain the nature of the Con-A binding which causes agglutination of synaptosomes.

Lectins are not only an interesting tool for separating classes of glycoproteins (31) and glycopeptides, but may also help us in understanding the functions of glycoproteins of cell surface membranes.

For instance there are some indications that the lectin-binding glycoproteins can influence cell-division and growth control (32). Binding of Con-A to lymphocyte surface glycoproteins is followed by mitosis (33) of these cells, and the addition of phytohaemagglutinin to neuroblastoma cells in culture stimulates the growth of processes (34).

Con-A stimulates, within a narrow concentration range, the growth of cellular processes of spinal ganglion cells in culture

(35). This effect is inhibited by α-methyl mannoside (35). These data indicate that in both spinal ganglion cells and synaptosomal plasma membranes there are glycoprotein "receptors" for Con-A.

Although the functional significance of these results is not clear, we can speculate that cell surface glycoproteins are important as receptors for signals mediated by direct contact with cells or through extracellular compounds. It is evident that precise knowledge of the structure of plasma membrane glycoproteins is essential for the understanding of these phenomena.

ACKNOWLEDGEMENTS

This work was in part supported by a grant from the "Institut National de la Santé et de la Recherche Médicale (contract 71 11698)" and from the "Fondation pour la Recherche Médicale Française".

REFERENCES

(1) Evans, W.H., Biochim. Biophys. Acta 211: 578 (1970).

(2) Henning, R., Kaulen, H.D. and Stoffel, W., Hoppe Seyler's Z. Physiol. Chem. 351: 1191 (1970).

(3) Shimizu, S. and Kunakoshi, I., Biochim. Biophys. Acta 203: 167 (1970).

(4) Glick, M.C., Comstock, C.A., Cohen, M.A. and Warren, L., Biochim. Biophys. Acta 233: 247 (1971).

(5) Rambourg, A. and Leblond, C.P., J. Cell Biol. 32: 27 (1967).

(6) Brunngraber, E.G., Dekirmenjian, H. and Brown, B.D., Biochem. J. 103: 73 (1967).

(7) Dekirmenjian, H. and Brunngraber, E.G., Biochim. Biophys. Acta 177: 1 (1969).

(8) Dekirmenjian, H., Brunngraber, E.G., Lemkey-Johnston, N. and Larramendi, L.M.H., Exptl. Brain Res. 8: 97 (1969).

(9) Morgan, I.G., Wolfe, L.S., Mandel, P. and Gombos, G., Biochim. Biophys. Acta 241: 737 (1971).

(10) Morgan, I.G., Reith, M., Marinari, U., Breckenridge, W.C. and Gombos, G., these Proceedings.

(11) Di Benedetta, C., Brunngraber, E.G., Whitney, G., Brown, B.D. and Aro, A., Arch. Biochem. Biophys. 131: 404 (1969).

(12) Whittaker, V.P., Michaelson, I.A. and Kirkland, R.J.A., Biochem. J. 90: 293 (1964).

(13) Suzuki, K., Poduslo, S.E. and Norton, W.T., Biochim. Biophys. Acta 144: 375 (1967).

(14) Norton, W.T. and Poduslo, S.E., J. Lipid Res. 12: 84 (1971).

(15) Lowden, J.A. and Wolfe, L.S., Canad. J. Biochem. 42: 1587 (1965).

(16) Derry, D.M. and Wolfe, L.S., Science 158: 1450 (1967).

(17) Breckenridge, W.C., Breckenridge, J.E. and Morgan, I.G., in "Structural and Functional Proteins of the Nervous System", Ed. A.N. Davison, I.G. Morgan and P. Mandel, Plenum Press, New York, in press.

(18) Waehneldt, T.V., Morgan, I.G. and Gombos, G., Brain Res. 34: 403 (1971).

(19) Spiro, R.G., Ann. Rev. Biochem. 39: 599 (1970).

(20) Margolis, R.U. and Margolis, R.K., in Abstracts 3rd Intern. Meeting I.S.N., Ed. J. Domonkos, A. Fonyó, I. Huszák and J. Szentágothai, Akadémiai Kiadó, Budapest, p. 276 (1971).

(21) Miyajima, N., Kawasaki, T. and Yamashina, I., FEBS Letters 11: 29 (1970).

(22) Fukuda, M. and Egami, F., Biochem. J. 123: 415 (1971).

(23) Barondes, S.H., J. Neurochem. 15: 699 (1968).

(24) Barondes, S.H. and Dutton, G.R., J. Neurobiol. 1: 99 (1969).

(25) Bosmann, H.B. and Hemsworth, B.A., J. Biol. Chem. 245: 363 (1970).

(26) Festoff, B.W., Appel, S.H. and Day, E., Fed. Proc. 28: 734 (1969).

(27) Broquet, P. and Louisot, P., Biochimie, in press.

(28) Montgomery, R., Wu, Y.C. and Lee, Y.C., Biochemistry 4: 578 (1965).

(29) Goldstein, I.J., Hollerman, C.E. and Smith, E.E., Biochemistry 4: 876 (1965).

(30) Goldstein, I.J., Hollerman, C.E. and Merrick, J.M., Biochim. Biophys. Acta 97: 68 (1965).

(31) Avrameas, S. and Guilbert, B., Biochimie 53: 603 (1971).

(32) Burger, M.M. and Noonan, K.D., Nature 288: 512 (1970).

(33) Powell, A.E. and Leon, M.A., Exptl. Cell Res. 62: 315 (1970).

(34) Pyke, K.W. and Dent, P.B., Fed. Proc. 30: 398 Abs (1971).

(35) Treska-Ciesielski, J., Gombos, G. and Morgan, I.G., C.R. Acad. Sci. Paris 273: 1041 (1971).

(36) Moczar, E., Moczar, M., Schillinger, G. and Robert, L., J. Chromatog. 31: 561 (1967).

(37) Dische, Z. and Schettles, L.B., J. Biol. Chem. 175: 595 (1948).

(38) Warren, L., J. Biol. Chem. 234: 1971 (1959).
(39) Svennerholm, L., Biochim. Biophys. Acta 24: 604 (1957).

GLYCOPROTEINS DURING THE DEVELOPMENT OF THE RAT BRAIN (°)

C. DI BENEDETTA and L. A. CIOFFI

Institute of Physiology, Medical School

University of Napoli, Italy

The rat brain development is sufficiently known so that different stages in its growth can be distinguished, and the most peculiar features occurring within each step can be specified. The scheme given by McIlwain (1) is still valid since all the most recent data tend to support it (2, 3).

At birth the rat brain has reached the 93-95% of its final number of neurons which, in the second stage, begin to give the dendritic and axonal outgrowth, making their connections. It is accepted that this phenomenon lasts the first 10-15 days after birth. Peculiar feature of this period is the increase of the amount of gangliosides (4, 5) and of the activity of enzymes associated with their synthesis (6, 7). It is described also during this period the changing morphology of the glial components which progress to the third stage of development when the oligodendroglial cells actively partecipate to the myelination, completed around the 40th day, as it has been pointed out from biochemical (8) and morphological studies (9).

(°) Part of this study was communicated at the 3rd International Meeting of the International Society for Neurochemistry, Budapest, 5-9 July, 1971.

Abbreviations used: NaNa = N-acetylneuraminic acid; HX = Hexose; HA = Hexosamine.

From a functional point of view the first signs of electrical activity (10, 11) appear in the second stage of growth of the rat brain, in good agreement with the morphological as well as the biochemical evidences of new synapse formations.

The location of the gangliosides, on the other hand, has been demonstrated in the neuronal membranes (3, 12), although a small amount of gangliosides is present in the myelin (13). From the literature it can be inferred, also, that NaNa containing glycoproteins are located on the external surface of the axonal and synaptosomal membranes, although their presence in the glial processes can not be ruled out (14, 15).

The glycoproteins have been regarded by many authors as informational macromolecules (16), due to their morphological location (17), peculiar metabolism (18) and large heterogeneity (19).

The glycopeptides from the whole rat brain are, indeed, very heterogeneous either in the nondialyzable or in the dialyzable fraction (20). It has been also demonstrated that the s.c. soluble glycoproteins of the rat brain in the prealbumin region show some fractions which can not be detected in the liver extracts and in the blood serum. Furthermore the glycopeptides isolated from the soluble fraction are different for chemical composition and biophysical characteristics from the glycopeptides of the insoluble glycoproteins (21).

The aim of the present research is to follow the developmental pattern of the sugars in the soluble and insoluble fractions of the rat brain glycoproteins, in order to correlate their changes to the morphological and, possibly, to the functional modifications of the nervous tissue.

MATERIAL AND METHODS

Sprague Dawley rats have been randomly distributed in litter of 9 animals. At birth three litters for each age were simultaneously reared. At different ages the rats were killed by decapitation; the brain was rapidly removed and quickly frozen. The brains from each group of animals were homogeneized in TRIS-HCl buffer pH 8.0. The supernatant was set aside and the residue reextracted with water. The two supernatants were combined. This fraction contained the s.c.

soluble proteins. The residue was extracted with C:M 2:1 and 1:2 and the gangliosides were prepared (22). The residue contained the insoluble proteins.

The NaNa was determined by the method of Warren (23); the hexoses by that of Dische, Shettles and Osnos (24); the hexosamines by the Boas' method (25); the protein by the method of Lowry et al. (26) and by the biuret method (35).

RESULTS

Table I reports the changes with the age of insoluble proteins and of NaNa, hexose and hexosamines in the rat brain. The values are expressed per gram of wet tissue. The concentration of all the components increases steadily with age, as it could be expected, but the rate of increase is not the same for all the components, since the proteins show a three-fold increase, the NaNa almost a 5-fold, while the hexoses at 26 days almost double the 5 days concentration. The hexosamines show a 3-fold increase, almost as much as the proteins, with the highest rate around 17-19 days of life.

Table I - Protein, NaNa, hexose and hexosamine content of insoluble glycoproteins.

Age (days)	Protein (mg/g w. t.)	NaNa	Hexose (μg/g w. t.)	Hexosamine
5	12.42	84	126	89
7	19.00	159	152	119
11	19.20	195	167	130
13	29.62	290	182	168
15	29.50	314	151	167
17	36.86	382	202	186
19	34.38	328	210	259
21	36.10	325	235	255
26	41.08	375	238	293

Table II - Protein, NaNa, hexose and hexosamine content of soluble glycoproteins.

Age (days)	Protein (mg/g w. t.)	NaNa	Hexose (μg/g w. t.)	Hexosamine
7	27.82	176	387	167
13	27.53	136	355	160
15	30.18	121	375	218
17	28.50	122	320	237
19	30.50	138	323	198
21	29.70	130	364	221
26	35.18	122	414	239
31	33.78	121	-	284

Table II reports the changes of proteins, NaNa, hexoses and hexosamines in the soluble fraction. In this case the protein concentration changes very little, confirming our previous data (27), while that of the NaNa decreases with the increase of the age. Changes are not evident in the soluble fraction of the hexoses. A different trend is evidentiated for the hexosamines which increase in concentration from the 7th up to the 31st day, showing the maximum rate around 15 days of life.

To evidentiate the differences among the specific concentrations of the sugars, the ratio of the NaNa, hexose and hexosamine to the proteins have been calculated. In the insoluble fraction (Fig. 1) it is clear that NaNa/proteins ratio increases steadily, while that of the hexose/proteins decreases with the age and the hexosamine/proteins ratio remains almost the same. It is worth to note that the ratio of the sum of the carbohydrates to the proteins does not change during this period.

Fig. 2 reports the same ratios in the soluble fraction. The ratio of the sum of the carbohydrates to proteins does not change appreciably; the NaNa/proteins ratio decreases, while that of the hexosamine/proteins increases steadily with age; the hexose/proteins ratio seems to be held constant.

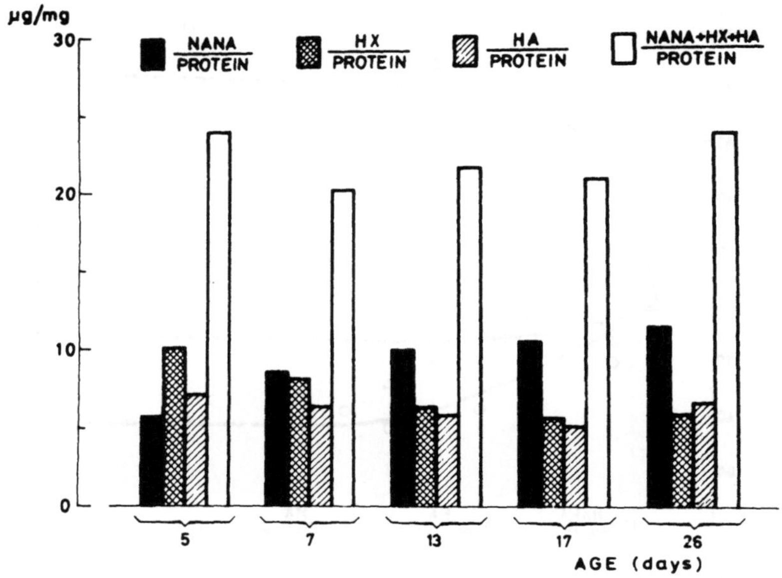

Fig. 1. NaNa, hexose and hexosamine specific concentration of the insoluble glycoproteins during the development of the rat brain.

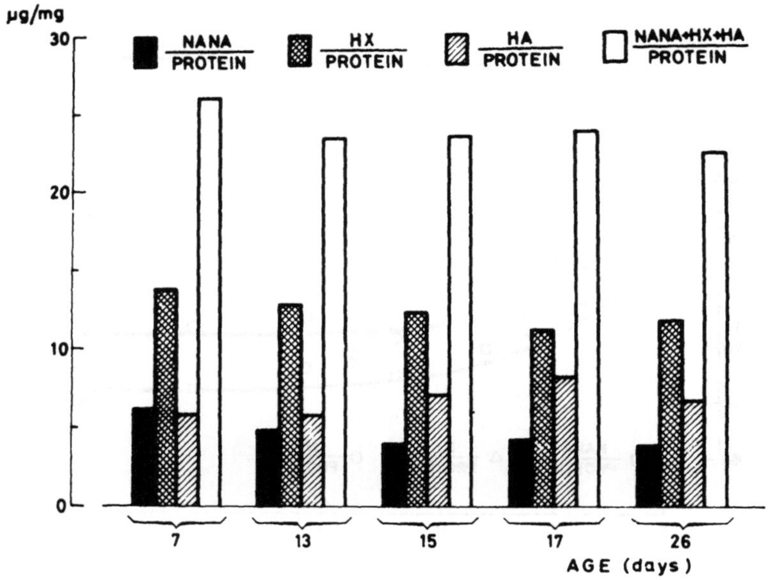

Fig. 2. NaNa, hexose and hexosamine specific concentration of the soluble glycoproteins during the development of the rat brain.

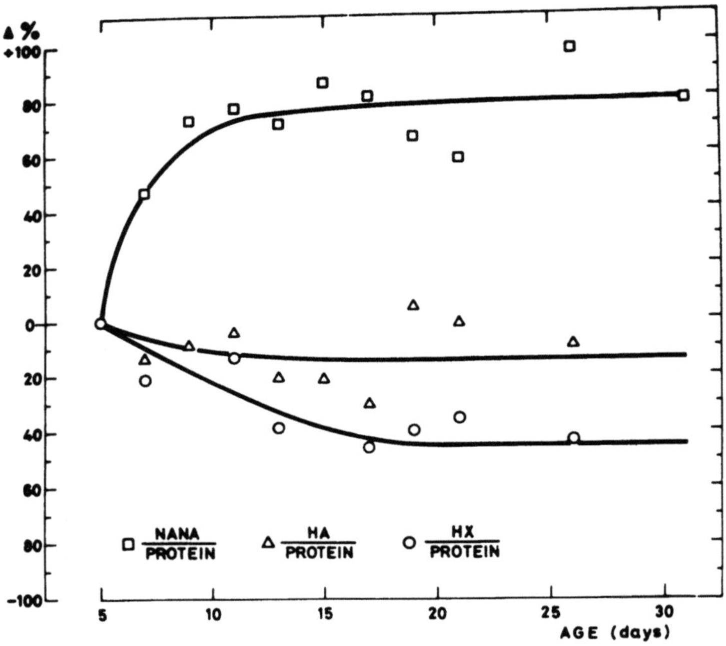

Fig. 3. Percent difference of the NaNa/protein, HX/protein and HA/protein ratios in the insoluble glycoproteins during the rat brain development.

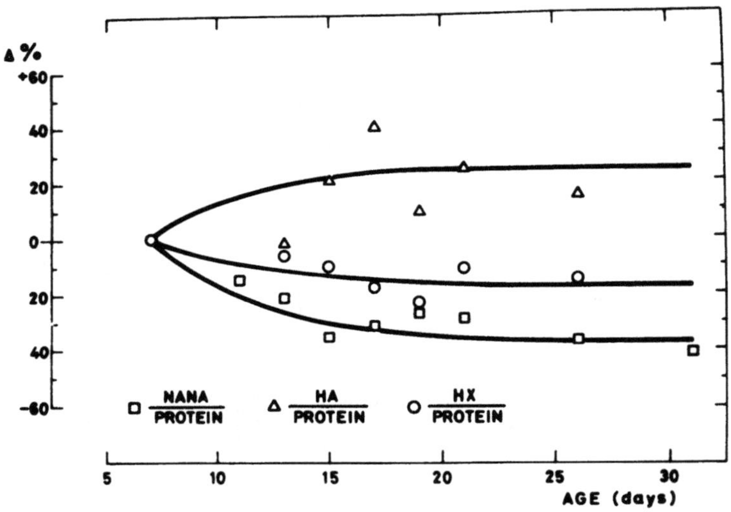

Fig. 4. Percent difference of the NaNa/protein, HX/protein and HA/protein ratios in the soluble glycoproteins during the rat brain development.

To compare the changes of the specific activity of the different sugars, their developmental trend in the soluble and insoluble fractions has been expressed as percent difference of their initial value.

The values obtained for the insoluble fraction (Fig. 3) are related to their own 5 days ratios. The NaNa/proteins ratio substantially increases, reaching a difference of + 80% at 31 days of life. The changes which occur in the hexoses and hexosamines are less marked.

The ratios of the sugars in the soluble glycoproteins are reported in Fig. 4 as percent difference of the 7 days ratios. It is evident that the NaNa/proteins ratio decreases to - 40% at 31st day while the hexosamines/proteins ratio slightly increases, and the hexoses ratio decreases.

DISCUSSION

The comparison of the developmental patterns of the carbohydrates of soluble and insoluble fractions indicates that the changes of the insoluble are different from those of the soluble fraction.

The ratios of the sum of the carbohydrates to protein content of both soluble and insoluble fractions show values very close to each other and constant throughout this period. Therefore the quantitative evaluation of the total carbohydrate content of the rat brain glycoproteins can not evidentiate significant changes during the development. A more detailed analysis of the different sugars allows to detect peculiar patterns for some of them. The opposite trends of NaNa in the soluble and insoluble glycoproteins suggest its different role in the development of different brain glycoproteins.

Some of the previous results indicated that the concentration of the soluble proteins does not change with age, but that there is a continuous heterogeneization of the soluble proteins and glycoproteins (27) depending on different amount of sialic acid residues on the molecule. Although we do not yet have evidence for this hypothesis, the decrease of the relative concentration of the NaNa to the proteins in the soluble fraction could account for this phenomenon, which occurs around the 10th day of life, indicating perhaps a need for higher degree of specificity.

The alternative is that the heterogeneization is due to the synthesis of a class of molecules different for the protein and for the sugar moieties. This phenomenon has been observed also for the acetylcholinesterase (28) and monoamine oxidase (29).

The developmental pattern of the NaNa of the insoluble glycoproteins resembles very closely what has been described for gangliosides and for their synthesizing enzymes (6, 7), indicating a discrete parallelism between the phenomena related to the two groups of substances both located in plasma membrane. The increase of NaNa concentration is highest between the 5th and 11th day, just about or shorter before the onset of detectable electrical activity in the central nervous system (10, 11).

From the present data it is possible to notice the striking difference between the pattern of the NaNa and those of the hexoses and hexosamines, the variations of which are less marked in both soluble and insoluble fractions. This observation points to a peculiar role of the sialic acid, supporting the hypothesis that the NaNa, the most reactive and the most externally located molecule in the gangliosides and in the glycoproteins, could be more interactive with the microenvironment (30).

At present the different developmental trends of the sugars concentration in the soluble and insoluble fractions are difficult to explain, but agree with the observation that in the adult animals the two fractions are different for chemical and physical characteristics (21).

Some data would indicate that the soluble glycoproteins have a great tendency to aggregate so that they show very large molecular weight by Sepharose gel filtration (31). This behaviour suggests that a similar phenomenon could occur to the soluble glycoproteins during the maturation of the brain.

Very recently it was reported that over or undernutrition, applied during the fast growth of the rat brain, modifies the correct developmental pattern of the NaNa of the soluble and insoluble glycoproteins (32) and of the gangliosides (33), impairing the function of the central nervous system (34). These data indicate that any modification in the correct timing of the brain development affects the integrated function of the organ, involving the gangliosides and NaNa containing glycoproteins, components of the neural and, possibly, of the glial membranes.

AKNOWLEDGEMENTS

The authors wish to thank Mr. Antonio Manfredi for his skillful technical assistance.

REFERENCES

(1) McIlwain, H., Biochemistry and the Central Nervous System, J. A. Churchill Ltd. (1966).
(2) Wells, M. A. and Dittmer, J. C., Biochemistry 6: 3169 (1967).
(3) Winick, M. and Noble, A., Develop. Biol. 12: 451 (1965).
(4) Spence, M. W. and Wolfe, L. S., Canad. J. Biochem. 45: 671 (1967).
(5) Rubiolo de Maccioni, A. H. and Caputto, R., J. Neurochem. 15: 1257 (1968).
(6) Hildebrand, J., Stoffyn, P. and Hauser, G., J. Neurochem. 17: 403 (1970).
(7) Shah, S. N., J. Neurochem. 18: 395 (1971).
(8) Bass, N. H., Netsky, M. G. and Young, E., Neurology 19: 405 (1969).
(9) Bunge, R. P., Physiol. Rev. 48: 197 (1968).
(10) Crain, S. M., Proc. Soc. Exp. Biol. Med. 81: 49 (1952).
(11) Blozovski, M., J. Physiol. 55: 202 (1963).
(12) Lapetina, E. G., Soto, E. F. and de Robertis, E., Biochim. Biophys. Acta 135: 33 (1967).
(13) Suzuki, K., Poduslo, S. E. and Norton, W. T., Biochim. Biophys. Acta 144: 375 (1967).
(14) Dekirmenjian, H., Brunngraber, E. G., Lemkey Johnston, N. and Larramendi, L. M. H., Exp. Brain Res. 8: 97 (1969).
(15) Dekirmenjian, H. and Brunngraber, E. G., Biochim. Biophys. Acta 177: 1 (1969).
(16) Dische, Z., Prot. Biol. Fluids 13: 1 (1966).
(17) Rambourg, A. and Leblond, C. P., J. Cell. Biol. 32: 27 (1967).
(18) Bosman, H. B., Hagopian, A. and Eylar, E. H., Arch. Biochem. Biophys. 130: 573 (1969).
(19) Schmid, K., Binette, J. P., Kamiyama, S., Pfister, V. and Takahashi, S., Biochemistry 1: 959 (1962).
(20) Di Benedetta, C., Brunngraber, E. G., Whitney, G., Brown, B. D. and Aro, A., Arch. Biochem. Biophys. 131: 404 (1969).

(21) Di Benedetta, C., Chang, I. and Brunngraber, E. G., Ital. J. Biochem. 20: (3) in press.
(22) Folch, J., Lees, M. and Sloane Stanley, H., J. Biol. Chem. 226: 497 (1957).
(23) Warren, L., J. Biol. Chem. 234: 1971 (1959).
(24) Dische, Z., Shettles, L. B. and Osnos, M., Arch. Biochem. 22: 169 (1949).
(25) Boas, N. F., J. Biol. Chem. 204: 553 (1953).
(26) Lowry, O. H., Rosebrough, G. H., Farr, A. L. and Randall, R. J., J. Biol. Chem. 193: 263 (1951).
(27) Di Benedetta, C., De Luca, B. and Cioffi, L. A., Prot. Biol. Fluids 18: 181 (1970).
(28) Iqbal, Z. and Talwar, G. P., J. Neurochem. 18: 1261 (1971).
(29) Shih, J. -H. C. and Eiduson, S., J. Neurochem. 18: 1221 (1971).
(30) Lehninger, A. L., Proc. Nat. Acad. Sci. (Washington) 60: 1069 (1968).
(31) DI Benedetta, C. and Brunngraber, E. G., data to be published.
(32) Cioffi, L. A., De Luca, B. and Di Benedetta, C., Prot. Biol. Fluids 18: 185 (1970).
(33) Di Benedetta, C. and Cioffi, L. A., XIII Int. Congress of Pediatrics (Vienna) 17 (1971), in press.
(34) Di Benedetta, C. and Cioffi, L. A., Biblioteca Nutritio et Dieta 17: (1971) in press.
(35) Gornall, A. G., Bardawill, C. W. and David, M. M., J. Biol. Chem. 177: 751 (1949).

SECTION II

CHEMISTRY AND METABOLISM OF GLYCOLIPIDS OF THE NERVOUS SYSTEM

RECENT ADVANCES ON THE CHEMISTRY AND LOCALISATION OF BRAIN GANGLIOSIDES AND RELATED GLYCOSPHINGOLIPIDS [1]

H. WIEGANDT

Institut für Physiologische Chemie

Philipps-Universität, Marburg, B. D. R.

Still very little is known about the biological function of the glycosphingolipids. Various experimental approaches were considered in the past by a great number of working groups to learn more of the possible physiological significances of these substances, in particular of their sialic acid containing representatives, the gangliosides. Hereby studies to localise more accurately these lipids in the organs at a cellular and subcellular level played a prominent role. A further attempt was to observe possible variations in the pattern of the glycosphingolipids under normal or pathological conditions. In fact, the quantitative distribution of these substances is not constant, but may change e. g. during the development of an organ (2) or a cellculture (3). Some such changes may partly also be sexdependent or hormoneinduced (4). Great emphasis has been placed lately by various investigators on the elucidation of changes in the pattern of glycosphingolipids during malignant cell transformation. It is of great importance of course for all such studies to

---

[1] Abbreviations: Gangliosides are designated according to principles outlined earlier (1). G stands for ganglioside. An index gives the neutral carbohydrate moiety e. g. G_{Lac}. To this is added the number (and if necessary the nature) of the sialic acid residues, a, b, etc. distinguishing sialic acid position isomers (e. g. G_{Gtet}3b). neuNAc = N-acetylneuraminic acid, neuNGl = N-glycolylneuraminic acid, Lac = lactose, Gtri = gangliotriose, Gtet = gangliotetraose, Lntet = lacto-neotetraose, Gfpt = ganglio-fucopentaose.

know exactly the chemical nature of all the glycosphingolipids occurring normally in the tissue and their quantitative as well as qualitative distribution. This in mind, I would like with the present contribution to report about more recent experiments, which were carried out in collaboration with Prof. Dr. Rahmann and Dr. Rösner, Zoolog. Inst. University of Münster, Germany, aiming at a further localisation of the gangliosides in the central nervous system of fish. The brain ganglioside of fish is very rich in polysialolipids, i.e. tri-, tetra- and pentasialogangliosides (5). In collaboration with Dr. Ishizuka we have elucidated the chemical structures of these gangliosides. In the final portion of this lecture I would like to discuss newly discovered extraneural gangliosides from human as well as bovine spleen, kidney and liver.

Even the crude quantitative comparison of the brain ganglioside distribution between gray and white matter is an indication of their main occurence in ganglion cells. Hence their name "gangliosides". Upon cellular fractionation experiments the highest concentration of brain gangliosides was found by Derry and Wolfe (6) in the nerve fibre terminations, whereas Dekirmenjian (7) observed higher sialic acid to protein ratios in the fraction consisting mainly of axons rather than those containing nerve endings. Tamai et al. (8) concluded from their findings, that the gangliosides were more or less evenly distributed over the whole neuronal membrane.

In the present investigation an attempt was made to localize the gangliosides in the brain of fish, in particular the crucian carp (carassius carassius), because the brain anatomy here is much simpler than in higher vertebrates. Moreover, the localisation and the sites of synthesis of important compounds of basic metabolism, such as RNA, proteins, polysaccharides and some lipids were already determined by H. Rahmann and coworkers in similar experiments (9, 10, 11). For the <u>in vivo</u> labelling of the brain gangliosides radioactive N-acetylmannosamine was chosen, because as far as glycosphingolipids were concerned, this sugar was reported by Brady (12) to be incorporated specifically and exclusively into the sialic acid moiety of these lipids. N-acetylmannosamine was injected intracranially and the incorporation of label followed by autoradiography as well as chemical analysis. The autoradiographic determinations were carried out on the tectum opticum, because of its rather clear structural anatomy. This consists mainly of two strata; one, periventricular layer is composed of perikarya and the other layer is

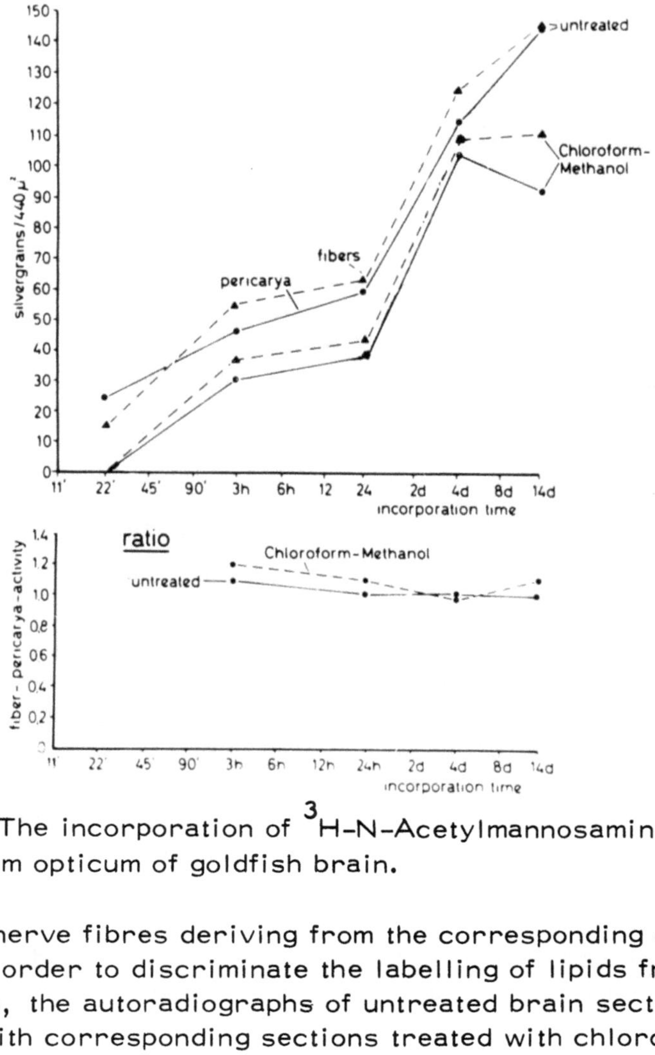

Fig. 1. The incorporation of ^{3}H-N-Acetylmannosamine into the tectum opticum of goldfish brain.

made up of nerve fibres deriving from the corresponding cell bodies beneath. In order to discriminate the labelling of lipids from that of the proteins, the autoradiographs of untreated brain sections were compared with corresponding sections treated with chloroform-methanol. Fig. 1 illustrates the time dependent incorporation of the tracer into the perikarya and fibres of the tectum. It shows the steady increase of labelling in the fibres as well as in the cell bodies over the whole period of incorporation. Unlike in the case of other compounds such as proteins, RNA or polysaccharides, no significant differences in the labelling of fibres and cell bodies could be detected. The same was found irrespective of whether gangliosides had previously been extracted from the tissue-slides or not. For the chemical localisation of the radioactive label the whole brains were extracted with chloroform-methanol, and the gangliosides partitioned into an aqueous phase. After freeze drying and reextrac-

sequence of preparation	I	II	III	IV
A	incubation with neuraminidase	incubation with inactivated neuraminidase	ganglioside extraction	ganglioside extraction
B	dialysis	dialysis	incubation with neuraminidase	incubation with inactivated neuraminidase
C	ganglioside extraction	ganglioside extraction	dialysis	dialysis
D	estimation of ganglioside - neuNAc	estimation of ganglioside - neuNAc	estimation of ganglioside - neuNAc	estimation of ganglioside - neuNAc
µg neuNAc pro mg protein	26.4 ± 2.0	47.1 ± 4.0	27.2 ± 2.4	50.8 ± 6.5

Fig. 2. Estimation of the amount of ganglioside-sialic acid from a microsomal fraction of bovine brain gray matter hydrolysed by neuraminidase. (Sialic acid was measured by thiobarbituric acid method after hydrolysis of the gangliosides with mineral acid).

tion with organic solvent the gangliosides were determined fluorometrically according to Hess and Rolde (13). Protein in the extraction residue was estimated by Lowry's method and the radioactivity measured by liquid scintillation counting. The ratio of labelling of the gangliosides compared to the proteins was time dependent and closely paralleled the data obtained by the method of autoradiography.

The gangliosides of the central nervous system are constituents of the cell membranes. To learn more about their arrangement, i. e. whether they were exposed within these membranes directly to the outer environment, we treated an isolated gangliosiderich membrane fraction of bovine brain with neuraminidase (vibr. col.) and determined the amount of sialic acid split off by the enzyme. A parallel experiment had to be performed, in which at first the gangliosides were extracted from another portion of the same membrane preparation and subsequently treated with the enzyme. This was necessary in order to establish the amount of the ganglioside-sialic acid, from gangliosides $G_{Gtri}1$ and $G_{Gtet}1$, which can not be split by neurami-

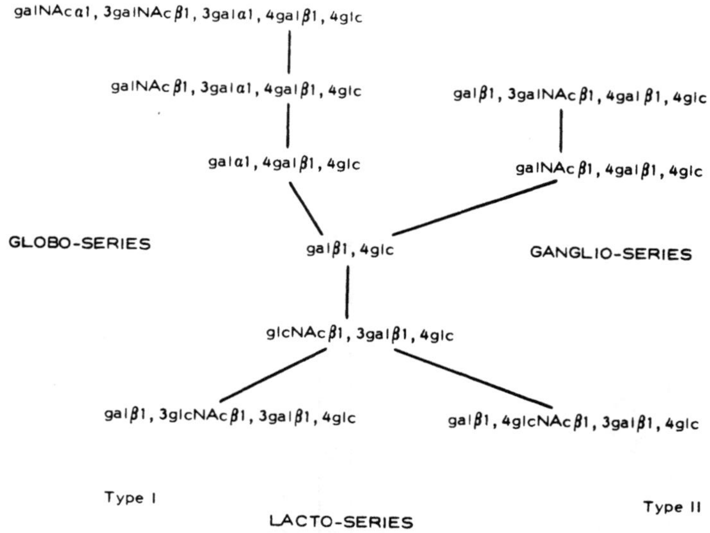

Fig. 3. Oligosaccharide series, from which glycosphingolipids may be derived.

dase. The results, shown in Fig. 2 indicate, that all neuraminidase-labile ganglioside within the membranes was accessible for the enzyme. May I now draw the attention to considerations concerning the chemical structures of these lipids.

The carbohydrate moiety of most of the glycosphingolipids occuring in animal tissue is derived from one of the four oligosaccharide series described in Fig. 3, namely the globo-, ganglio-series or the lacto-series of type I or type II (neo-).

The gangliosides, which have been found in the central nervous system of animals, all belong - with the only exception of ganglioside $G_{Gal}1$ neuNAc (14) - to the ganglio-series. Analysis of the brain gangliosides of <u>fishes</u> confirmed this and we have obtained no indication that, in this case, except for the occurence of 8-0, N-diacetyl-neuraminic acid, the structural design is in any way different from those of mammalian or reptile (crocodile) brain. But there are drastic differences in the quantitative distribution as shown in Fig. 4. We have now elucidated the structures of the tetra- and penta-sialoganglioside and in addition the one of a newly discovered isomere of ganglioside $G_{Gtet}3a$. We did not find this ganglioside in mammalian brain as yet.

Upon treatment of ganglioside $G_{Gtet}5$ with *vibrio cholerae* neura-

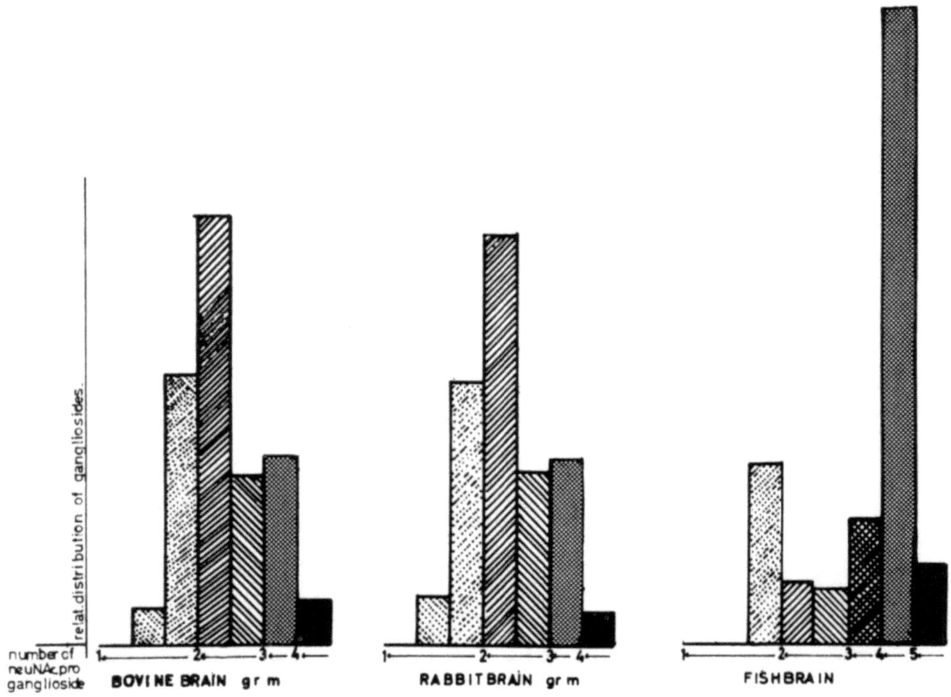

Fig. 4.

minidase and the same by mild hydrolysis with mineral acid, sialic acid residues are cleaved sequentially from the molecule and the following ganglioside transitions were observed:

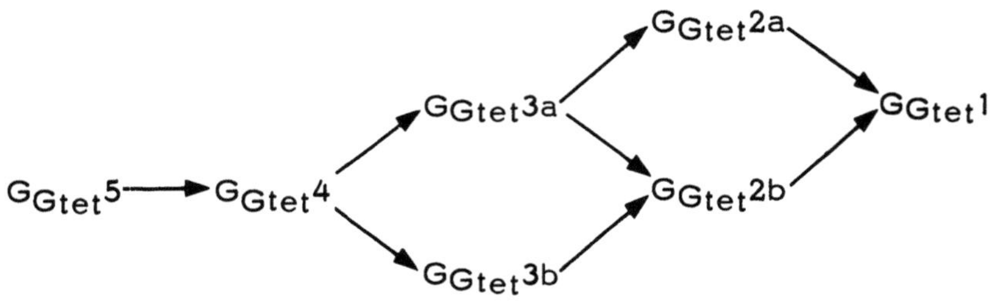

In contrast to this, ganglioside $G_{Gtet}5$ was not degraded by the neuraminidase of the chicken foul plaque virus. This enzyme is known to hydrolyse readily only $\alpha 2 \longrightarrow 3$ and $\alpha 2 \longrightarrow 6$ ketosidically linked sialic acid and leave the neuNAc $\alpha 2, 8$ neuNAc-grouping intact. Ganglioside $G_{Gtet}4$ was split by this neuraminidase to ganglioside $G_{Gtet}3b$ only, which again was resistant to the enzyme:

$$G_{Gtet}5 \not\longrightarrow G_{Gtet}4 \longrightarrow G_{Gtet}3b \not\longrightarrow G_{Gtet}2$$

Periodate oxidation followed by borohydride reduction of ganglioside $G_{Gtet}3b$ was performed and it was found that by this treatment the molecule had lost one residue of galactose and that of the three sialic acid residues one was converted to the 7C-analogue. By the same reaction ganglioside $G_{Gtet}3a$ had not lost any galactose, but two out of three sialic acid residues were oxidized to 7C-sialic acid. Also in gangliosides $G_{Gtet}4$ and $G_{Gtet}5$ both galactose residues were protected against oxidation by periodate. Furthermore in ganglioside $G_{Gtet}4$ two and in ganglioside $G_{Gtet}5$ three sialic acid residues could not be degraded by periodate, because of their substitution by other neuNAc-molecules in the C-8-position. These findings prove the following chemical structures of the fish brain gangliosides:

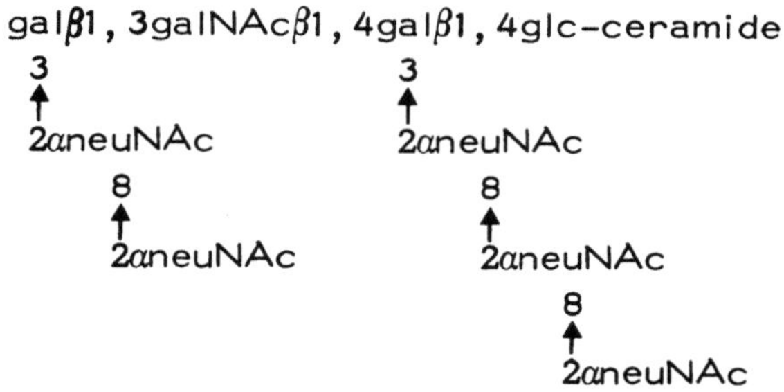

Ganglioside $G_{Gtet}5$

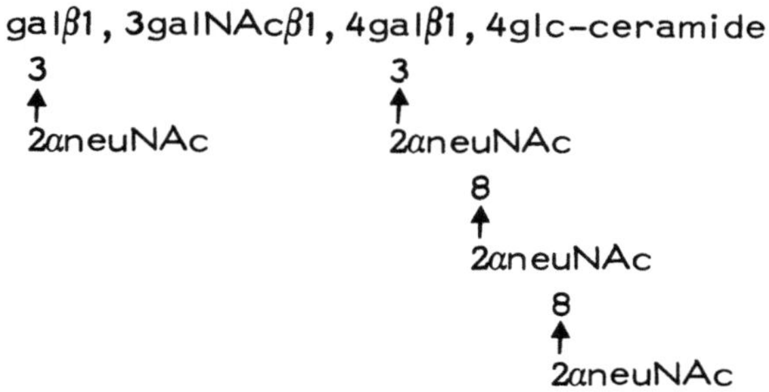

Ganglioside $G_{Gtet}4b$

```
galβ1,3galNAcβ1,4galβ1,4glc-ceramide
                  3
                  ↑
                  2αneuNAc
                     8
                     ↑
                     2αneuNAc
                        8
                        ↑
                        2αneuNAc
```

Ganglioside 3b

```
galβ1,3galNAcβ1,4galβ1,4glc-ceramide
 3                3
 ↑                ↑
 2αneuNAc         2αneuNAc
                     8
                     ↑
                     2αneuNAc
```

Ganglioside $G_{Gtet}3a$

Different from the brain, the carbohydrate moiety of gangliosides of extraneural origin may, in addition to those of the ganglioseries, also be derived from both of the lacto-series (cf. Fig. 3). The gangliosides of bovine spleen as compared to bovine kidney were found to be - at least up to "tetraose" - gangliosides qualitatively very similar. A closer analysis revealed, that these gangliosides were also similar to those isolated from human spleen and kidney with the only exception of the additional occurence of N-glycolylneuraminic acid in the bovine material, which may replace N-acetylneuraminic acid at any of its positions on the neutral carbohydrate backbone. The following gangliosides were found in spleen and kidney:

Gangliosides G_{Lac}

```
galβ1,4glc-ceramide                 galβ1,4glc-ceramide
 3                                   3
 ↑              G_Lac1               ↑              G_Lac2
 2αneuNAc                            2αneuNAc
                                        8
                                        ↑
                                        2αneuNAc
```

Gangliosides G_{Gtri}

galNAcβ1, 4galβ1, 4glc-ceramide
3
↑
2αneuNAc — G_{Gtri}^{1}

Gangliosides G_{Gtet}

galβ1, 3galNAcβ1, 4β1, 4galβ1, 4glc-ceramide
3
↑
2αneuNAc — G_{Gtet}^{1}

galβ1, 3galNAcβ1, 4galβ1, 4glc-ceramide
3 ↑ 2αneuNAc (on galβ1) ; 3 ↑ 2αneuNAc (on galβ1, 4glc) — G_{Gtet}^{2a}

Gangliosides G_{Lntet}

galβ1, 4glcNAcβ1, 3galβ1, 4glc-ceramide
3
↑
2αneuNAc — G_{Lntet}^{1a}

galβ1, 4glcNAcβ1, 3galβ1, 4glc-ceramide
6
↑
2αneuNAc — G_{Lntet}^{1b}

Within the higher gangliosides of spleen and kidney the preponderance of compounds derived from the lacto-neotetraose series was always observed, only trace amounts of gangliosides of the lacto-series of type I were detected. The structure of ganglioside G_{Lntet}^{1b} was found to be identical with the pentasaccharide C of human milk (1) by analysis of the ratio of hexose : hexosamine :

---

(1) In an earlier publication (15) erroneously this ganglioside (Nr. 5 of human spleen) was thought to contain two residues of sialic acid, because of its rather low migration velocity on paper chromatography. This of course may now be explained by the very elongated form of the molecule, when the sialic acid residue stands at the C-6-position of the terminal galactose of the lacto-neo-tetraose.

#	Glc	Gal	GlcNAc	NANA	
3	1	2	1	1	Lac-ntet
5	1	2	1	1	
6	1	3	2	1	Lac-ntet + GlcNAc + Gal
7	1	3	2	1	
8	1	4	2	1	

De-sialo-3 = De-sialo-5

De-sialo-6 = De-sialo-7

Fig. 5. Human spleen gangliosides.

sialic acid, partial hydrolysis and periodate oxidation experiments.

More recently we have isolated some of the higher gangliosides of human spleen (1) in sufficient quantity for analysis of their basic carbohydrate constituents. The data obtained are given in Fig. 5.

The gangliosides Nr. 6 and Nr. 7 were shown to be derived from the lacto-neotetraose series and to be sialic acid position isomers.

A comparison of the sialo-oligosaccharide moieties of the gangliosides of bovine liver with those of bovine spleen done by paper chromatography revealed a drastic difference in the pattern between these two organs (Fig. 6). So far only gangliosides G_{Lac} and the gangliosides $G_{Gtet}1$ and $G_{Gtet}2a$ were found to be present in both organs, spleen and liver. A major difference appeared to be the virtual absence in liver of the in spleen predominating gangliosides of the lacto-neotetraose-series.

In quantity next to the gangliosides G_{Lac} we found in bovine liver a ganglioside, which contains in addition to one mole sialic acid one mole of fucose. This ganglioside could not be split by neuraminidase. The ratio of its basic carbohydrate constituents as obtained after total hydrolysis and gas-liquid-chromatography was found to be glucose : galactose : galactosamine : fucose : N-glycolyl-neuraminic acid 1:2:1:1:1.

(1) For the numbering of these gangliosides confer (15).

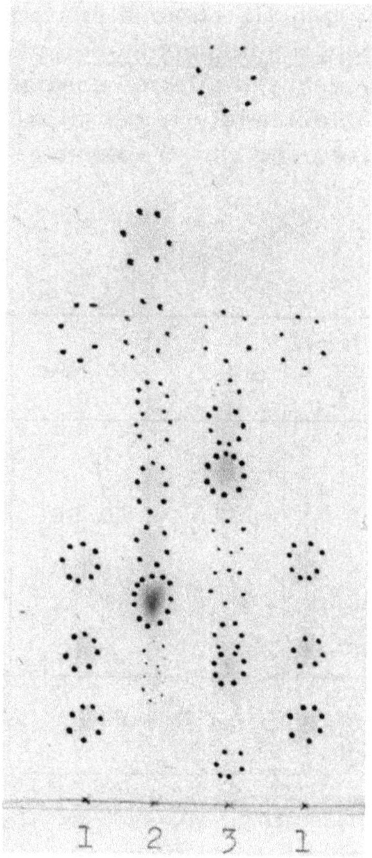

Fig. 6. Paperchromatographic comparison of the sialo-oligosaccharide moieties of the gangliosides of bovine spleen and liver.

1. Bovine brain-des-sphingosine-gangliosides.
   From above: Des-Sph-G_{Gtet}1neuNAc
   Des-Sph-G_{Gtet}2aneuNAc
   Des-Sph-G_{Gtet}2bneuNAc
   Des-Sph-G_{Gtet}3neuNAc.
2. Sialo-oligosaccharides from liver.
3. Sialo-oligosaccharides from spleen.

Mild partial hydrolysis of the monosialo-fuco-pentaose obtained from the liver ganglioside yielded three oligosaccharides. The smallest of them was ganglio-tetraose (identified by 1. ratio of constituent monosaccharides (see Fig. 7), 2. paper chromatography of the tetraose and also of its degradation products after partial hydrolysis or treatment with alcali). The other two were a pentasaccharide containing gangliotetraose substituted either with one mole of fucose, the ganglio-fucopentaose, or one mole sialic acid. Upon paper chromatography the latter compound comigrated with the sialo-oligosaccharide moiety of ganglioside $G_{Gtet}1neuNGl$ isolated from bovine spleen and was the same as this resistant against neuraminidase.

Compound	glc	gal	galNAc	fuc	
Gtpt	1	2	1	1	calc.
	1	2, 03	0, 98	0, 72 (")	found.
Gtet (')	1	2	1	-	calc.
	1	2, 00	1, 01	-	found.

(') From ganglioside $G_{Gfpt}1neuNGl$
(") Uncorrected.

Fig. 7. Carbohydrate analysis (gas-liquid chromatography) of ganglio-fucopentaose and ganglio-tetraose obtained from the liver ganglioside $G_{Gfpt}1neuGl$.

These results indicate that, except for neuNGl instead of neuNAc, the ganglioside $G_{Gfpt}1neuNGl$ of bovine liver resembles the monosialoganglioside first described in the central nervous system with an additional molecule of fucose in a still unknown position.

The majority of the glycosphingolipids of animal origin are with regard to their carbohydrate moiety structurally derived from the lactose molecule. This feature they have in common with the free, reducing oligosaccharides found in secretions, e. g. milk and urine and indeed a great number of such sugars have been found to be structurally completely identical with the carbohydrate portion of some glycosphingolipids. In a final figure all such lactose derived oligosaccharides known up to date, found free in secretions

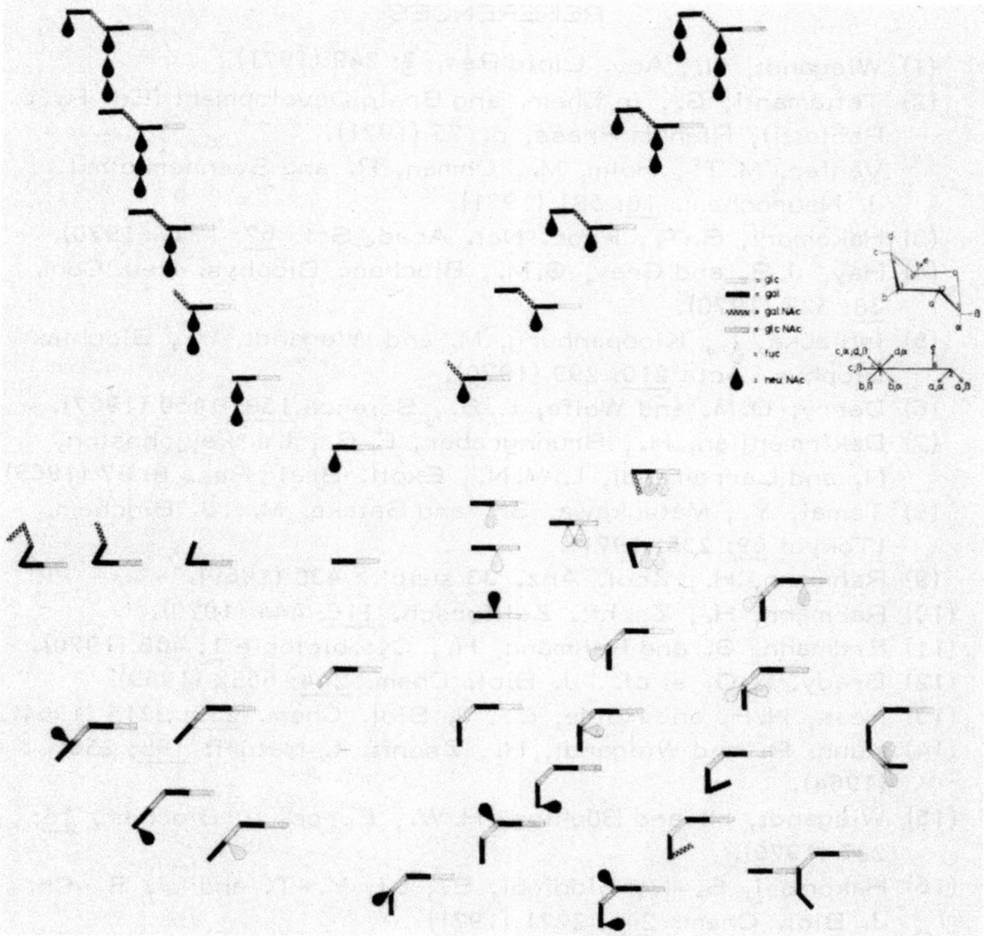

Fig. 8. Schematic representation of the oligosaccharides found in glycosphingolipids and as free reducing sugars in secretions (milk, urine), which are derived from lactose.

or as constituents of lipids have been summarized (Fig. 8). Of all the series only the globosaccharides have never been found to contain branchings of fucose or sialic acid. This may have its reason by the deviation of these sugars from the form of a more elongated molecular chain caused by the introduction of α-glycosidic linkages (16). Comparing the sources and the specificities of all these oligosaccharides it may seem, that on the first level we first observe organ-specificity and, second to this, species-specificity and that mainly in the regions of the higher oligosaccharides specificity of individuality might be expressed.

REFERENCES

(1) Wiegandt, H., Adv. Lipid Res. 9: 249 (1971).

(2) Tettamanti, G., in Chem. and Brain Development (Ed. R. Paoletti), Plenum Press, p. 75 (1971).
Vanier, M.T., Holm, M., Ohman, R. and Svennerholm L., J. Neurochem. 18: 581 (1971).

(3) Hakomori, S.T., Proc. Nat. Acad. Sci. 67: 1741 (1970).

(4) Hay, J.B. and Gray, G.M., Biochem. Biophys. Res. Com. 38: 527 (1970).

(5) Ishizuka, I., Kloppenburg, M. and Wiegandt, H., Biochim. Biophys. Acta 210: 299 (1970).

(6) Derry, D.M. and Wolfe, L.S., Science 158: 1450 (1967).

(7) Dekirmenjian, H., Brunngraber, E.G., Lemkeyjohnston, N. and Larramendi, L.M.N., Exptl. Brain Res. 8: 97 (1969).

(8) Tamai, Y., Matsukawa, Sh. and Satake, M., J. Biochem. (Tokyo) 69: 235 (1971).

(9) Rahmann, H., Zool. Anz. 33 suppl.: 430 (1969).

(10) Rahmann, H., Zschft. Zellforsch. 110: 444 (1970).

(11) Erdmann, G. and Rahmann, H., Cytobiologie 1: 468 (1970).

(12) Brady, R.O. et al., J. Biol. Chem. 244: 6552 (1969).

(13) Hess, H.H. and Rolde, E., J. Biol. Chem. 239: 3215 (1964).

(14) Kuhn, R. and Wiegandt, H., Zschft. f. Naturf. 19b: 256 (1964).

(15) Wiegandt, H. and Bücking, H.W., Europ. J. Biochem. 15: 287 (1970).

(16) Hakomori, S.-I., Siddiqui, B., Li, Y.-T. and Li, S.-Ch., J. Biol. Chem. 246: 2271 (1971).

ENZYMATIC ASPECTS OF SPHINGOLIPID METABOLISM

S. GATT

Department of Biochemistry, The Hebrew University

Hadassah Medical School, Jerusalem, Israel

In the last few years my coworkers and I have been engaged in an investigation of enzymes which hydrolyze the complex lipids. This resulted in the identification, isolation from brain tissue and partial purification of most enzymes which degrade the phospho- and glycosphingolipids, as well as several brain phospholipases. Two review articles summarized parts of this work (1, 2). Recently, a major part of our research efforts has been directed to studies on the interaction of enzymes with their respective lipid substrates. The results of these investigations will be reported elsewhere (3).

During these studies a multitude of data has accumulated on the properties of the above enzymes, though much more is still unknown than has been revealed. As in any new field, errors were made during these investigations: fortunately, most of them were corrected prior to reaching a scientific journal. Some were however published, and I consider it instructive to point out two errors made by us. Both pertain to the enzymology of the lipidoses.

Tay Sachs' disease (Infantile amaurotic familial idiocy; GM_2-gangliosidosis) is characterized, chemically, by the accumulation of a mono-sialo ganglioside (ceramide-glucose-galactose (N-acetyl neuraminic acid) - N-acetyl galactosamine) and of lesser quantities of its asialo derivative (ceramide-Glc-Gal-GalNAc). In 1968, data were already available which suggested that several sphingolipidoses might be caused by a deficiency of hydrolytic enzymes. Since the two compounds which accumulate in Tay Sachs' disease have a terminal β-N-acetyl galactosaminidic linkage, we, as well as other investigators, suspected that this disease might be caused by a deficiency in the enzyme N-acetyl hexosaminidase. We designed experiments to compare the levels of this enzyme in normal human brain tissue and in that of patients who died of Tay Sachs' disease. We have previously found (4) that the enzyme hydrolyzed p-nitrophenyl N-acetyl glucosaminide, p-nitrophenyl galactosaminide, the asialo Tay Sachs' ganglioside, and globoside from human erythrocyte stroma. The GM_2-ganglioside itself was however hydrolyzed very poorly; 24 hr. incubations were required for this purpose. The above comparison of the enzyme levels was therefore done using the p-nitrophenyl hexosaminides and the asialo ganglioside as substrates. With either of these substrates a deficiency of the enzyme in the pathological tissue could not be detected. We therefore concluded that the disease is not caused by a hexosaminidase deficiency and suggested the following alternate hypothesis (2). The rates of all brain sphingolipid hydrolases are about the same magnitude, except for the hydrolysis of the GM_2-ganglioside. The hydrolysis of the terminal N-acetyl galactosaminyl linkages of this compound is about 100-fold slower than that of each of the other hydrolytic steps. This is therefore the rate-limiting step of the total degradation of the brain gangliosides. I suggested (2) that in Tay Sachs' disease the gangliosides might be somewhat less accessible to the hexosaminidase. This would result in a further decrease of the rate of this limiting step, thus unbalancing the metabolic steady state of biosynthesis and degradation, and causing an accumulation of the GM_2-ganglioside. Although this suggestion has not been proven wrong, it is now well established that Tay Sachs' disease is characterized by a deficiency of one isoenzyme, hexosaminidase A, while the second isoenzyme, B, is overcompensated and is present in quantities which exceed those of normal tissue.

The second example is that of Fabry's disease, which is characterized, chemically, by the accumulation of a trihexoside, ceramide-glucose-galactose-galactose. This compound might be derived from globoside (Ceramide-Glc-Gal-Gal-GalNAc) which was considered to have all β-linkages. The trihexoside was therefore assumed to have a terminal β-galactosidic linkage, and an enzyme, considered specific for this bond only, was isolated (5). In a paper on the substrate specificity of brain β-galactosidase (6) I showed that this enzyme hydrolyzed either of the two galactosidic linkages of trihexosides which were isolated from Fabry tissue or from normal human erythrocytes. In a former paper (7) we found that this preparation of β-galactosidase had none or only negligible activity with melibiose as substrate. We therefore assumed that the enzyme was free of α-galactosidase. Hydrolysis of the trihexosyl ceramide by this enzyme therefore further confirmed the presumed all β-linkages of this glycolipid. When, following Kint's report (8) it became clear that the terminal linkage was α, rather than β-galactosidic, we rechecked our β-galactosidase. We found that in spite of the low utilization of melibiose, its p-nitrophenyl α-galactosidase activity was about the same as that of the p-nitrophenyl β-galactosidase. The hydrolysis of the two galactosidic linkages can therefore be accounted for by the combined action of the α- and β-galactosidases.

I would now like to present data on a contribution to the chemical pathology of sphingolipid metabolism. A great volume of work on the metabolic basis of the sphingolipidoses has been published in the last few years. In many of these diseases, the excessive accumulation of a deposited lipid has been attributed to an enzymatic defect of the metabolism of this compound. In all cases investigated to date, the enzymatic defect was on the degradative rather than on the biosynthetic route, and was characterized by a diminished activity or even total deletion of an enzyme or one of its isoenzymes (for review, see 9).

We have recently investigated a "sphingolipidosis" in a yeast. Stodola and Wickerham (10) showed that one strain (NRRL Y-1031, F-60-10) of the yeast Hansenula ciferri produced large quantities of phytosphingosine. This was not stored in the cell, but was secreted into the extracellular medium, where it accumulated not as a free, but as partially or fully acetylated phytosphingosine base (10, 11). Greene et al. (11) have shown that, similar to mammalian systems, this yeast utilizes serine and palmitate for the biosynthesis of the sphingosine bases and several investigators have since utilized this yeast strain to investigate the metabolism of these bases. We were interested in the acetylating enzyme. Two enzymes which acylate sphingosine bases have been described. One utilized free, long-chain fatty acids, in a reversal of the hydrolytic reaction catalyzed by the enzyme ceramidase (12, 13). The second was a microsomal system which utilized long-chain fatty acyl coenzyme A derivatives (14, 15). However, in neither case was acetic acid or acetyl coenzyme A utilized for the formation of O- or N-acetyl derivatives of the sphingosine bases. Using the method of Braun and Snell (16), we disrupted the yeast using a French press (17) and prepared subcellular fractions of this yeast extract. The enzyme (long-chain base acetyl coenzyme A acetyl transferase) was located in the microsomal fraction (17). An extensive investigation of this enzyme (17, 18) revealed several interesting properties which may be summarized as follows:

a. The enzyme acetylated the three main sphingosine bases (i.e., sphingosine, dihydrosphingosine and phytosphingosine) and their N-acetylated derivatives, but not other derivatives of these bases.

b. The bases were acetylated at either their hydroxyl or amino groups; these two activities could not be separated from each other and kinetic experiments could also not distinguish between the two. We therefore concluded that both the O- and N-acetylation are catalyzed by one enzymatic protein. This is of interest since all transacetylases described to date are specific for either O- or N-acetylation. This is the first enzyme which catalyzes the transfer of the acetyl group of acetyl coenzyme A to both hydroxyl and amino groups.

c. Primary, normal amines of 6-18 carbon atoms were substrates for the enzyme; long-chain primary alcohols were not.

d. The enzyme catalyzes a bisubstrate reaction in which one substrate (acetyl CoA) is water soluble while the second (the base) is a lipid. The V/S curves were hyperbolic when the base was maintained at a fixed concentration and that of the acetyl CoA was varied. However, when the acetyl CoA concentration was fixed and that of the base was varied, the curves were non-symmetrical sigmoids. The sigmoidal curves could be converted into rectangular hyperbolas by the addition of serum albumin. These findings were explained by assuming that the enzyme acted on micellar dispersions of the bases but not on their monomers in solution. Albumin-bound monomers could however be utilized by the enzyme. Measurement of the critical micellar concentrations of the bases added support to this theory.

We then turned to the study of the metabolic lesion which results in the excessive accumulation of the acetylated phytosphingosine in this yeast. Dr. Wickerham supplied us with 25 yeast strains, all of the genus Hansenula. We grew them for 4 days and determined their content of long-chain bases. On the basis of this parameter, the strains could be divided into three subgroups: 1. "Low producers", which had 2-5 μmoles bases per liter of growth medium. 2. "Intermediate producers", which had 10-40 μmoles per liter of growth medium. 3. "High producers", which had 120-260 μmoles of long-chain bases per liter of growth medium. All "intermediate and high producers" belonged to strain H. ciferri, NRRL Y-1031. The low producers contained all species other than H. ciferri and also several variants of this species.

We then compared representative strains of each of the above subgroups, and found no correlation between base production and growth rate, generation time or DNA content (Table I).

Table I - Production of Sphingosine Bases and Parameters of Growth.

Experimental property	H. anomala and H. saturnus	H. ciferri NRRL Y-1031		
		Low	Intermediate	High
Long chain base (μmoles/L)	4	10	60	350
Generation time (min)	105	150	135	150
DNA (mg/g packed cells)	1.26	1.14	1.31	1.56
Packed cells (g/L)	29.1	17.0	22.6	20.0

Neither was there any correlation with several general metabolic properties, such as oxygen uptake, CO_2 evolution, incorporation of radioactively-labelled amino acids into cell proteins and incorporation of glucose into the total fatty acids. Radioactively labelled serine, glucose, palmitate or acetate were each incubated for 1 hr. with cell suspensions of the above representative yeast strains. Incorporation of the labelled substrates into the sphingosine bases paralelled the quantities of these bases as enumerated above. Representative experiments, using serine and glucose, are

Table II - Incorporation of ^{14}C-glucose and ^{3}H-serine into Sphingosine Bases.

Experimental property	H. anomala and H. saturnus	H. ciferri NRRL Y-1031		
		Low	Inter.	High
U-^{14}C-glucose incorporated into:				
N-acetyl dihydrosphingosine	-	< 0.1	0.6	16.5
N-acetyl phytosphingosine	-	< 0.1	12.0	58.0
^{3}H-serine incorporated into:				
N-acetyl dihydrosphingosine	0.28	0.16	2.00	3.17
N-acetyl phytosphingosine	0.15	0.13	0.63	2.98
Dihydrosphingosine	0.06	0.03	0.31	0.75

All values are expressed as nmoles of radioactive compound incorporated in one hr by 1 gr of cells, except dihydrosphingosine where the values are calculated per 1 mg microsomes.

shown in Table II. The incorporation into the bases of the "high producers" was 15-500 times greater than into those of the "low producers". In another experiment (shown in the last line of Table II), microsomes prepared from the above strains were incubated for 1 hr. with tritium labelled serine in the presence of palmitate and the necessary cofactors. Dihydrosphingosine was then isolated and its radioactivity determined. The incorporation into this base using the microsomes of the "high producers" was about 20 times greater than with those of the "low producers". We would have preferred measuring the incorporation into phytosphingosine rather than dihydrosphingosine. However, no *in vitro* system for the synthesis of phytosphingosine has been described to date.

To ascertain that the yeast "disease" is not caused by a defect in the degradation of the bases, tritium-labelled dihydrosphingosine or phytosphingosine were incubated with cell suspensions of representative yeast strains. After four days the cells were harvested, the lipids were extracted and the fatty acids were isolated by saponification. No clear cut differences could be detected in the radioactivity content of the normal or hydroxy fatty acids of the various subgroups.

The above results suggest that the "high" and "intermediate producers" have increased rates of biosynthesis of the sphingosine bases, as compared with those of the "low producers". This is most likely the reason for the excessive accumulation of the bases in the afflicted strains. This yeast "lipidosis" is thus a class in itself and differs from the human sphingolipidoses, which are caused by deficiencies of degradative enzymes.

Table III points out one more striking difference between the three yeast subgroups.

Table III - LCB - Acetyl CoA Acetyltransferase Activity.

Reaction catalyzed	H. anomala and H. saturnus	H. ciferri NRRL Y-1031		
		Low	Inter.	High
O-acetylation of N-acetyl phytosphingosine	0.04	0.06	6.08	11.60
O- and N-acetylation of phytosphingosine	0.04	0.13	2.85	3.30
N-acetylation of phytosphingosine	0.02	0.06	1.26	1.59

All values are expressed as nmoles ^{3}H-acetate transferred by 1 mg microsomes in 1 hr.

The enzyme which transfers the acetyl group of acetyl CoA to the bases was present in the "high" and "intermediate producers" but was practically absent from the strains subgrouped as "low producers". This evokes some thoughts on the possible role of this enzyme in the excessive accumulation of the bases. We cannot decide to date whether the enzyme is responsible for the survival of the "sick" yeast or, conversely, has actually caused the "disease" An argument can be presented for either of these two possibilities. If, because of some genetic defect, biosynthetic rates increased, the elevated levels of the bases will probably be toxic to the cell and these must be "detoxified". It is thus probable that only those yeast strains which adapted to this situation and produced an enzyme which neutralized the amino group through acetylation, survived. Furthermore, the fact that the base in the outer medium is not free but acetylated, suggests that probably only the acetylated base can be secreted. Thus, production of the acetylating enzyme not only relieved the cell of a toxic base, but also enabled it to secrete the excess through the cell wall.

The second argument is that the enzyme might actually have caused the disease. It is possible that the free base exerts a feed-back inhibition or repression on its own biosynthesis. Those cells which have the enzyme, maintain a very low level of the free base, thus eliminating this control of biosynthesis. This results in increased biosynthetic rates, and in excessive accumulation of the acetylated bases. A point of interest to this discussion is the probability that the acetylation is irreversible. We could not demonstrate splitting of the N-acetyl linkage of the acetylated dihydrosphingosine by the yeast extracts. Further experiments such as elucidation of the genetics of the Hansenulas or induction of the acetylating enzyme in the "low producers", might assist in clarifying the role of the acetylating enzyme, as discussed above. Experiments are currently in progress to determine which of the enzymatic steps on the biosynthetic route of the sphingosine bases is the one whose rate is elevated.

My last subject is a brief progress report on a topic which I have frequently discussed, but supported by rather few experiments. All the sphingolipid hydrolases of brain tissue reside on one subcellular fraction which is obtained when rat brain particles are subjected to sonic irradiations. This, as well as theoretical considerations, have suggested the possibility that these enzymes might be grouped in a multi-enzyme complex (2). We have recently initiated the following experiments to investigate this. All the sphingolipid hydrolases required addition of detergents for activity (1). However, use of these detergents disrupted the membrane holding these enzymes (i. e., the proposed multi-enzyme complex) and solubilized them. We are attempting to disperse these complex lipids to enable them to interact with the membranous enzymes in the absence of detergents; these experiments were mostly unsuccessful to date. Experiments are in progress to further purify the subcellular membranous fraction which contains the bound sphingolipid hydrolases. Fig. 1 shows the results of a subfractionation of rat brain particles in a discontinuous sucrose density gradient. By

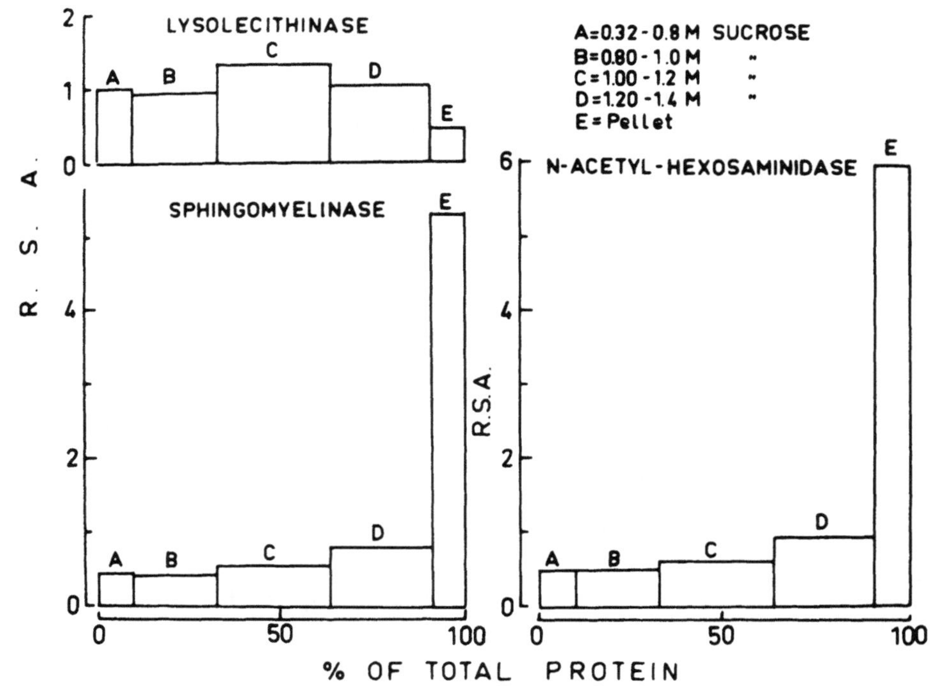

Fig. 1. Subcellular localization of sphingomyelinase, β-N-acetyl hexosaminidase and lysolecithinase in rat brain.

using Koenig's fractionation (19) the relative specific activity was greatly increased in the pellet-E, which sedimented below 1.4M sucrose. The similarity of the distribution of sphingomyelinase and N-acetyl hexosaminidase strongly suggests a localization in lysosomes or a lysosome-like particle in brain (for comparison, the dissimilar distribution of the non-lysosomal enzyme lysolecithinase is also shown in Fig. 1). Similar to the behaviour of the intact brain particles, the partially purified preparation could be subjected to sonic irradiation or hypotonic disruption with only partial loss of enzymatic activity.

ACKNOWLEDGEMENTS

The investigations on the metabolism of H. ciferri and on the acetylating enzyme were made by Y. Barenholz and I. Edelman. The experiments on the subcellular fractionation were performed by A. Herzl. All the investigations here reported were supported by a U.S. Public Health Service Grant (NS-02967).

REFERENCES

(1) Gatt, S., in Lipids, Methods in Enzymology, vol. 14 (Ed. J.M. Lowenstein), Academic Press, p. 134 (1969).
(2) Gatt, S., Chem. Phys. Lipids 5: 235 (1970).
(3) Gatt, S., Barenholz, Y., Borkovski, I. and Leibovitz-Ben Gershon, Z., in "Sphingolipids, Sphingolipidoses and Allied Diseases" (Ed. B.W. Volk), Plenum Press, to be published.
(4) Frohwein, Y.Z. and Gatt, S., Biochemistry 6: 2783 (1967).
(5) Brady, R.O., Gal, A.E., Bradley, R.M. and Martensson, E., J. Biol. Chem. 242: 1021 (1967).
(6) Gatt, S., Biochim. Biophys. Acta 137: 192 (1967).
(7) Gatt, S. and Rapport, M.M., Biochem. J. 101: 680 (1966).
(8) Kint, J.A., Science 167: 1268 (1970).
(9) Brady, R.O., Ann. Rev. Med. 21: 317 (1970).
(10) Wickerham, L.J. and Stodola, F.H., J. Bacteriol. 80: 484 (1960).
(11) Greene, M.L., Kaneshiro, T. and Law, J.H., Biochim. Biophys. Acta 98: 582 (1965).
(12) Gatt, S., J. Biol. Chem. 241: 3724 (1966).
(13) Yavin, E. and Gatt, S., Biochemistry 8: 1692 (1969).
(14) Sribney, M., Biochim. Biophys. Acta 125: 542 (1966).
(15) Morell, P. and Radin, N.S., J. Biol. Chem. 245: 342 (1970).
(16) Braun, P.E. and Snell, E.E., Proc. Natl. Acad. Sc. 58: 298 (1967).
(17) Barenholz, Y. and Gatt, S., Biochem. Biophys. Res. Comm. 35: 676 (1969).
(18) Barenholz, Y. and Gatt, S., submitted for publication.
(19) Koenig, H., Gaines, D., McDonald, T., Gray, R. and Scott, J., J. Neurochem. 11: 729 (1964).

RECENT STUDIES ON THE ENZYMES THAT SYNTHESIZE BRAIN GANGLIOSIDES (°)

J. A. DAIN, J. L. DI CESARE, M. C. M. YIP

Dept. of Biochemistry, University of Rhode Island,
Kingston, Rhode Island 02881

and H. WEICKER

S. J. Thannhauser-Abteilung fur Stoffwechseluntersuchungen, Medizinische-Univ-Poliklinik, Heidelberg

INTRODUCTION

The enzymic synthesis of brain gangliosides proceeds via a stepwise addition of monosaccharide units from sugar nucleotides to glycosphingolipid acceptors. Several possible pathways for ganglioside biosynthesis have been proposed (s. Scheme) utilizing this mechanism. Preliminary reports by Steigerwald et al. (1) and Kaufman et al. (2) have described a particulate embryonic chick brain preparation which would catalize the synthesis of gangliosides via the sequence of reactions 2, 3, and 4. The demonstrations in brain of reaction 3 in the frog (3) and rat (4, 5), reaction 4 in the frog (3, 6, 7) and rat (5, 8, 9), and reaction 2 in frog (Table I) suggests that this pathway is probably universal in vertebrates.

(°) Abbreviations: the abbreviations for individual gangliosides are those proposed by Svennerholm (19). Gp_1 is a tetrasialoganglioside and Gp_2 is an unknown polysialoganglioside in frog brain.

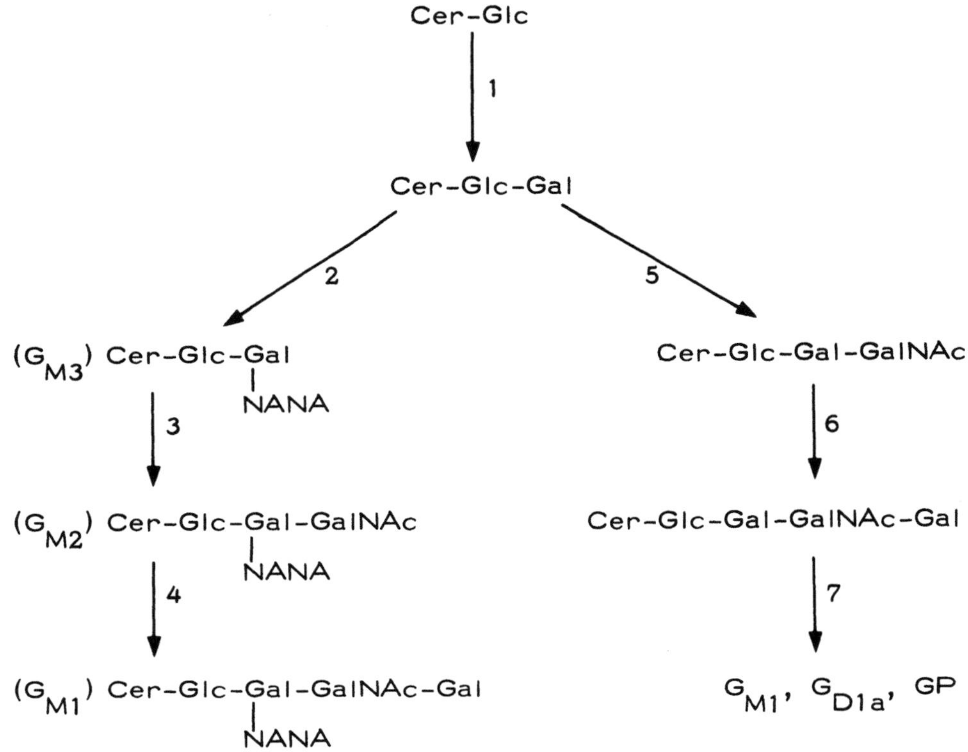

THE ALTERNATE PATHWAY FOR GANGLIOSIDE BIOSYNTHESIS

Experimental data from our laboratory have led us to suggest (7, 10) that ganglioside biosynthesis can also proceed by an alternate pathway (reactions 5, 6, and 7). The first indication of an alternate pathway for ganglioside biosynthesis was from our studies with the *in vivo* incorporation of glucosamine 1-^{14}C into frog brain gangliosides (Figure 1).

The rate of the incorporation of glucosamine-1-^{14}C into the individual gangliosides (Fig. 1) shows a sharp increase in radioactivity into the polysialogangliosides (G_{P1} and G_{P2}), reaching a maximum between 24 and 50 hours after injection. During this time, the ^{14}C incorporation into the other major gangliosides (G_{M2}, G_{M1}, and G_{D1a}) was relatively constant and at a considerably lower level. This *in vivo* incorporation study suggested a complete ganglioside synthesizing system. However, the much

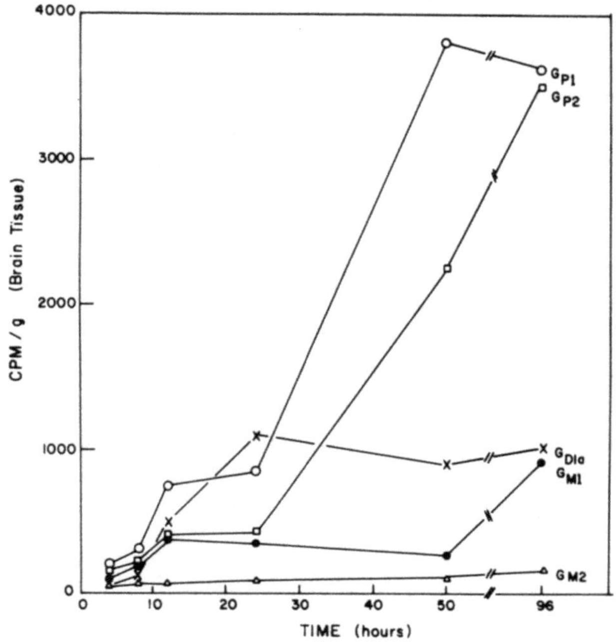

Fig. 1. The in vivo incorporation of D-glucosamine-1-^{14}C into gangliosides of adult frog brains (Rana pipiens) as a function of time (11, 12). D-glucosamine-1-^{14}C (0.5 $\mu c/g$ body weight) was injected into adult male frogs which were subsequently sacrificed at various time intervals after injection and the brains removed. The brain tissue was extracted using a "Folch" (13) procedure with 20 volumes of chloroform-methanol (2:1 v/v) and partitioned with 0.2 volumes of 0.1 N KCl. The upper phase was removed and the lower phase washed twice with an equal volume of theoretical upper phase containing 0.1 N KCl and once without KCl. The upper phases were combined, dialized and lyophilized. The incorporation into crude gangliosides increased linearly up to 50 hours (11). To remove radioactive nucleotide sugars, which have been reported (14) to contaminate the ganglioside preparations, the crude gangliosides were treated with phosphodiesterase and alkaline phosphatase. Control experiments indicated that this treatment removed less than 10% of the radioactivity in the crude ganglioside preparations. It should be mentioned, that a significant amount of radioactivity was lost after the above enzymatically treated preparation was further purified (11) by TLC with n-propanol: water, 3:1 as the developing solvent.

more rapid labeling of the polysialogangliosides (G_{P1} and G_{P2}) after 25 hours suggested to us that they are synthesized by a route other than that of G_{M2} (3, 11) and GM_1 (7) whose synthesis we could demonstrate in vitro.

The evidence for this alternate pathway was further augmented by our discovery (15) of an enzymic activity in tadpole, adult frog, 8-day old and adult rat brain, which catalyzed the transfer of galactose from UDP-Gal to GalNAc-Gal-Glc-Ceramide (Reactio 6). This finding, together with the report by Handa and Burton (16) which was later confirmed by our laboratory (4, 5) of GalNAc-Gal-Glc-Ceramide (Reaction 5) and the further reports of a CMP-NANA: Gal-GalNAc-Gal-Glc-Ceramide transferase activity (Reaction 7) in embryonic chick (2) and young rat (17) brain led us to examine the activity of CMP-NANA glycolipid transferase(s) with various substrates in a frog brain homogenate (Table I).

Table I - Substrate Specificity of CMP-NANA: Glycolipid Transferase (11).

Adult frog brains were homogenized in 0.25M sucrose solution containing 0.11% 2-mercaptoethanol. The incubation mixtures and assay procedures have been described (2). CMP-NANA (Ac^{14}C), 50 nmoles (6.92×10^5 CMP/nmoles) was a gift from Dr. J. Hickman. 1.7 mg protein was used for each assay. All glycolipids were present in the concentration of 100 nmoles.

Substrate		(^{14}C) NANA incorporated (counts/min)
None		116
Gangliosides:	G_{M3}	117
	G_{M2}	210
	G_{M1}	145
	G_{D1a}	210
	G_{T1}	205
Glc-Cer		225
Gal-Glc-Cer		321
Gal-Gal-Glc-Cer		495
Gal-GalNAc-Gal-Glc-Cer		7755

Gal-GalNAc-Gal-Glc-Ceramide (Reaction 7) was by far the best substrate under the assay conditions employed. Gal-Gal-Glc-Ceramide, which should not be a substrate in any proposed ganglioside biosynthetic pathway, was, in fact, the second best acceptor! Better in fact than Gal-Glc-Ceramide (Reaction 2), a proposed intermediate in the originally suggested ganglioside biosynthetic pathway (2) (Reaction 2, 3, and 4). Even with G_{M1}, G_{D1a} and G_{T1}, which would presumably be substrates for the formation of G_{P1} and G_{P2}, only a minimal transfer of NANA from CMP-NANA was observed. Since labelled G_{D1a}, G_{P1} and G_{P2} were formed from endogenous acceptors (Fig. 1) a specific transferase(s) must be present in frog brain for their synthesis. The deduction from the experimental data above led us to conclude that reactions 5, 6, and 7 represent another major pathway for ganglioside biosynthesis (15).

The enzymes catalyzing the different reaction in each pathway have been studied by various investigators using different systems and it has yet to be determined just how many enzymes are involved in ganglioside biosynthesis. Substrate competition studies (8) with the galactosyl transferases catalyzing reactions 1 and 4 have shown they are different enzymes. Competition studies in our laboratory (5) between the substrates G_{M2} and GalNAc-Gal-Glc-Ceramide in reactions 4 and 6 respectively as shown in Figure 2, demonstrated a competitive type inhibition. The same type of competitive inhibition curve was obtained using the reverse conditions (5), that is, varying the GalNAc-Gal-Glc-Ceramide concentration against a constant concentration of G_{M2}. This experimental test suggests that reactions 4 and 6 are catalyzed by the same galactosyltransferase. When the same type of competition experiments were performed on the N-acetylgalactosaminyl transferase(s) in reactions 3 and 5 using G_{M3} and Gal-Glc-Ceramide as substrate, no competition was observed, strongly suggesting that the N-acetylgalactosaminyl transferase involved in reaction 3 is different from the one involved in reaction 5. The nonidentity of the N-acetylgalactosaminyl transferases, which are essentially branch point enzymes between the two pathways, further indicated the importance of 2 separate pathways for the biosynthesis of gangliosides. The subcellular localization according to the methods of Lapetina et al. (18) of the transferases involved in reactions 3 and 4 are reported elsewhere (5) and indicate that these two enzymes are closely associated with gangliosides and with acetylcholinesterase on the synaptic membranes of 7-day old rat brain.

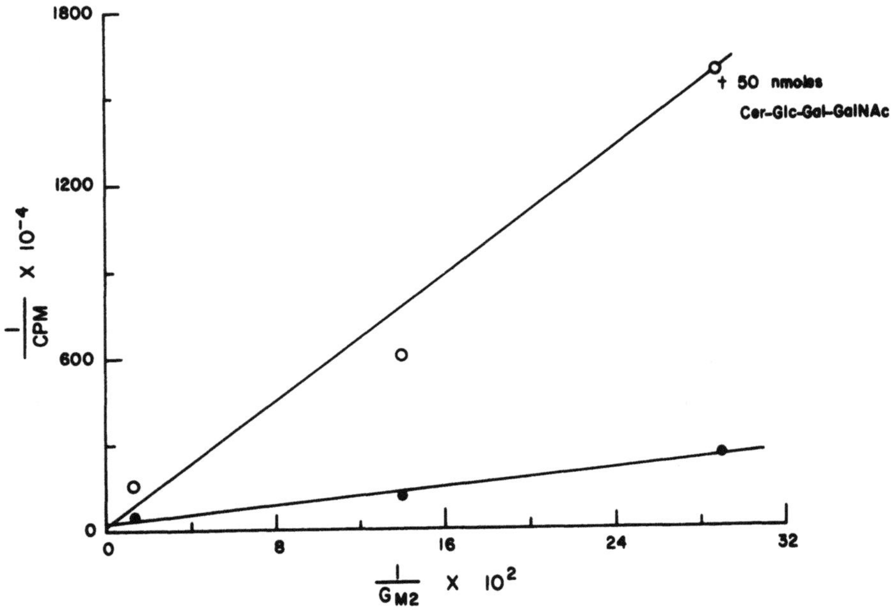

Fig. 2. Competition study where the G_{M2} concentration is varied against a fixed concentration of GalNAc-Gal-Glc-Ceramide.

PRODUCT INHIBITION OF GANGLIOSIDE BIOSYNTHESIS IN THE RAT

The observations that product gangliosides can inhibit their own biosynthetic transferase enzymes (Reactions 3 and 4) by complexing with the Mn^{+2} which activated these enzymes in a frog brain preparation has been described (3, 10). These studies have been extended using young rat brain. Table II demonstrates that increasing concentrations of G_{M2}, the product of reaction 3, increasingly inhibits this reaction.

The same type of inhibition was observed under similar experimental conditions for the galactosyl transferase of both reaction 4 and 6 where increasing concentrations of either G_{M1} or mixed brain gangliosides were added. This study suggests that the same type of feedback inhibition mechanism originally suggested (7, 10) for frog brain ganglioside synthesis in reaction 4 can be postulated in rat brain for reaction 3 and 4 and also for the first time for reaction 6 of the alternate pathway.

Table II - Inhibition of the N-Acetylgalactosaminyl Transferase in Reaction 3.

The incubation mixtures and assay procedures for this type of rat brain preparation have been described (4). The incubation time was 4 hours. Each tube contained 0.17 mg microsomal protein and 50 nmoles G_{M2} as substrate.

Product Addition	% Inhibition
NONE	0
100 nmoles G_{M2}	27
200 nmoles G_{M2}	51
300 nmoles G_{M2}	74
400 nmoles G_{M2}	81

UDP-GALACTOSE: G_{M2} GALACTOSYL TRANSFERASE IN FROG RED CELL STROMA (REACTION 4)

The galactosyl transferase in reaction 4 can be demonstrated (14) in the red cell stroma from *Rana temporaria* and *Rana pipiens*. However, reaction 3, which is the preceeding reaction in the ganglioside biosynthetic pathway, could not be demonstrated in the stroma nor could G_{M1} or any higher gangliosides be isolated from the frog red cell stroma. Furthermore, attempts to demonstrate reaction 4, the red cell stroma of other species (chicken, rabbit, rat, human, and fetal human) proved unsuccessful. The reason for this activity in frog red cell stroma remains a mystery.

SUMMARY

(1). The evidences for two ganglioside biosynthetic pathways are detailed.

(2). Competition studies suggest that the N-acetylgalactosaminyl transferase involved in the formation of G_{M2} from G_{M3} is different from the N-acetylgalactosaminyl transferase involved in the formation of GalNAc-Gal-Glc-Ceramide from Gal-Glc-

Ceramide , while, in contrast, both the conversion of G_{M1} from G_{M2} and of Gal-GalNAc-Gal-Glc-Ceramide from GalNAc-Gal-Glc-Ceramide appear to be catalyzed by the same galactosyl transferase.

(3). A feedback inhibition mechanism for ganglioside biosynthesis in rat brain is proposed.

(4). The finding of UDP-Galactose: G_{M2} Galactosyl transferase activity in frog but not chicken, rabbit, rat, human, and fetal human red cell stroma is presented.

ACKNOWLEDGEMENTS

The work was supported in part by grants from the National Institute of Health, USPHS and the Deutschen Forschungsgemeinschaft.

REFERENCES

(1) Steigerwald, J. C., Kaufman, B., Basu, S. and Roseman, S., Fed. Proc. 25: 587 (1966).
(2) Kaufman, B., Basu, S. and Roseman, S., in Aronson, S. M. and Volk, B. W., Proc. 3rd Intern. Symp. Cerebral Sphingolipidoses, Pergamon Press, New York, p. 193 (1966).
(3) Dain, J. A., Mark, M., Yip, M. C. M., Yiamouyiannis, J. and Y-Cha, 154th Am. Chem. Soc. Meeting, Chicago, 69 c (1967).
(4) Di Cesare, J. L. and Dain, J. A., Biochim. Biophys. Acta 231: 385 (1971).
(5) Di Cesare, J. L. and Dain, J. A., J. Neurochem., in press, (1972).
(6) Yiamouyiannis, J. A. and Dain, J. A., Lipids 3: 378 (1968).
(7) Yip, M. C. M. and Dain, J. A., Biochem. J. 118: 247 (1970).
(8) Hildebrand, J., Stoffyn, P. and Hauser, G., J. Neurochem. 17: 403 (1970).
(9) Yip, G. B. and Dain, J. A., Biochim. Biophys. Acta 206: 252 (1970).
(10) Dain, J. A. and Yip, M. C. M., Metabolismo 5: 129 (1969).
(11) Yip, M. C. M., Studies on the Biosynthesis of Gangliosides in Rana Pipiens, Ph. D. Thesis, University of Rhode Island (1968).

(12) Di Cesare, J. L. and Dain, J. A., unpublished data.
(13) Folch, J., Lees, M. and Sloane-Stanley, G. H., J. Biol. Chem. 226: 497 (1957).
(14) Kanfer, J. N. and Brady, R. O., in Aronson, S. M. and Volk, B. W., Proc. 3rd Intern. Symp. Cerebral Sphingolipidoses, Pergamon Press, New York, p. 187 (1968).
(15) Yip, M. C. M. and Dain, J. A., Lipids 4: 270 (1969).
(16) Handa, I. S. and Burton, R. M., Lipids 4: 589 (1969).
(17) Arce, A., Maccioni, H. F. and Capputto, R., Arch. Biochem. Biophys. 116: 52 (1966).
(18) Lapetina, E. G., Soto, E. F. and De Robertis, E., Biochim. Biophys. Acta 135: 33 (1967).
(19) Svennerholm, L., J. Lipid Res. 5: 145 (1964).

BRAIN NEURAMINIDASES

G. TETTAMANTI, B. VENERANDO, A. PRETI,

A. LOMBARDO and V. ZAMBOTTI

Institute of Biological Chemistry, Medical School

University of Milan, Italy

INTRODUCTION

The presence in the brain of a neuraminidase (sialidase; N-acetyl-neuraminylhydrolase, E.C., 3.2.1.18), reported by Carubelli et al. in 1962 (1), was later confirmed by many investigators (2, 3, 4, 5, 6).

The first detailed description of a mammalian brain neuraminidase was given by Leibovitz and Gatt (7) in 1968. The enzyme, prepared from calf brain, was reported to be particle-bound and specific for gangliosides. The finding was confirmed by Di Donato and Tettamanti (8) in pig and by Preti et al. (9) in rat brain.

Conversely Tettamanti and Zambotti (10) revealed and purified from pig brain a soluble neuraminidase which had no specificity for gangliosides. A soluble neuraminidase was also described in rat brain by Carubelli (11) and in human brain by Ohman et al. (12).

Notwithstanding the numerous investigations carried out on the topic, the relationship between particle bound and soluble neuraminidase, the substrate specificity and the physiological involvement of

---

Abbreviations used: NANA = N-acetylneuraminic acid; NGNA = N--glycolylneuraminic acid.

the two enzymes have not yet been established.

In this report the available data on brain neuraminidases will be critically reviewed and some recent results presented.

SEPARATION OF SOLUBLE AND PARTICULATE NEURAMINIDASES

The presence in brain of particle bound and soluble neuraminidase has been definitely ascertained. The separation of the two enzymes is attained by brain homogenization in isotonic buffered sucrose solution (0.32 M sucrose, 10^{-6} EDTA M, 10^{-6} M $MgSO_4$, 5.10^{-5} M Tris, final pH 7.2) under the mild conditions recommended by Gray and Whittaker (13) and De Robertis et al. (14) (glass homogenizer provided with a teflon pestle, clearance 0.25 mm, rotation at 800 r.p.m.; two homogenization periods, 1 min and 5 up and down strokes each, with a refrigeration interval of 1 min in an ice bath) and centrifugation at 105.000 x g for 1 hour (15). The pellet and the supernatant obtained constitute respectively the "crude preparation of particulate neuraminidase" and the "crude preparation of soluble neuraminidase".

ASSAY OF BRAIN NEURAMINIDASES (°)

1) The endogenous substrates of brain neuraminidase

The crude preparation of brain particulate neuraminidase contains practically all the sialocompounds (sialoglycolipids – gangliosides–, and sialoglycoproteins) present in brain (16, 17).

In 1963 Morgan and Laurell (2) provided evidence, later confirmed by several authors (7, 18), that NANA could be released from brain particulate material by the action of neuraminidase. More recently the features of the hydrolysis of endogenous substrates by particulate neuraminidase have been elucidated (12, 19, 20).

The most important aspects of this phenomenon, as occurring in the brain of different animals, are presented in Fig. 1, 2, 3 and 4.

(°) 1 Unit of neuraminidase is defined as the amount of enzyme which liberates, under the given conditions, 1 nMole NANA/min.

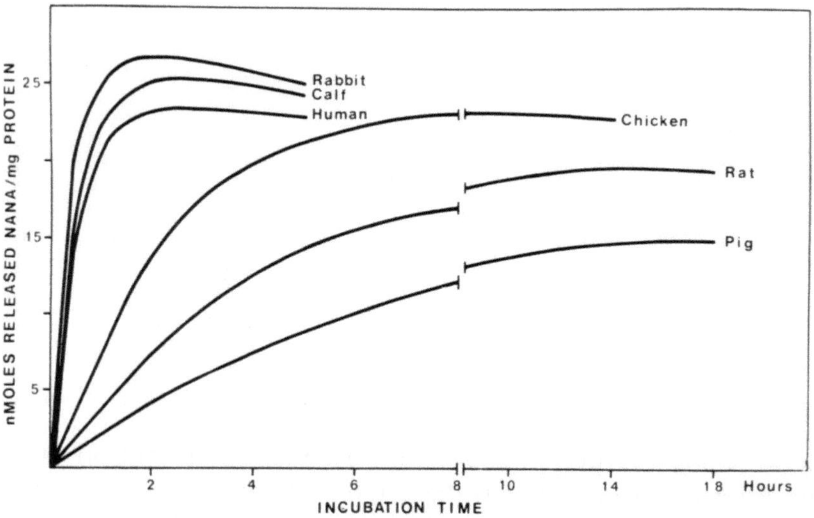

Fig. 1. Effect of time on the hydrolysis of endogenous substrates by brain particulate neuraminidase from different animals. The experiments were performed at 37°C in the presence of Triton X-100 at the optimal final concentration.

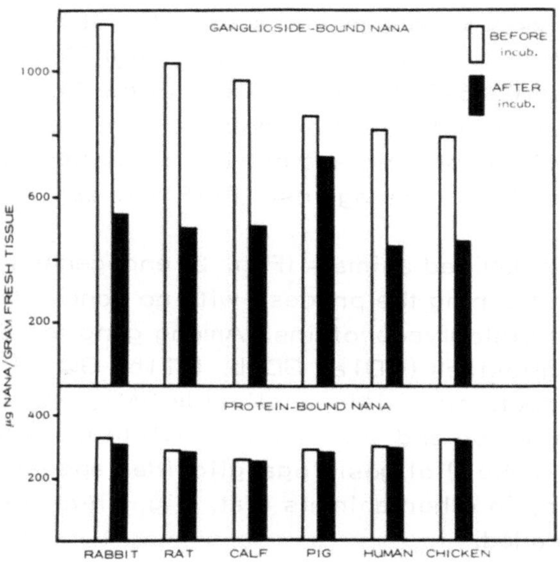

Fig. 2. Ganglioside-bound and protein-bound NANA present in the crude preparation of brain particulate neuraminidase from different animals before and after incubation under the conditions of maximal enzymatic hydrolysis of endogenous substrates.

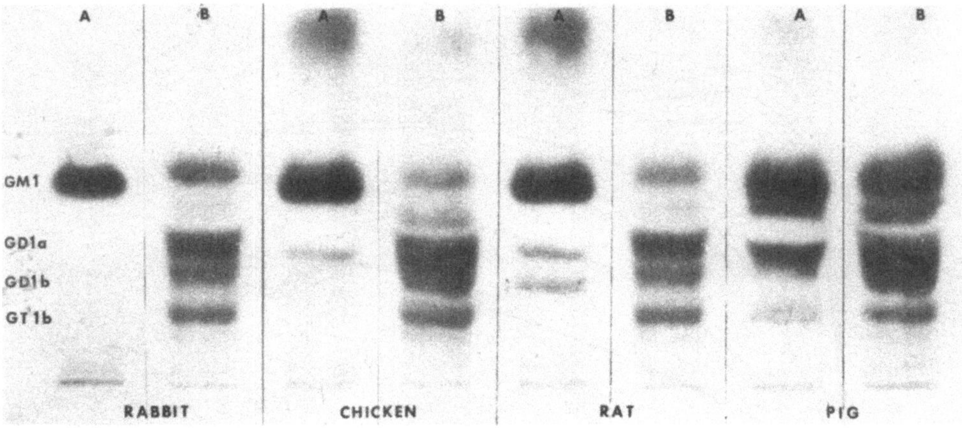

Fig. 3. Gangliosides present in the crude preparation of brain particulate neuraminidase from different animals before (B) and after (A) incubation under the conditions for the maximum hydrolysis of neuraminidase endogenous substrates. The situation of human and calf brain is similar to that depicted for rabbit brain.

In the brain of some animals (human, rabbit, calf) (Fig. 1) the rate of hydrolysis of endogenous substrates at 37° is much higher than that of other animals (chicken, rat, pig) (0. 6-0. 8 Units/mg protein against 0. 05-0. 2 Units). Also, in the former group of animals the maximum level of hydrolysis is higher (23-27 nMoles released NANA/mg protein against 15-19 nMoles) and reached in a shorter time (1-2 hours against 10-18 hours).

In all the examined animals (Fig. 2) endogenous gangliosides are hydrolyzed during the process with no concurrent degradation of endogenous sialoglycoproteins. Among gangliosides (Fig. 3) oligosialogangliosides (GD1a, GD1b, GT1b, GQ1) (°) are degraded to a final product, monosialoganglioside GM1, which is resistant to the enzyme action and accumulates. While in some animals (human, calf, rabbit) oligosialogangliosides appear to be completely exhausted, in other animals (rat, pig, chicken) they are only partially degraded.

The kinetics of the hydrolysis of endogenous oligosialogangliosides by particulate neuraminidase from rabbit brain is represent-

(°) The ganglioside nomenclature according to Svennerholm has been adopted (21).

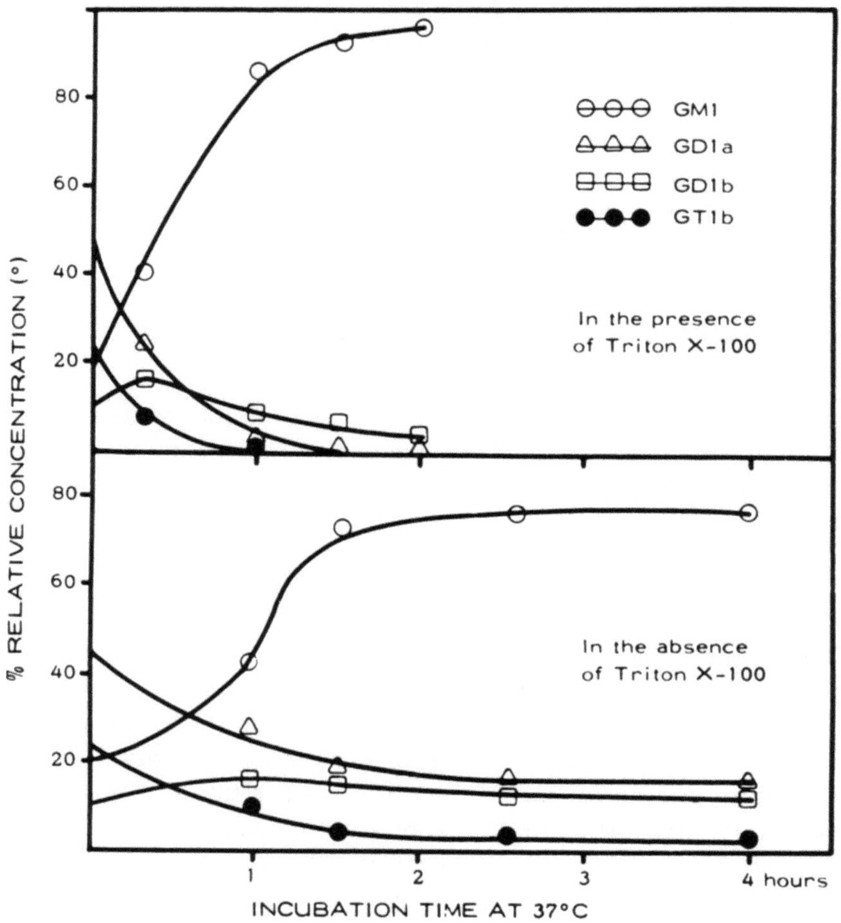

(°) The relative concentration of each ganglioside is expressed as % of the NANA content of the individual ganglioside on total ganglioside NANA.

Fig. 4. Action of the particulate neuraminidase from rabbit brain on the different gangliosides present in the crude enzyme preparation. The incubation were performed at 37°, for different times, in the presence (0.3% final concentration) or absence of Triton X-100.

ed in Fig. 4. The rate and the final level of hydrolysis of the different gangliosides is highly enhanced by Triton X-100 (optimum final concentration: 0.3%). Disialoganglioside GD1a, trisialoganglioside GT1b and tetrasialoganglioside GQ1 appear to be hydrolyzed at higher rate than disialoganglioside GD1b, which actually tends to accumulate when the incubation is carried out in the absence of Triton X-100.

Thus the general conclusions which can be drawn from these studies are as follows:

a) oligosialogangliosides act as the endogenous substrates of the enzyme (monosialoganglioside GM1 and sialoglycoproteins being resistant);

b) the complete autohydrolysis of endogenous substrates cannot be achieved in all animals. This second evidence must be taken in mind in the choice of criteria for the assay of brain particulate neuraminidase.

2) Assay of brain particulate neuraminidase

Conventionally a meaningful enzyme assay requires the absence of endogenous substrates from the enzyme preparation and the use of a proper substrate added in sufficient amounts to saturate the enzyme. The assay of brain particulate neuraminidase can hardly fullfil these requirements. In fact only in the case of some animals (human, calf, rabbit) the endogenous substrates of particulate neuraminidase can be completely exhausted by autolysis or other means.

Furthermore even if the present evidences indicate oligosialogangliosides as the best substrates for the enzyme, either the capacity of neuraminidase to act as well on other sialocompounds, or the presence in the crude enzyme preparation of additional enzymes specific for other sialocompounds cannot be ruled out.

Thus the approach to the determination of brain particulate neuraminidase could be provisionally based on the following criteria (see also Roukema et al., 18): a) to establish the maximum hydrolysis rate of endogenous substrates (which are oligosialogangliosides); b) to add, if necessary, adequate amounts of a possible physiological substrate (for instance disialoganglioside GD1a or tri-

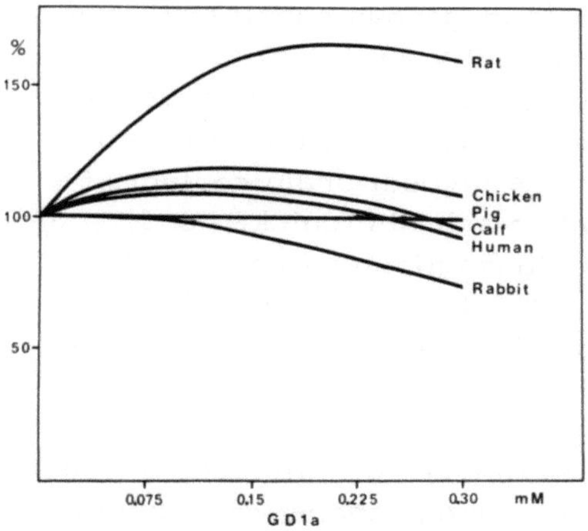

Fig. 5. Effect of the addition of disialoganglioside GD1a on the activity of brain particulate neuraminidase contained in the crude enzyme preparation from different animals. The effect is represented as % modification of the activity on endogenous substrates, assayed under optimum conditions.

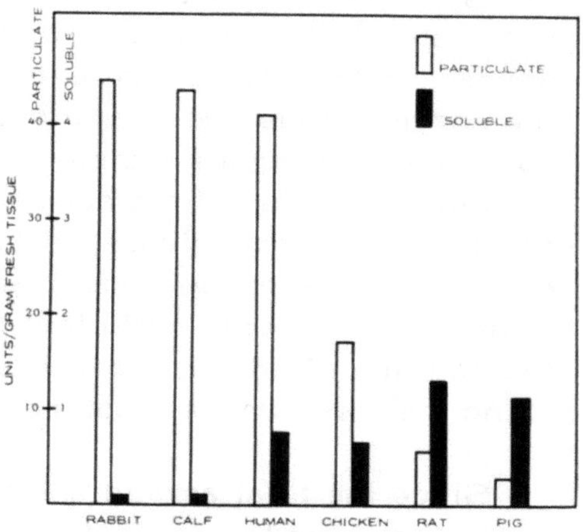

Fig. 6. Content of particulate neuraminidase in the brain of different animals. Both enzymes were assayed at 37°C under optimum conditions, GD1a being used (if necessary for reading maximum activities, in the case of particulate neuraminidase) as added substrate.

sialoganglioside GT1b) till the maximum reaction rate is reached. Practically the activity on endogenous substrates is included in the assay. Accordingly, the following assay procedure, used in our laboratory, could be suggested: 0.5-1 mg (as protein) of 0-105.000 x g pellet (washed once in 0.1 M sucrose, 15) is incubated at 37° in 0.15 M Na acetate buffer, in the presence of 0.3% (final concentration) Triton X-100, at optimum pH and convenient time (30-60 min) with the addition (if necessary) of proper amounts of GD1a. Released NANA is directly determined by the Warren's procedure (22, 19). A blank prepared with boiled (10 min) enzyme is employed.

The graphs reported in Fig. 5 show the changes of the activity of brain particulate neuraminidase from different animals caused by the addition of GD1a. In Fig. 6 the data concerning the content of particulate neuraminidase in the brain of different animals are presented. Two situations, merging from the above figures, are worth being commented: a) in the brain of some animals (human, calf, pig, rabbit) oligosialogangliosides (endogenous substrates) are present in "available" amounts practically sufficient to saturate the enzyme, thus the rate of hydrolysis of endogenous substrates approximately equals the maximum reaction rate; the same animals, with the exception of pig have also the highest brain content of particulate neuraminidase; b) in other animals (chicken and, specially, rat) the endogenous oligosialogangliosides are not present in available amounts sufficient to saturate the enzyme, therefore the addition of adequate amounts of GD1a is necessary for reaching the maximum reaction rate.

Since the total amount and the pattern of gangliosides present in the brain (and, parallelly, in the crude preparation of particulate neuraminidase) of the examined animals, are nearly the same, it must be concluded that the actual availability of endogenous oligosialogangliosides for neuraminidase greatly changes with species. The problem of the physiological interaction between neuraminidase and ganglioside seems to be very intriguing.

3) Assay of brain soluble neuraminidase

Various conditions have been described for the assay of brain soluble neuraminidase (10, 23). At the present time no definite evidences are available concerning: a) the possible presence of different forms of soluble neuraminidases; b) the substrate specificity of the enzyme; c) the occurrence of endogenous substrate in the crude preparation of the enzyme.

The following procedure, set up in our laboratory, could be used for the assay of soluble neuraminidase: 10–15 mg (as protein) of the 105.000 x g supernatant of brain homogenate in 0.32 M sucrose are incubated at 37° in 0.15 M Na acetate buffer in the presence of saturating amounts of GD1a (generally 0.1–0.2 mM) at optimum pH for 1–2 hours.

A blank with boiled enzyme (10 min) is concurrently run. The reaction mixture, stopped by dropwise addition of 1 N HCl to pH 1.0, is centrifuged at 100.000 x g for 30 min. The supernatant is poured on a column of Dowex 2x8, 0.5x6 cm, prepared and processed according to Svennerholm (24). The liberated NANA, recovered in the eluate (5 ml), is determined according to Warren (22).

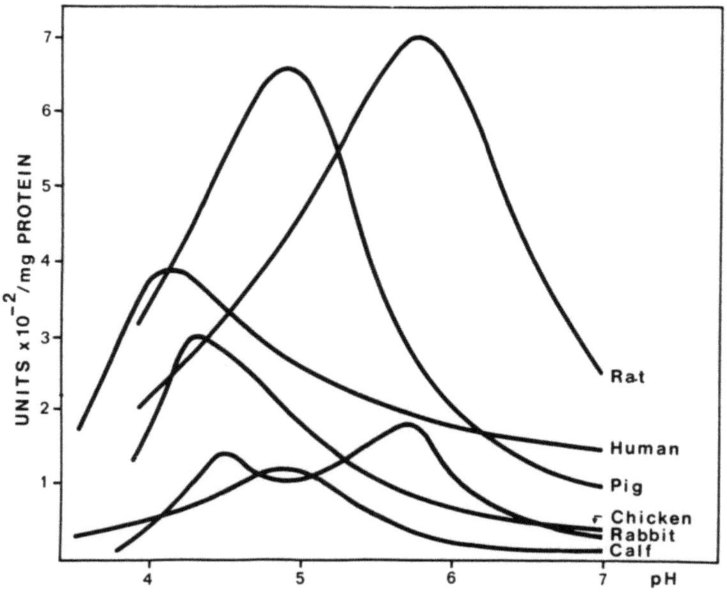

Fig. 7. Effect of pH on the activity of soluble neuraminidase from the brains of different animals.

The influence of pH on the activity of brain soluble neuraminidase from different animals is graphically represented in Fig. 7.

The optimum pH of the enzyme varies from 4.1 in human to 5.8 in rat brain. The enzyme preparation from rabbit brain provides two optimum pH values: 4.6 and 5.8 (two enzymes?).

The total content of soluble neuraminidase in the brain of different animals is reported in Fig. 6. It's worth pointing out that the animals, having in brain the highest amount of particulate neuraminidase, have the lowest value of the soluble enzyme and vice-versa.

PROPERTIES OF BRAIN NEURAMINIDASES

a) Particulate neuraminidase

The general properties of the particulate neuraminidase devoid of endogenous substrates, prepared from rabbit brain, are exposed in Fig. 8 a, b, c, d. The enzyme activity, assayed with GD1a as the substrate, has optimum pH 4.2, is linear with time till 90 min and

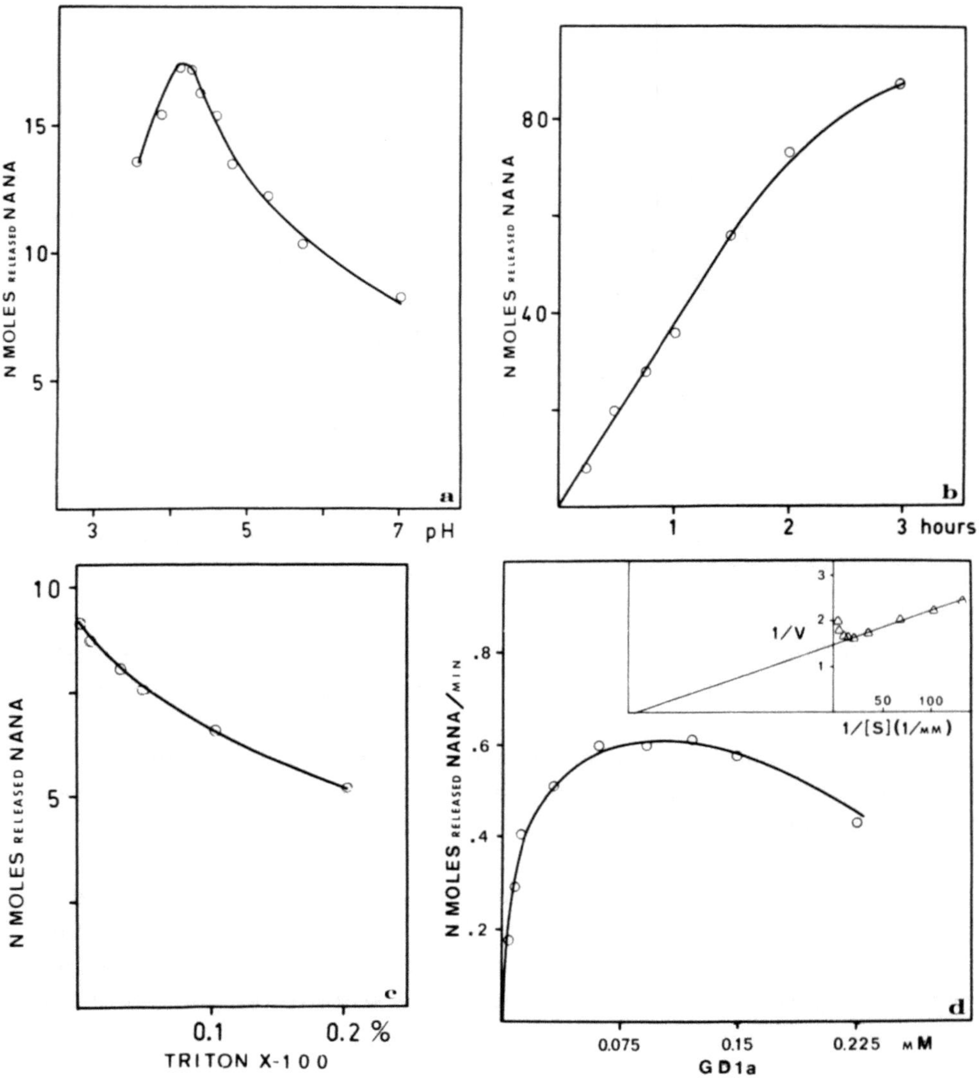

Fig. 8. Effect of pH (a), incubation time (b), Triton X-100 (c) and substrate (GD1a) concentration (d), on the activity of the particulate neuraminidase, devoid of endogenous substrate, prepared from rabbit brain.

markedly inhibited by Triton X-100 (and other unionic and anionic detergents). The inhibitory effect of Triton X-100 is probably due to the absence of endogenous substrates since the detergent is known to activate the enzyme in both crude (19) or purified (7) preparations containing endogenous substrates.

A typical substrate (GD1a)-velocity curve is demonstrated in Fig. 8d; the calculated value (from the Lineweaver and Burk plot, 25) for Km is $4.8 \cdot 10^{-6}$ M and for Vmax 1.6 nMoles released NANA/min. At high substrate concentration a pronounced fall in velocity is observed: a 50% decrease from the highest observed velocity occurs at 0.5 mM GD1a concentration.

The optimum temperature of the enzyme, ascertained by the approach indicated in Fig. 9a, is unusually high: 70° (Fig. 9b).

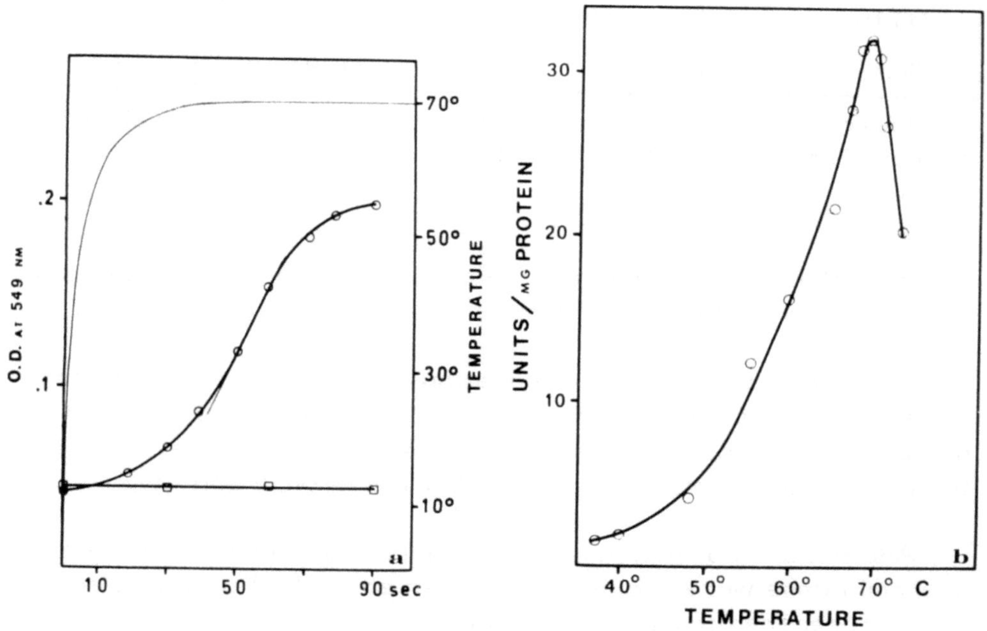

Fig. 9. a) Time course of the activity of brain particulate neuraminidase at 69°. Thick lines: enzyme activity curves (-□-□-□- : complete assay mixture; -o-o-o- : assay mixture with boiled enzyme); thin line: temperature curve. b) Effect of temperature on the activity of brain particulate neuraminidase. The enzyme devoid of endogenous substrates, prepared from rabbit brain, was employed.

Table I - Kinetic constants obtained for the action of brain particulate and soluble neuraminidase on different ganglioside substrates.

Ganglioside substrate	Particulate neuraminidase (')		Soluble neuraminidase ('')	
	Km (M)	Vmax (°)	Km (M)	Vmax (°)
GM3-NANA	1.0×10^{-5}	33	-	-
GM3-NGNA	2.6×10^{-5}	19	-	-
GD1a	2.8×10^{-5}	38	2.9×10^{-4}	51
GD1b	1.0×10^{-4}	26	3.7×10^{-4}	17
GT1b	1.2×10^{-5}	35	1.1×10^{-4}	56
GQ1	-	-	1.9×10^{-4}	62

(') Human brain - Data from Ohman et al. (12). ('') Pig brain - Data from Penati et al. (27). (°) nMoles of released NANA (or NGNA) /hour.

The resistance of the enzyme to high temperature was proved to be strictly dependent on the presence of the substrate (GD1a).

Storage of the enzyme preparation rid of endogenous substrates at 0-4° for 6 days and frozen at -20° for 3 months resulted in no appreciable changes of neuraminidase activity. Freezing and thawing up to 10 times does not modify the enzyme activity.

Additional informations on the properties of brain particulate neuraminidase were provided by Ohman et al. (12) who studied the enzyme prepared from human brain.

The Km and Vmax values of this enzyme for different gangliosides (GT1b, GD1a, GD1b, GM3-NANA, GM3-NGNA) are represented in Table I. A lower affinity is provided for GD1b and different Km and Vmax values correspond to the forms of GM3 containing N-acetylneuraminic acid (GM3-NANA) or N-glycolylneuraminic acid (GM3-NGNA). A competition has been clearly ascertained between GD1a and GT1b, GD1a and GM3, this indicating a single neuraminidase activity on all three gangliosides.

Sialyllactose, human transferrin and ovine submaxillary mucin release NANA by enzyme action more slowly than disialogangliosi-de GD1a, regardless of pH. Except for GM3 no NANA is released from monosialogangliosides (GM2, GM1).

EDTA at a concentration of 0.05 M is not inhibitory, but at this concentration, Cu^{++}, Cd^{++}, or Ag^{+} practically abolish activity, and Zn^{++} or Fe^{++} cause more than a 50% reduction. Hg^{++} causes no inhibition and Ca^{++}, Mg^{++} and Mn^{++} are without effect, as are also the monovalent alkali metals Li^{+}, Na^{+} and K^{+}. Sulfate depressed activity some 60% but not phosphate, nitrate, acetate, chloride, bromide or iodide.

Recently Kolodny et al. (26) reported the existence in brain of a particulate neuraminidase able to remove sialic acid from labelled monosialoganglioside GM2. The few available data on the properties and substrate specificity of this enzyme are consistent with the hypothesis that it is quite different from the "major" neuraminidase described above.

b) Soluble neuraminidase

Some properties of a 600-fold purified soluble neuraminidase from pig brain were described by Tettamanti and Zambotti in 1968 (10). At that time the possible occurrence of solubilization of the particulate enzyme during brain homogenization in 0.156 M KCl was pointed out. Recent experiments (28) showed no difference in the total amount and basic properties of the enzyme when brain homogenization was carried either in isotonic KCl or isotonic sucrose. Thus, it is now conceivable that the enzyme described in that paper in true soluble neuraminidase.

The optimum pH of the enzyme in 0.03 M diNa phosphate-citric acid buffer, is 4.9 for gangliosides (oligosialogangliosides and monosialoganglioside GM3), 4.7 for sialyllactose, 4.4 for ovine submaxillary mucin.

The Km and Vmax values for oligosialogangliosides (27), presented in Table I, clearly show the different kinetics toward the same substrates of the soluble and the particulate enzyme.

The enzyme, which is strongly inhibited by unionic and anionic detergents, maintains the full activity for 2 days at 0-2° and for 4 months when stored at -20° (28). On the contrary freezing and thawing for 3-4 times results in a complete inactivation of the enzyme (28).

The soluble enzyme from rat brain was reported to have different properties than the pig enzyme (23). The relative reaction rates toward a number of substrates are different (higher activity on sialyllactose than on gangliosides); Triton X-100 is ineffective, etc. Species differencies (see Fig. 6) and the inadeguacy of the enzyme assay (for instance the used concentration of gangliosides) may be responsible for the above differences. Some indications concerning the effect of ions on the activity of soluble neuraminidase were provided by Carubelli and Tulsiani (23). Ca^{++} and Mg^{++} (as their chlorides) give a very slight stimulation (up to 8%), while Cu^{++}, Hg^{++} and Zn^{++} at concentrations 1.10^{-4} M and above cause marked inhibition (over 85-90%). Li^{+}, Na^{+}, or K^{+} (as their chlorides) at concentrations up to 2.10^{-1} do not inhibit the enzyme.

REGIONAL DISTRIBUTION OF BRAIN NEURAMINIDASES

Gielen and Harpprecht (29) reported the occurrence of a higher activity of particulate neuraminidase in grey matter than in white matter. The evidence was confirmed by Roukema et al. (18) and better specified by Ohman (30). While the enzyme activity in the cerebral cortex is about 10-15 times higher than in white matter, no relevant differences in activity are recorded in the various cortical areas. The neuraminidase activity of the cerebellar cortex is significantly higher than that of frontal cerebral cortex. Among the studied nuclei the enzyme activity in the caudate nucleus tends to be higher than in the lenticular nucleus and the talamus. So far, studies on the regional distribution of soluble neuraminidase have not been reported.

DEVELOPMENTAL PROFILES OF BRAIN NEURAMINIDASE

The early appearance of neuraminidase in chick embryo brain has been recently studied by Schengrund and Rosenberg (31). The enzyme activity toward endogenous substrates, which starts being detectable in the 5-day embryo, undergoes a sharp increase (9-fold) from that time till hatching and a further increase (12-fold) through the adult.

Carubelli and Tulsiani (23) reported the pattern of soluble and particulate brain neuraminidase during the pre- and the postnatal development of the rat. The activity of the soluble enzyme, very low before birth (7 days) and at birth, increases gradually till reaching, at the 20th day after birth, a maximum value which is later maintained. The increase is about 12-fold.

On the contrary, the activity of the particulate enzyme undergoes no changes from the 7th day before birth through the 30th day of postnatal age. However the evidence provided by Roukema et al. (32) that a 3-fold increase of the activity of the particulate neuraminidase occurs from birth to the 30-35-day old animal seems to be much more reliable. An increase of the activity of particle-bound neuraminidase from birth to the 10th year of age was also observed in human brain by Ohman and Svennerholm (33). It's interesting to emphasize the observation (31, 33) that the developmental profile of neuraminidase (at least of the particulate enzyme) strictly parallels that of the sialo-derivatives, specially gangliosides, present in brain.

SUBCELLULAR LOCATION OF BRAIN NEURAMINIDASE

At the present time there is enough evidence that brain "soluble" neuraminidase (soluble from a procedural point of view) is a cytosol enzyme (23, 28).

Much work has been carried out in our laboratory on the subcellular location of particulate neuraminidase since 1968 (34). The evidences provided by the initial experiments (34, 35, 36, 37), are only indicative, since the assay method used didn't take into proper account endogenous substrates. According to those evidences, anyway, a neuraminidase activity is present in the "myelin", "synaptosomal" and "microsomal" brain subcellular fractions. Ohman (38), who studied the human brain enzyme (rid of endogenous substrates) was recently able to confirm this general pattern.

At present, the situation is being reinvestigated on rabbit brain. When brain homogenate is divided into the conventional "nuclear", "crude mitochondrial" and "microsomal" primary fractions, according to Gray and Whittaker (13), the activity of particulate neuraminidase is almost completely recovered in the "crude mitochondrial" and "microsomal" fractions and equally distributed between them.

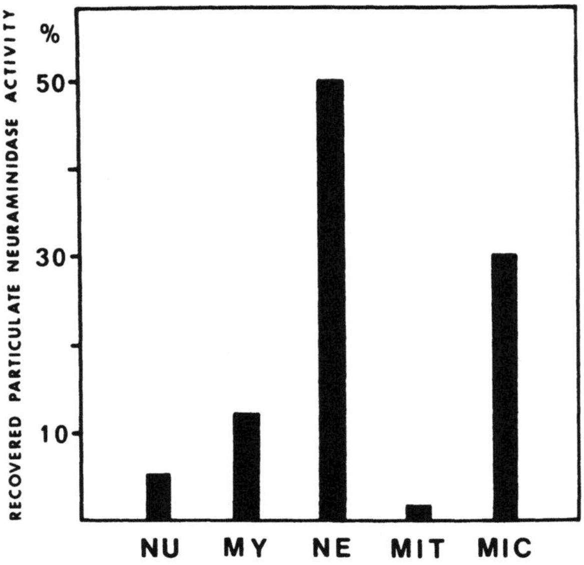

Fig. 10. Distribution of recovered particulate neuraminidase activity in different subcellular fractions of rabbit brain. NU = nuclear fraction; MY = myelin fraction; NE = nerve ending fraction; MIT = mitochondrial fraction; MIC = pure microsomal fraction.

The "microsomal" fraction, from which nerve endings have been removed, contains approximately 30% of the neuraminidase activity initially present in brain (Fig. 10).

After subfractionation of the crude "mitochondrial" fraction on a sucrose density gradient (0.8-1.0-1.2-1.4 M) according to De Robertis et al. (39), 80% of the activity is recovered in bands C and D rich in nerve endings, the remainder being present in the myelin subfraction (band A). No appreciable activity is detectable in the pellet, which contains mitochondria and large lysosomes. In conclusion, the activity of particulate neuraminidase in brain is distributed 50% in the "nerve ending" subfraction, 30% in the microsomal fraction, 12% in the "myelin" subfraction (Fig. 10).

Very recent experiments showed that after rupturing nerve endings by ipo-osmotic shock and preparing synaptosomal plasma membranes according to Morgan et al. (40), the activity of neuraminidase present in nerve endings remains linked to membranes.

The specific activity of neuraminidase in the purest (light) plasma membrane fraction undergoes the same enrichment, (10-12 fold) with respect to the starting brain homogenate, than gangliosides and K^+ and Na^+ stimulated ATPase, considered valuable markers of neuronal plasma membranes.

The available data strongly suggest the hypothesis that brain particulate neuraminidase acting on gangliosides is linked to nerve tissue membranes (synaptosomal plasma membranes, myelin or axonal membranes, perikarya plasma membrane, dendryte membranes).

Conversely no convincing evidences have been provided (34, 35, 36, 37, 38) for the presence or absence of neuraminidase in brain lysosomes. However only intensive investigations on the action of brain particulate neuraminidase on different sialoderivatives will be able to give a definite answer to this problem. In this respect the approach recently introduced by Kolodny et al. (26), seems to be extremely promising.

PHYSIOLOGICAL ROLE OF BRAIN NEURAMINIDASE

The present uncertainity on the enzyme assay and the poor information on the kinetics, substrate specificity, subcellular site of the enzyme(s) render merely speculative any attempts to device a physiological role for brain neuraminidases.

The possible involvements of neuraminidases in brain metabolism and function may be seen in the following directions:

a) Degradation of gangliosides. Oligosialogangliosides appear to act as the physiological substrates for "major" particulate and soluble neuraminidase. Both enzymes have a preferential action on GQ1, GT1b and GD1a than on GD1b. Also, the enzyme degradation of GT1b occurs almost completely via GD1b. So the initial steps of oligosialoganglioside degradation follow this pattern (7, 10, 12):

GM1, as well GM2, are resistant to the enzyme action, apparently by a steric hindrance due to the hexosamine contained in their molecule.

Another enzyme is involved, according to Kolodny et al. (26), in the degradation of the hexosamine containing monosialogangliosides, at least GM2. The hexosamine free monosialoganglioside GM3 is again hydrolyzed by the former neuraminidase(s). Thus, by the action of the different neuraminidases, all NANA is removed from gangliosides, the neutral oligosaccharide portion of their molecule being available for further degradation.

b) Regulation of the negative charge of neuronal membranes. It is well known that gangliosides are specific components of nervous membranes (41, 42). They contribute, by the dissociation of sialic acid residues, to the net negative charge of the membrane. The release of NANA from their molecule, catalyzed by neuraminidase, causes a decrease of the negative charge of the membrane; conversely the increase of the same charge follows the action of sialyltransferase which restores the content of ganglioside NANA. Now, both neuraminidase (34, 36) and sialyltransferase (43) are located in the brain particulate fractions richest in gangliosides. This finding suggests the hypothesis that the two enzymes play a role in the modulation of the net negative charge of the ganglioside containing membranes and are therefore involved in the correlated functional events.

c) Intermittent modification of neuronal membrane permeability. Gammack (44) observed that in aqueous media oligosialogangliosides form micelles with higher size than monosialogangliosides. Considering that gangliosides are fundamental constituents of some nervous membranes, the observed *in vivo* degradation of oligosialo- into monosialoganglioside, should be followed by dramatic changes in the architecture and, consequently, in the permeability of the membranes. To be involved in higher specialized brain functions (for instance, nervous transmission) these modifications should be reversible, intermittent and adequately fast.

The Km values of brain particulate neuraminidase for most oligosialogangliosides (order of magnitude: 10^{-5}–10^{-6} M) indicate a high affinity of the enzyme for these "physiological substrates". But, in consideration of the pronounced inhibitory effect of substrate at high concentration, it can be assumed that particulate neuraminidase, in the physiological milieu (ganglioside concentration range of the order 10^{-3}–10^{-4} M) is under constant inhibition. The actual pH value (the optimum pH of the enzyme is around 5.0) is an additional inhibitory factor.

Consequently either the enzyme "never" has the possibility to act or it is "allowed" to act under certain circumstances and at the proper time, that is its action is "intermittent".

Nothing is known on the specific mechanisms by which brain neuraminidase starts actually acting.

Of course further and more detailed investigations are needed in order to assess the validity of the above hypotheses.

REFERENCES

(1) Carubelli, R., Trucco, R.E. and Caputto, R., Biochim. Biophys. Acta 60: 196 (1962).
(2) Morgan, E.N. and Laurell, C.B., Nature 197: 921 (1963).
(3) Korey, S. and Stein, A., in Brain Lipids and Lipoproteins and the Leucodystrophies, Ed. Folch-Pi, J. and Bauer, H., Elsevier Publishing Co., Amsterdam, p. 71 (1963).
(4) Zambotti, V., Tettamanti, G. and Berra, B., Abstracts Congress Fed. Europ. Biochem. Soc., Vienna, A 236 (1965).
(5) Kelly, R.T. and Greiff, D., Biochim. Biophys. Acta 110: 548 (1965).
(6) Svennerholm, L., in Inborn Disorders of Sphingolipid Metabolism, Ed. Aronson, S.A. and Volk, B.M., Pergamon Press, Oxford, p. 169 (1967).
(7) Leibovitz, Z. and Gatt, S., Biochim. Biophys. Acta 152: 136 (1968).
(8) Di Donato, S. and Tettamanti, G., Boll. Soc. It. Biol. Sper. 43: 1032 (1967).
(9) Preti, A., Tettamanti, G. and Di Donato, S., Boll. Soc. It. Biol. Sper. 44: 1143 (1968).
(10) Tettamanti, G. and Zambotti, V., Enzymologia 35: 61 (1968).
(11) Carubelli, R., Nature 219: 955 (1968).
(12) Ohman, R., Rosenberg, A. and Svennerholm, L., Biochemistry 9: 3774 (1970).
(13) Gray, E.G. and Whittaker, V.P., J. Anat. (London) 96: 79 (1962).
(14) De Robertis, E., Salganicoff, L., Pellegrino de Iraldi, A. and Rodriguez de Lores Arnaiz, G., J. Neurochem. 9: 23 (1962).
(15) Tettamanti, G., Preti, A., Lombardo, A., Gasparini, M. and Zambotti, V., Biochim. Biophys. Acta 258: 228 (1972).

(16) Wolfe, L.S., Biochem. J. 79: 348 (1961).
(17) Brunngraber, E.G., Dekirmenjian, H. and Brown, B.D., Biochem. J. 103: 73 (1967).
(18) Roukema, P.A. and Heijlman, J., J. Neurochem. 17: 773 (1970).
(19) Preti, A., Lombardo, A. and Tettamanti, G., Italian J. Biochem. 19: 371 (1970).
(20) Lombardo, A., Preti, A. and Tettamanti, G., Italian J. Biochem. 19: 386 (1970).
(21) Svennerholm, L., in Comprehensive Biochemistry - Ed. Florkin M. and Stotz E.H., Pergamon Press, Oxford, 18: 1691 (1969).
(22) Warren, L., J. Biol. Chem. 234: 1971 (1959).
(23) Carubelli, R. and Tulsiani, D.R.P., Biochim. Biophys. Acta 237: 78 (1971).
(24) Svennerholm, L., Biochim. Biophys. Acta 24: 604 (1957).
(25) Lineweaver, H. and Burk, D., J. Amer. Chem. Soc. 56: 659 (1934).
(26) Kolodny, E.H., Kanfer, J., Quirk, J.M. and Brady, R.O., J. Biol. Chem. 246: 1426 (1971).
(27) Penati, M.R., Tettamanti, G., Bolognani, L. and Venerando, B., Abstracts Meeting Soc. It. Biol. Sper., S. Vincent (1967).
(28) Tettamanti, G. and Venerando, B., in preparation.
(29) Gielen, W. and Harpprecht, V., Z. Physiol. Chem. 350: 201 (1969).
(30) Ohman, R., J. Neurochem. 18: 531 (1971).
(31) Schengrund, C.L. and Rosenberg, A., Biochemistry 10: 24 (1971).
(32) Roukema, P.A., Van den Eijnden, D.H. and Van der Berg, G., FEBS Letters 9: 267 (1970).
(33) Ohman, R. and Svennerholm, L., J. Neurochem. 18: 79 (1971).
(34) Tettamanti, G., Preti, A., Di Donato, S. and Marconi, M., Rend. Ist. Lomb. Sc. Lett. (B) 102: 292 (1968).
(35) Preti, A., Tettamanti, G. and Di Donato, S., Boll. Soc. It. Biol. Sper. 44: 1143 (1968).
(36) Tettamanti, G., Preti, A., Di Donato, S., Bassi, M. and Zambotti, V., Metabolismo 5: 207 (1969).
(37) Tettamanti, G., Preti, A., Lombardo, A. and Zambotti, V., Com. 12° Int. Conference Biochemistry Lipids, Atene (1969).
(38) Ohman, R., J. Neurochem. 18: 89 (1971).
(39) De Robertis, E., Alberici, M., Rodriguez de Lores Arnaiz, G. and Azcurra, T., Life Sciences 5: 577 (1966).

(40) Morgan, I.G., Wolfe, L.S., Mandel, P. and Gombos, G., Biochim. Biophys. Acta 241: 737 (1971).

(41) Lapetina, E.G., Soto, E.S. and De Robertis, E., Biochim. Biophys. Acta 135: 33 (1967).

(42) Wiegandt, H., J. Neurochem. 14: 671 (1967).

(43) Den, H., Kaufman, B. and Roseman, S., J. Biol. Chem. 245: 6607 (1970).

(44) Gammack, D.B., Biochem. J. 88: 373 (1963).

INFLUENCE OF THYROID UPON LIPID CONTENT IN DEVELOPING CEREBRAL CORTEX AND CEREBELLUM OF THE RAT

N. E. GHITTONI and I. F. de RAVEGLIA

Departamento de Química Biológica

Facultad de Farmacia y Bioquímica

Universidad de Buenos Aires, Argentina

Thyroid deficiency arising early in life has long been known to produce a severe and frequently irreversible damage of the nervous system. In the rat brain, the most remarkable histological alteration observed in this condition is a marked hypoplasia of both the axonal and dendritic components of the neuropil (1) and a defective development of Purkinje cells and cerebellar glomeruli (2).

Biochemical studies have demonstrated that the lack of thyroid function from birth causes several alterations of the nervous system of the rat, related to the metabolism of carbohydrates (3), of amino acids (4, 5), of nucleic acids (6, 7, 8, 9), of proteins (10, 11, 12), and of enzymes (13, 14, 15). Furthermore, Balázs et al. (16) have reported a defective myelination due to thyroid deficiency and similar results on formation of myelin lipids were obtained by Walravens and Chase (17).

In the present work, we have studied the changes in lipid content during postnatal maturation of the cerebral cortex and cerebellum at four stages of development, both in normal and thyroidectomized rats at birth.

EXPERIMENTAL

Experiments were carried out on rats of either sex, of a highly inbred Wistar strain. Entire litters, each reduced to 8 rats at birth were used as experimental or normal control groups. Each rat of the experimental group was made hypothyroid within two hours of birth by a single intraperitoneal injection of 120 μCi of ^{131}I, according to Goldberg and Chaikoff (18). Groups of 3 to 16 rats (the larger number representing the earlier stages of development) were used at 5, 10, 20 and 30 days of age. The animals were weighed, killed by decapitation and cerebral cortex and cerebellum rapidly excised and weighed.

Lipid fractions were obtained by extraction with chloroform : methanol (2:1) and partition (19), and repartition according to Suzuki and Chen (20). After the standard double extraction procedure, gangliosides were purified by Sephadex G-100 column chromatography (21). Gangliosides were separated by horizontal TLC, with propanol : butanol : water (65:10:25) as previously reported (21), and quantitation was done according to MacMillan and Wherrett (22).

Aliquots from the combined lower phases were used for further analytical procedures. Phospholipids were separated by TLC, according to Neskovic and Kostic (23), and phosphorus content of the various fractions was determined by the procedure of Chen et al. (24). Galactose from cerebrosides and cerebroside-sulphatides, was measured by the orcinol-sulphuric acid method of Hess and Lewin (25). Proteolipid proteins were determined with the method of Lowry et al. (26), as modified by Hess and Lewin (25). Cholesterol was estimated by the method of Searcy and Bergquist (27). DNA was determined by the procedure of Munro and Fleck (28), slightly modified by Pasquini et al. (6).

All results are presented on a DNA basis (6) as the means $\pm$ SEM of the number of individual experiments indicated. Values from normal controls and hypothyroid rats were compared by Student's t test and differences were considered statistically significant when P values were less than 0.05.

RESULTS

Our results on DNA changes in both normal and hypothyroid developing rat cerebral cortex and cerebellum are in good agreement with those found by Pasquini et al. (6) (Table I).

Table I - Effect of Neonatal Thyroidectomy on the DNA Content in the Cerebral Cortex and the Cerebellum of Rats.

Age (days)	Normal	Thyroidectomized
	CEREBRAL CORTEX	
5	1.71 ± 0.10	1.96 ± 0.15
10	1.44 ± 0.08	1.58 ± 0.10
20	0.98 ± 0.10	1.33 ± 0.09 (¹)
30	1.03 ± 0.07	1.35 ± 0.10 (¹)
	CEREBELLUM	
5	5.51 ± 0.62	5.26 ± 0.39
10	6.79 ± 0.52	6.49 ± 0.38
20	6.97 ± 0.49	7.21 ± 0.38
30	7.06 ± 0.40	7.35 ± 0.63

The results are presented as mg DNA/g fresh tissue and they are the means ± SEM of 10 individual experiments for each group and age. (¹) $P < 0.02$. Technical details are given in the text.

Cerebral cortex. During normal development there was a progressive increase in the lipid content of the cerebral cortex (Fig. 1, a). This increase was most remarkable for the cerebrosides. The results are in good agreement with the literature (29, 30, 31, 32, 33).

In the cerebral cortex of thyroidectomized rats, the patterns of changes with age were similar to controls, but the increase in lipids between 10 and 20 days was less noticeable. Furthermore, between 5 and 30 days, the cerebroside content of cretin cerebral cortex increased about 20-fold, while during the same period, controls had a 55-fold rise (Fig. 1, b).

Table II shows the effect of neonatal thyroidectomy upon the lipid content. At 20 and 30 days of age, this experimental condition led to a highly significant decrease in all the lipids studied.

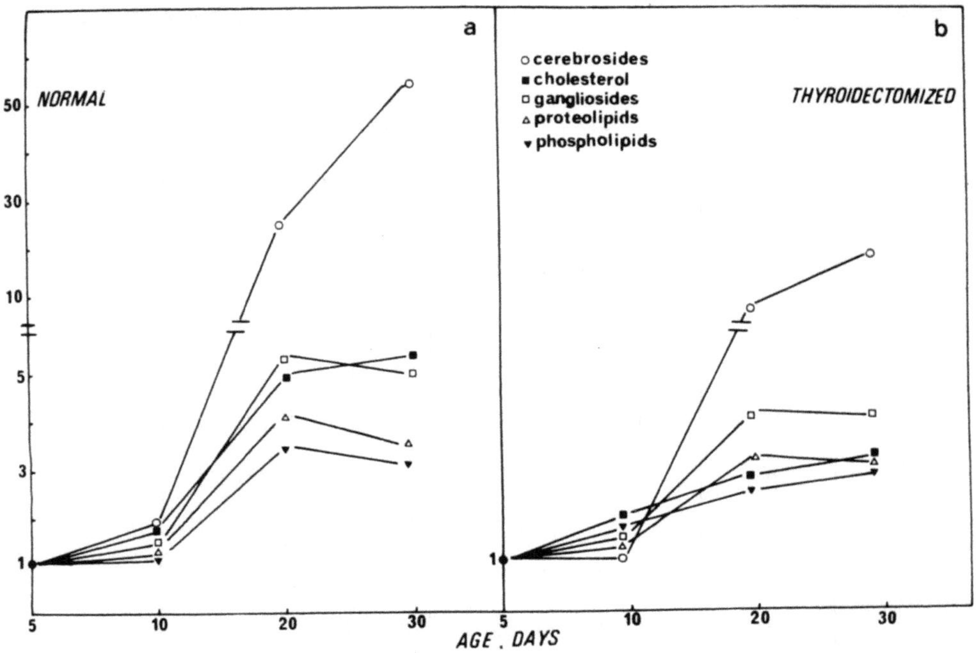

Fig. 1. Changes in the lipid content during postnatal development of cerebral cortex of normal and thyroidectomized rats. Each point represents the mean of 3-9 individual experiments. Results are expressed on a DNA basis and all they are given as the ratio to 5 days old values. See text for methods.

Table II - Effect of Neonatal Thyroidectomy on the Deposition of Lipids in the Cerebral Cortex of the Rat.

AGE (days)	GANGLIOSIDES µg NANA/mg DNA	PHOSPHOLIPIDS µg P/mg DNA	CHOLESTEROL mg/mg DNA	CEREBROSIDES mg/mg DNA	PROTEOLIPIDS µg protein/mg DNA
5 N	146 ± 8 (5)	469 ± 33 (4)	2.13 ± 0.19 (5)	0.057 ± 0.008 (4)	360 ± 28 (4)
T	96 ± 15 (3)	385 ± 10 (3)	2.24 ± 0.08 (3)	0.063 ± 0.022 (3)	314 ± 25 (3)
10 N	178 ± 6 (4)	697 ± 38 (6)	3.94 ± 0.10 (6)	0.107 ± 0.009 (7)	540 ± 26 (5)
T	156 ± 32 (4)	679 ± 10 (4)	4.33 ± 0.50 (4)	0.079 ± 0.011 (4)	431 ± 28 (4)
					P < .025
20 N	627 ± 40 (4)	1564 ± 61 (6)	10.74 ± 0.57 (6)	1.470 ± 0.062 (6)	1503 ± 93 (5)
T	390 ± 41 (8)	943 ± 35 (7)	6.48 ± 0.19 (6)	0.820 ± 0.102 (6)	989 ± 52 (9)
	P < .005	P < .001	P < .001	P < .001	P < .001
30 N	590 ± 35 (5)	1461 ± 68 (5)	11.33 ± 0.65 (3)	3.210 ± 0.187 (5)	1262 ± 93 (5)
T	385 ± 24 (5)	1085 ± 62 (4)	6.88 ± 0.25	1.224 ± 0.115 (5)	951 ± 62 (6)
	P < .005	P < .01	P < .001	P < .001	P < .02

The number of determinations are given in brackets. The value of P is indicated when the difference between the experimental and control groups were significant. The details of the experimental conditions and methods are given in the text.

N : normal controls; T : thyroidectomized rats.

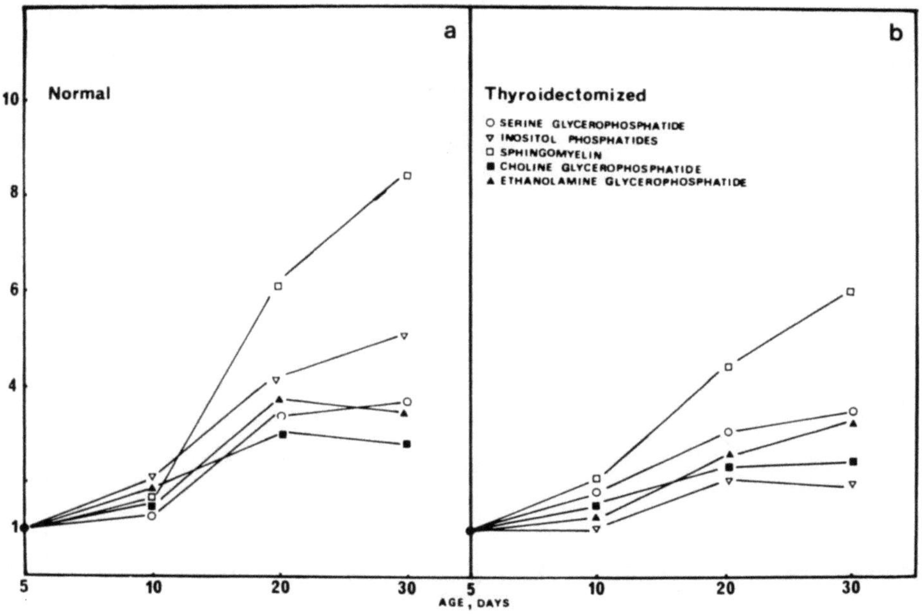

Fig. 2. Changes in the phospholipid content during postnatal development of cerebral cortex of normal and thyroidectomized rats. The results are presented as in Fig. 1.

Table III - Effect of Neonatal Thyroidectomy on the Deposition of Phospholipids in the Cerebral Cortex of the Rat.

Age (days)		S. G.	I. P.	Sph.	C. G.	E. G.	
5	N	44 ± 2	12 ± 2	10 ± 1	195 ± 7	129 ± 4	(4)
	T	38 ± 4	18 ± 5	10 ± 1	180 ± 10	109 ± 12	(3)
10	N	65 ± 6	24 ± 2	16 ± 1	301 ± 17	220 ± 17	(6)
	T	60 ± 3	21 ± 1	20 ± 3	272 ± 25	160 ± 17	(4)
						P < .05	
20	N	150 ± 8	51 ± 5	62 ± 4	594 ± 27	484 ± 28	(6)
	T	121 ± 11	38 ± 6	46 ± 3	393 ± 10	292 ± 24	(7)
				P < .02	P < .001	P < .001	
30	N	158 ± 17	63 ± 10	84 ± 4	543 ± 43	445 ± 49	(5)
	T	133 ± 12	36 ± 4	62 ± 5	451 ± 45	372 ± 30	(4)
			P < .05	P < .02			

The results are expressed as μg P/mg DNA, and they are presented as in Table II. S. G. : serine glycerophosphatide; I. P. : inositol phosphatides; Sph. : sphingomyelin; C. G. : choline glycerophosphatide; E. G. : ethanolamine glycerophosphatide.

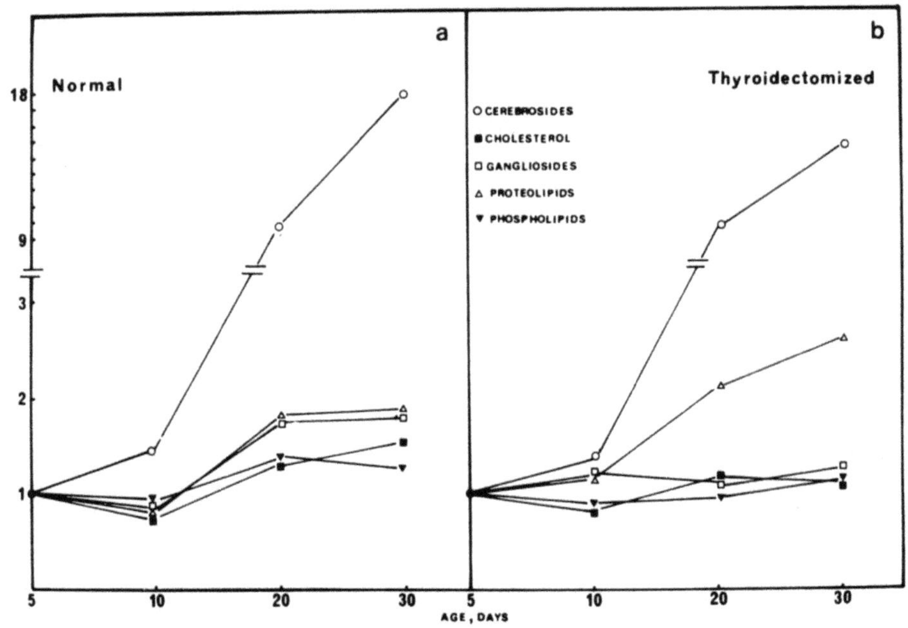

Fig. 3. Changes in the lipid content during postnatal development of cerebellum of normal and thyroidectomized rats. The results are presented as in Fig. 1. Technical details are given in the text.

Table IV - Effect of Neonatal Thyroidectomy on the Deposition of Gangliosides in the Cerebral Cortex of the Rat.

AGE (days)	G_O	%	G_1 (G_{T1})	%	G_2 (G_{D1b})	%	G_3 (G_{D1a})	%	G_4 (G_{M1})	%	G_5 (G_{M2})	%	
5 N	18 ± 4	14 ± 3	27 ± 2	22 ± 2	13 ± 1	11 ± 1	41 ± 3	33 ± 2	20 ± 2	16 ± 2	5 ± 1	4 ± 1	(5)
T	10 ± 2	10 ± 2	23 ± 3	23 ± 3	12 ± 1	12 ± 1	31 ± 3	31 ± 3	17 ± 5	17 ± 5	7 ± 2	7 ± 2	(3)
10 N	28 ± 13	14 ± 7	38 ± 2	19 ± 1	29 ± 6	15 ± 3	73 ± 6	37 ± 3	27 ± 2	14 ± 1	5 ± 1	3 ± 1	(4)
T	17 ± 3	10 ± 2	26 ± 5	15 ± 2	21 ± 3	12 ± 2	68 ± 13	40 ± 8	29 ± 4	17 ± 2	8 ± 1	5 ± 1	(4)
20 N	96 ± 13	15 ± 2	101 ± 8	16 ± 1	80 ± 4	13 ± 1	213 ± 15	33 ± 2	127 ± 13	20 ± 2	22 ± 7	3 ± 1	(4)
T	39 ± 6	10 ± 2	84 ± 11	21 ± 3	50 ± 4	12 ± 1	148 ± 14	37 ± 3	68 ± 8	17 ± 2	13 ± 3	3 ± 1	(8)
	P <.001				P <.005		P <.02		P <.005				
30 N	65 ± 20	11 ± 4	77 ± 17	14 ± 3	78 ± 10	14 ± 2	194 ± 26	34 ± 5	132 ± 10	23 ± 2	23 ± 3	4 ± 1	(5)
T	45 ± 9	12 ± 2	59 ± 6	15 ± 2	49 ± 3	13 ± 1	144 ± 16	37 ± 4	71 ± 5	18 ± 1	21 ± 5	5 ± 1	(5)
					P <.05				P <.001				

The results are expressed as μg NANA/mg DNA, and they are presented as in Table II. Nomenclature of individual gangliosides is that of Korey and Gonatas (38) with the corresponding Svennerholm nomenclature (39) in parentheses. % : the distribution of individual gangliosides is expressed in percentage of the total gangliosides. The details of the experimental conditions and methods are given in the text.

Fig. 2 shows the changes of individual phospholipid content in developing cerebral cortex of normal and thyroidectomized rats.

Table V - Effect of Neonatal Thyroidectomy on the Deposition of Lipids in the Cerebellum of the Rat.

AGE (days)	GANGLIOSIDES µg NANA/mg DNA	PHOSPHOLIPIDS µg P/mg DNA	CHOLESTEROL mg/mg DNA	CEREBROSIDES mg/mg DNA	PROTEOLIPIDS µg protein/mg DNA
5 N	33.4 ± 6.0 (4)	140 ± 6 (4)	1.01 ± 0.05 (4)	0.061 ± 0.009 (2)	105 ± 21 (2)
T	33.8 ± 1.5 (3)	153 ± 32 (2)	1.03 ± 0.12 (2)	0.044 ± 0.022 (2)	86 ± 9 (2)
10 N	28.8 ± 0.9 (5)	135 ± 9 (7)	0.79 ± 0.06 (6)	0.087 ± 0.015 (6)	89 ± 6 (7)
T	39.7 ± 0.7 (4)	135 ± 5 (4)	0.86 ± 0.10 (4)	0.059 ± 0.003 (4)	104 ± 4 (4)
20 N	58.0 ± 4.4 (6)	196 ± 9 (5)	1.32 ± 0.06 (7)	0.661 ± 0.059 (5)	192 ± 17 (7)
T	36.5 ± 3.3 (6)	144 ± 8 (6)	1.17 ± 0.07 (6)	0.415 ± 0.043 (5)	180 ± 22 (7)
	P < .005	P < .005		P < .025	
30 N	58.9 ± 5.7 (4)	178 ± 7 (5)	1.56 ± 0.08 (5)	1.093 ± 0.087 (6)	196 ± 5 (4)
T	42.2 ± 3.4 (5)	173 ± 11 (5)	1.15 ± 0.06 (5)	0.645 ± 0.077 (4)	223 ± 15 (4)
	P < .05		P < .005	P < .01	

The number of determinations are given in brackets. The value of P is indicated when the difference between the experimental and control groups were significant. The details of the experimental conditions and methods are given in the text.
N : normal controls; T : thyroidectomized rats.

At 30 days of age, sphingomyelin and inositol phosphatides were the most affected by the lack of thyroid function (Table III).

Although the gangliosides concentration decreased in cerebral cortex of 20 and 30 days old cretin rats, the percent distribution of individual gangliosides did not change (Table IV).

Cerebellum. In normal rats, the content of cholesterol, gangliosides, proteolipids and phospholipids increased slightly between 10 and 20 days of age (Fig. 3, a). On the contrary, cerebrosides showed a marked increase during all the period studied. The lipid content of the cerebellum was slightly affected by neonatal thyroidectomy (Fig. 3, b). When the deposition of gangliosides and cerebrosides was calculated, it was evident that the amount was reduced both at 20 and 30 days of age. The amount of phospholipids was reduced only for a transient period (20 days of age), whereas the cholesterol content was decreased in the cerebellum of 30 day old hypothyroid rats (Table V).

DISCUSSION

The major defects in the composition of the brain attributable to neonatal thyroidectomy have proved to be evident mainly after 15 days of postnatal life (34, 35).

Our observations are in agreement with biochemical (4, 6, 7, 14) and also with histological findings (1, 2, 34, 35). We found a statistically significant decrease of all lipids studied in the cerebral cortex of 20 and 30 day old cretin rats.

Evidences point to a localization of the gangliosides in dendritic and axonal membranes as well as in synaptic junctions (36, 37). After neonatal thyroidectomy the arborization of dendrites is retarded and the axonal density in the cortex is reduced (1). The depression in the concentration of gangliosides from cerebral cortex is consistent with the histological alterations.

Our results showing that the lack of thyroid function produces a significant decrease of myelin lipids (i.e. cholesterol, cerebrosides, sphingomyelin, etc.) agree with previous reports of Balázs et al. (16), Dalal et al. (29) and Walravens et al. (17).

The cerebellar lipids seemed to be only partially affected by neonatal thyroidectomy. This might be due to the ontogenic heterogeneousness of the cerebellum, an organ in which the final assembly of cells occurs to a great extent after birth (2).

ACKNOWLEDGEMENTS

The authors are grateful to Dr.Carlos J. Gómez for his guidance and encouragement in this work, and to Dr. Eduardo F. Soto for his kind help during the correction of the English text. They also thank Miss María I. Rigo Ordóñez for her technical assistance. This work was supported by Grants of the Consejo Nacional de Investigaciones Científicas y Técnicas (1003 b) and the Instituto Nacional de Bromatología y Farmacología, Argentina.

REFERENCES

(1) Eayrs, J.T., in Regional Neurochemistry, Ed. Kety, S.S. and Elkes, J., Pergamon Press, New York, p. 423 (1961).
(2) Legrand, J., C.R. Acad. Sci. (Paris) 261: 544 (1965).
(3) Ghittoni, N.E. and Gómez, C.J., Life Sci. 3: 979 (1964).
(4) Ramirez de Guglielmone, A.E. and Gómez, C.J., J. Neurochem. 13: 1017 (1966).
(5) Gómez, C.J. and Ramirez de Guglielmone, A.E., J. Neurochem. 14: 1119 (1967).

(6) Pasquini, J.M., Kaplun, B., Garcia Argiz, C.A. and Gómez, C.J., Brain Res. 6: 621 (1967).
(7) Geel, S.E. and Timiras, P.S., Brain Res. 4: 135 (1967).
(8) Balázs, R., Kovacs, S., Teichgraber, P., Cocks, W.A. and Eayrs, J.T., J. Neurochem. 15: 1335 (1968).
(9) Ramirez de Guglielmone, A.E., Duvilanski de Feldguer, B. and Gómez, C.J., Congr. Arg. Cienc. Biol. Buenos Aires, p. 118 (1970).
(10) Geel, S.E., Valcana, T. and Timiras, P.S., Brain Res. 4: 143 (1967).
(11) Dainat, J., Rebiere, A. and Legrand, J., J. Neurochem. 17: 581 (1970).
(12) Szijan, I., Kalbermann, L.E. and Gómez, C.J., Brain Res. 27: 309 (1971).
(13) Geel, S.E. and Timiras, P.S., Endocrinology 80: 1069 (1967).
(14) Garcia Argiz, C.A., Pasquini, J.M., Kaplun, B. and Gómez, C.J., Brain Res. 6: 635 (1967).
(15) Szijan, I., Chepelinsky, A.B. and Piras, M.M., Brain Res. 20: 313 (1970).
(16) Balázs, R., Brooksbauk, W.L., Davison, A.N., Eayrs, J. T. and Wilson, D.A., Brain Res. 15: 219 (1969).
(17) Walravens, P. and Chase, H.P., J. Neurochem. 16: 1477 (1969).
(18) Goldberg, R.C. and Chaikoff, I.L., Endocrinology 45: 64 (1949).
(19) Folch-Pi, J., Lees, M. and Sloane-Stanley, G.H., J. Biol. Chem. 226: 497 (1957).
(20) Suzuki, K. and Chen, G., J. Neurochem. 14: 911 (1967).
(21) Faryna de Raveglia, I. and Ghittoni, N.E., J. Chromat. 58: 288 (1971).
(22) MacMillan, V.H. and Wherrett, J.R., J. Neurochem. 16: 1621 (1969).
(23) Neskovic, N.M. and Kostic, D.M., J. Chromat. 35: 297 (1968).
(24) Chen, P.S. (Jr.), Toribara, T.Y. and Warner, H., Anal. Chem. 28: 1756 (1956).
(25) Hess, H.H. and Lewin, E., J. Neurochem. 12: 205 (1965).
(26) Lowry, O.H., Rosebrough, N.J., Farr, A.L. and Randall, J.R., J. Biol. Chem. 193: 265 (1951).
(27) Searcy, R. and Bergquist, L.M., Clin. Chim. Acta 5: 192 (1960).
(28) Munro, H.N. and Fleck, A., Biochim. Biophys. Acta 55: 571 (1962).

(29) Dalal, K.B. and Roboz-Einstein, E., Brain Res. 16: 441 (1969).
(30) Sperry, W.M., in Biochemistry of Developing Nervous System, Ed. by Waelsch, H., Academic Press, New York, p. 261 (1955).
(31) Davison, A.N. and Wajda, M., J. Neurochem. 4: 353 (1959).
(32) Wender, M., Folia Biol. (Krakow) 13: 323 (1965).
(33) Galli, C. and Cecconi, D.R., Lipids 2: 76 (1967).
(34) Eayrs, J.T., J. Endocrin. 22: 409 (1961).
(35) Eayrs, J.T. and Levine, S., J. Endocrin. 25: 505 (1963).
(36) Seminario, L.M., Hren, N. and Gómez, C.J., J. Neurochem. 11: 197 (1964).
(37) Lapetina, E.G., Soto, E.F. and De Robertis, E., Biochim. Biophys. Acta 135: 33 (1967).
(38) Korey, S.R. and Gonatas, J., Life Sci. 2: 296 (1963).
(39) Svennerholm, L., J. Neurochem. 10: 613 (1963).

SECTION III

SEPARATION AND PURIFICATION OF THE SUBCELLULAR COMPONENTS AND PLASMA MEMBRANES OF THE NERVOUS SYSTEM

METHODS OF SEPARATING THE SUBCELLULAR COMPONENTS OF BRAIN TISSUE

S. Spanner

Department of Pharmacology (Preclinical)

The Medical School, Birmingham, B15 2TJ, U.K.

It was in the early 1950s that the first great advances were made in our knowledge of the structure of the cell. This came about as a result of the development of the electron microscope and of the high-speed refrigerated centrifuge. The former gave an insight into the high degree of complexity of the components of the cell, and the latter made possible the separation of some of these components from each other and consequently the study of their metabolism and interaction.

The mammalian brain is an organ made up of highly complex cell types. The glial cells, microglia, astroglia and oligodendroglia, surround the highly specialised neurones (nerve cells) with their branching processes and long axon, each of which ends in a series of swollen tips. The axons in the adult brain are surrounded by the complex and protective coils of myelin which are formed from the cell membranes of the oligodendroglia. It is this complexity of its cell structure that has made the brain so difficult an organ to study chemically at the subcellular and cellular level.

There are two possible approaches to this study. Either the cells can be dissected out and studied intact or the brain or cells from different parts of the brain can be disrupted and the cell components studied. The main weakness of the second approach is that, generally speaking, a wide variety of cells contribute to the pool of subcellular components. The first approach was pioneered by Hydén (1) who developed a highly sophisticated dissection technique. This technique gives isolated neurones, intact and viable in the sense that they respire, but the method is tedious and the yields are small. Another dissection technique was described by Roots and Johnston (2). Many attempts have been made to develop bulk preparative methods for the isolation of neurones and glia using slow speed centrifugation following a gentle disruption of the tissue. Korey (3), Rose (4, 5), Bloomstrand and Hamburger (6), Norton and Poduslo (7) and Giorgi (8) have all had a degree of success using this approach.

The first attempt to study the subcellular components of brain was by Brody and Bain (9) in 1952. Using a sucrose medium, they fractionated the subcellular components of the disrupted brain, by differential centrifugation. Four years later and using the same technique, Hebb and Smallman (10) demonstrated that the "mitochondrial fraction" contained choline acetyltransferase (EC 2.3.1.6), the enzyme responsible for the formation of acetylcholine, and that this enzyme was present in an inactive, latent or occluded form. In 1958, Hebb and Whittaker (11) demonstrated that the acetylcholine pool closely paralleled the distribution of the choline acetyltransferase and, though present in the mitochondrial fraction prepared by the method of Brody and Bain (9), could, in fact, be separated from the mitochondria. It was the evidence of this heterogeneity of the mitochondrial fraction combined with the furtherence of the study of acetylcholine and synaptic transmission, which stimulated the development of the technique of subcellular fractionation. Initially this was developed by Whittaker's group in England and by de Robertis in Argentina. For a review of this development the reader is referred to Whittaker (12).

Cell fractionation, according to de Duve (13), involves three successive steps. The first is the destruction of the existing structure to give the so-called homogenate. This is followed by a re-grouping, by centrifugation, of components, according to their physical properties of density and particle size. The third step is that of analysing the result, of reconstructing the tissue metabolism in the light of one's findings and bearing in mind the limitations of the methods employed. It was found, for example, by de Duve (14), that while mechanical disruption in a Waring blendor released into the homogenate the enzymic content of the lysosomes, more gentle homogenisation left the lysosomes intact and the enzymes could be released only after water shock or treatment with detergents.

The centrifugation techniques applied to the separation and isolation of brain subcellular components can be divided into two main classes. There is rate centrifugation suitable for the separation of heavy from light components (Fig. 1). Here the brain is homogenised in isotonic sucrose and centrifuged. A low gravitational force will bring down nuclei and whole cells (4×10^4 g min) while a force of 2.4×10^5 g min will bring down the remaining cell components apart from the soluble fraction of the cell and the microsomes from the broken endoplasmic reticulum. These remain in suspension. The cell sap and the microsomes require a force of 6×10^6 g min to separate them. As Anderson points out (15), it is only the fractions which are the last to come down in the centrifugation programme which are in any way pure when this method of separation is used. Another draw-back is that frequent sedimentation is necessary and this in itself is harmful since it necessitates the frequent bumping of organelles on the bottom of the centrifuge tube. However, it is upon the heterogenous fraction, sedimenting between 4×10^4 and 2.4×10^5 g min that most interest has been focussed and this fraction, the P_2 or crude mitochondrial, cannot be subdivided by simple rate centrifugation.

As has already been mentioned, Hebb and Whittaker (11) found that within the crude mitochondrial fraction were the organelles containing acetylcholine and that these were distinct from the mitochondria. The two fractions could be separated by equilibrium or isopycnic centrifugation in a sucrose gradient, the different components settling at the sucrose density equivalent to their own (Fig. 2). This first density centrifugation was on a simple discontinuous gradient of 0.32M on 0.8M on 1.2M sucrose. In this way the P_2 fraction was shown to be made up of three fractions, the mitochondria in the pellet, a medium fraction (0.8 - 1.2M) containing acetylcholine and a light fraction (0.32 - 0.8M).

It was at this stage that the tools of the cell biologist came once more to the aid of the biochemist. Examination of the fractions obtained by the above gradient by electron microscopy (16), showed that the medium fraction contained a high concen-

RATE SEPARATION

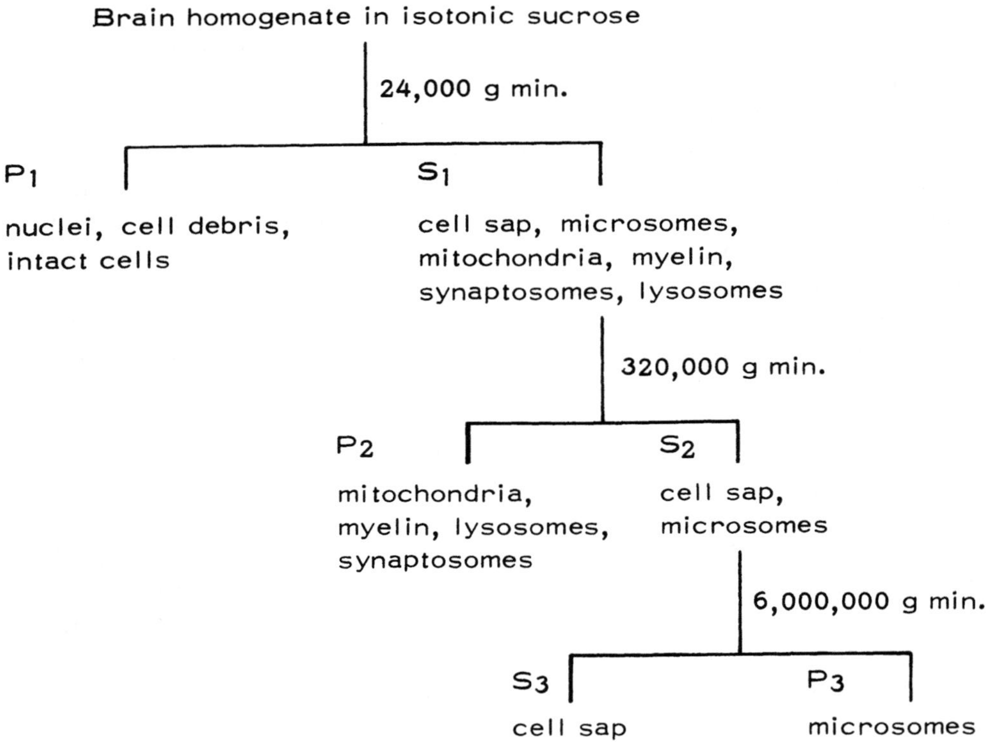

Fig. 1. Flow diagram of rate centrifugation of brain.

tration of particles surrounded by a thin membrane, and containing mitochondria and tiny vesicles. These were some of the characteristics of the nerve endings and were named synaptosomes by Whittaker.

The light fraction (0.32 - 0.8M) contained myelin fragments, the light myelin. There is also a considerable amount of myelin brought down in the nuclear fraction. Myelin may be isolated from these two crude fractions by further purification on sucrose gradients but a simpler method and one yielding myelin of a high degree of purity, was described by Autilio, Norton and Terry (17). They started with an homogenate of brain white matter and, by a series of gradients, produced two myelin populations; a light

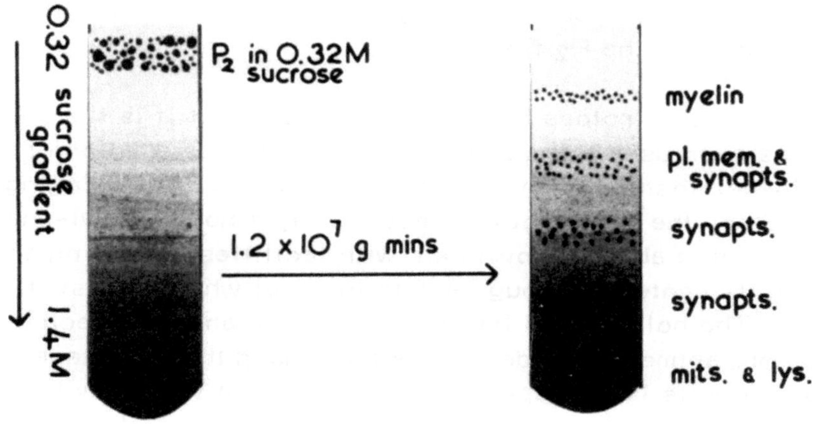

Fig. 2. Diagramatic representation of an isopycnic separation of brain tissue.

one sedimenting at 0.656 M and a heavy one at about 0.8 M sucrose.

Many workers have used a medium other than sucrose for preparing density gradients. Ficoll with sucrose has been extensively used, particularly in the fractionation of the P_2 fraction. Kurakawa, Sakamoto and Kato (18), Abdel-Latif (19) and Diamond and Kennedy (20) have used Ficoll in sucrose to prepare enriched synaptosomal fractions and Haga (21) uses a continuous gradient of Ficoll in sucrose to isolate fractions enriched in myelin, synaptosomes and mitochondria.

Caesium chloride has also been used, though this medium quickly destroys the aluminium caps of the tubes. Thompson, Goodwin and Cumings (22), for example, used a caesium chloride gradient for preparing myelin from brain white matter, and microsomes from brain cortex. Kornguth, Anderson and Scott (23) used caesium chloride to separate the components of the crude mitochondrial fraction.

Using finer stepwise gradients and continuous gradients, it has been possible to prepare fractions of far greater purity (24) and also to discover the presence of lysosomes in the brain (25). However, the small amount of material obtained and the relative difficulty of obtaining the synaptosomal fraction from the middle of the gradient, prompted us to apply the zonal rotor (B XIV) to the separation of the P_2 fraction.

Several zonal rotors have been developed but it is the B-type which has been used almost exclusively for the separation of brain tissue. Instead of the conventional tubes of the swinging bucket rotor, the zonal rotor is essentially a hollow bowl-shaped vessel spinning about its own axis with facilities for filling and removing its contents through a rotating seal while it is still spinning. The hollow bowl has a central core and four sector-shaped compartments divided by septa. Among the advantages of the rotor are its large capacity, the absence of wall effects, the stability of the gradient which is always under the influence of the centrifugal field and the ease of obtaining the fractions at the end of the run. The B XIV, aluminium rotor which is the one used in the work described in this paper, has a capacity of 650 ml, a maximum speed of 30,000 rpm giving a force of 6.7×10^4 g at the periphery. A full description of the rotor and its many uses is to be found in the following references (26, 27, 28).

As with the bucket rotor, the gradients used in the zonal rotor may be either continuous or discontinuous. In our early work with the P_2 fraction we used a simple linear gradient similar to the one used by Whittaker and his colleagues in their studies with tubes (24). This gave rather ill-defined fractions in the zone so a discontinuous gradient was adopted (29). The use of a stepwise gradient with shallow steps has made it possible to separate the components of the P_2 fraction into a number of decisive bands (Fig. 3). These components have been examined by electron microscopy and also studied using enzyme markers and their composition has been tentatively identified.

Peak A. This fraction is rich in myelin and also contains a high concentration of membranous material, partly plasma membranes and partly endoplasmic reticulum. This high membrane content is shown by the concentration of acetylcholinesterase, an enzyme absent from myelin. The myelin can be removed from the P_2 fraction in order to examine this membrane fraction on the zone or, the zonal centrifugation technique can be modified (Fig. 4). The zonal rotor was filled with a discontinuous gradient from 0.6 to 1.2M sucrose with a cushion of 1.6M sucrose. The P_2 fraction was applied in 0.6M sucrose and an overlay of 0.4M completed the gradient. This was centrifuged for 2 hr then the cushion was

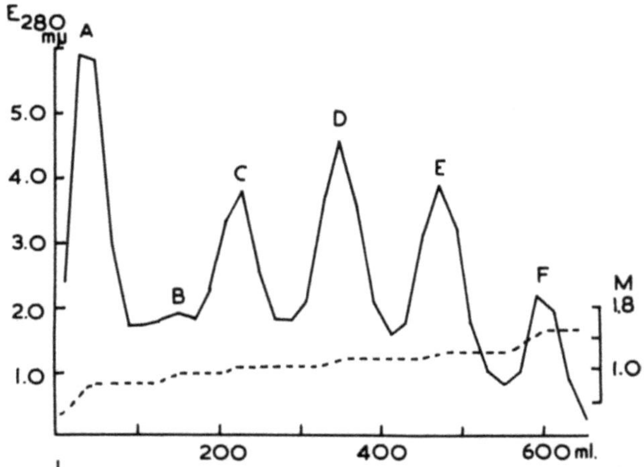

	A	B	C	D	E	F
AChE	45	17	22	10	4	2
occluded LDH	0	3	49	34	10	4
SDH	2	4	15	33	43	2
β-glucuronidase	5	5	12	14	12	52

Fig. 3. Graph of the protein concentration (O. D. 280 nm) of the fractions separated from the P_2 fraction by zonal centrifugation. The dotted line gives the sucrose concentration. Below the graph is a table giving the activity of the various enzyme markers. The results are expressed as a percentage of the total P_2 fraction.

pushed out. The first fractions were then pumped out in 20 ml samples using the second gradient from 1.2 M to 1.6 M sucrose. This was then centrifuged (2nd run) and then the fractions again collected. The result is seen in Fig. 4. The myelin is separated from the membranes (acetylcholinesterase-containing fraction). When white matter is fractionated by the zonal technique, two distinct myelin fractions are isolated, one between 0.4 and 0.6 M and the other between 0.6 and 0.8 M sucrose. This is in strict agreement with the findings, in tubes, of Autilio et al. (17).

Peak B. The small peak, also containing acetylcholinesterase activity, may well contain fragments from broken synaptosomes (c.f. Whittaker, 24) and also axonal fragments. Both Brunngraber (30) and Lenkey-Johnston (31) working with Dekirminjian, have shown that axonal fragments sediment at a sucrose concentration between 0.9 and 1.0 M.

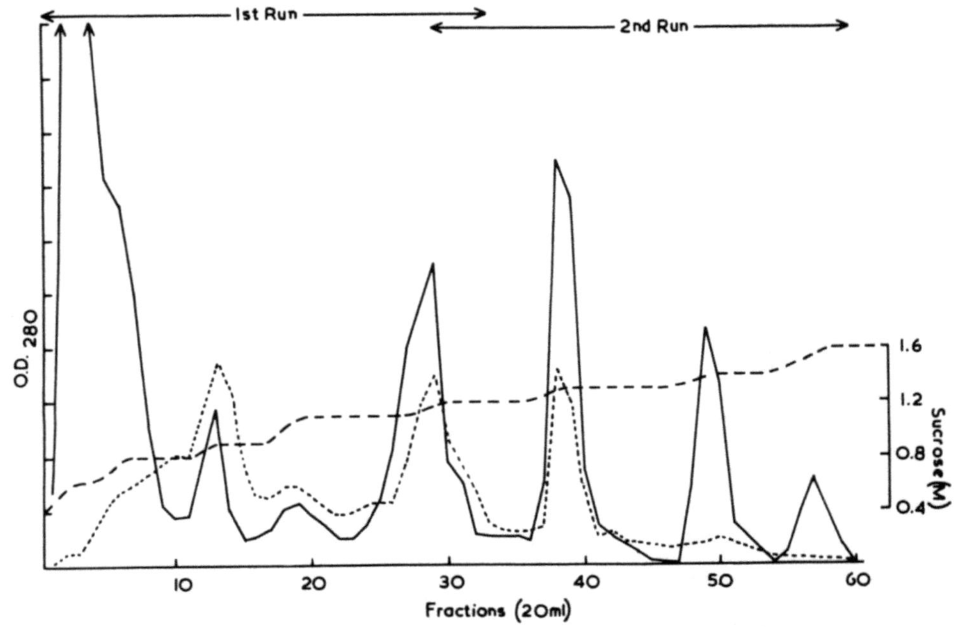

Fig. 4. The O. D. 280 nm (——) and the acetylcholinesterase activity (……) of the fractions obtained after two combined zonal fractionations of the P_2 fraction. For details see the text. The dashed line shows the sucrose concentration (----).

Peak E and F. By the low activity of acetylcholinesterase and β-glucuronidase and the high activity of succinic dehydrogenase and monoamine oxidase, it was concluded that fraction E was almost exclusively mitochondrial. It was well separated from the densest fraction (F) which contained a very high percentage of β-glucuronidase (Fig. 3 and 4). This was most certainly a lysosomal fraction and appeared to be fairly pure. When stained with acridine orange by the method of Koenig (32), no overlap of lysosomes with mitochondria could be detected.

Peak C and D. Fractions C and D are the two major synaptosomal peaks. Electron microscope pictures show both these fractions to contain the pinched off nerve endings described by Gray and Whittaker (16). The lactic dehydrogenase activity is all in an occluded form. This enzyme is a cytoplasmic marker and its presence in an occluded form is one indication that the synaptosomes are intact (33). Acetylcholinesterase activity is also high in these two fractions (Fig. 3 and 4). The presence of both succinic dehydrogenase and monoamine oxidase activity is due to the intraterminal mitochondria. It seemed worthwhile to examine these two fractions in greater detail to see if there were differences other than those of density, between the two.

In 1968 Iversen and Snyder (34) described some in vitro studies carried out with brain slices cut from the striatum and the hypothalamus. The slices were incubated in the presence of the radioactively labelled amines, noradrenalin and γ-aminobutyric acid. Subsequently the tissue was homogenised and the subcellular fractions prepared in tubes on a continuous sucrose gradient. The distribution of uptake within the synaptosomal band was not identical for the two amines. That for γ-aminobutyric acid had its peak between 0.9 and 1.0M while that for noradrenalin was between 1.1 and 1.2M.

Similarly Kuhar and his colleagues (35) studied the uptake of labelled γ-aminobutyric acid and 5-hydroxytryptamine into slices of the hypothalamus and mid-brain of rat. On fractionation in a sucrose gradient, these workers also found a difference in the peak of radioactivity of the two amines. That of γ-aminobutyric acid occurred in the same position as that of Iversen and Snyder (34) that is, a peak of radioactivity between 0.9 and 1.0M while that of 5-hydroxytryptamine lay between 1.0 and 1.1M.

Michaelson and Whittaker (36) showed that the maximum concentration of acetylcholine and 5-hydroxytryptamine did not completely correspond throughout the gradient and de Robertis and his colleagues (37) reported two populations of nerve endings, one cholinergic and the other non-cholinergic.

In view of all this evidence for the existance of at least two synaptosomal populations with different metabolic characteristics, it was decided to study, in more detail, the two fractions obtained by zonal centrifugation.

One approach was to study the uptake of α-methyl noradrenalin into the isolated synaptosomes. Each fraction was incubated for 30 min in Krebs' bicarbonate Ringer with the monoamine oxidase inhibitor pargylline and α-methyl noradrenalin. The synaptosomes were then centrifuged and the pellets dried, exposed to formaldehyde gas and mounted in Entellon. On examination by fluorescence microscopy it was seen that while the lighter fraction shows no evidence of α-methyl noradrenalin uptake, there is a considerable uptake into the heavier fraction. This finding of a preferential uptake is strictly comparable with the finding of Iversen and Snyder (34). So there would appear to be a direct relationship between the function and sedimentation density. It must, of course, be remembered that these synaptosomal fractions can be still further subdivided and work is proceeding along these lines.

Although there are still many improvements to be made in the zonal technique described, we are using the method to study some aspects of brain metabolism, particularly that of choline. It was shown (38) that [Me-^{14}C]choline, when injected intracerebrally, was rapidly metabolised by the brain. Recent studies with subcellular fractions have shown that over a five hour period after the injection of [Me-^{14}C]choline there was a rapid rise in the percentage of radioactivity in the P_2 fraction and an equally rapid fall in the soluble cell sap. The incorporation into the synaptosomes at all the times studied, was low, and this agrees with the studies in cats by Chakrin and Shideman (39).

Methods other than those based on centrifugation for the subcellular fractionation of brain tissue, have been described. One of these is based on electrophoresis. Sellinger and Borens (40) describe the treatment of a fraction containing synaptosomes, mitochondria and lysosomes by zonal density gradient electrophoresis. Runs were carried out at various different values of pH. The profiles obtained showed the presence of two and possibly three components. The subfraction of the crude mitochondrial fraction prepared on a three step sucrose gradient (A); 1.0, 1.3, 1.4M sucrose or (B); 1.1, 1.2, 1.3M sucrose in a refrigerated centrifuge were also subjected to the electrophoresis technique. This was used to study the distribution of membrane-bound acetylcholinesterase and glutamine synthetase. They confirmed the preferential association of acetylcholinesterase with the 1.0 to 1.1 M sucrose fraction.

The other method, that of ultrafiltration, has also been used to study the synaptosomal population of brain. Baldessarini and Vogt (41) have studied the uptake of noradrenalin by brain fractions. The crude mitochondrial fraction (P_2) was incubated with

L-[^{3}H]noradrenalin or the animals were injected with the labelled substrate and the brains subsequently fractionated. The suspensions were then poured through a millipore filter 25 mm in diameter with a pore size of 0.8 μm. This trapped the [^{3}H]noradrenalin-containing particles and the population of nerve endings thus isolated, would appear to correspond to the denser synaptosomal fraction described above.

It can be seen that methods for the subcellular fractionation of brain tissue have improved greatly since the pioneer work of Brody and Bain (9) about twenty years ago. Work is still in progress to prepare organelles of greater purity in large amounts, and for this, the zonal rotor is proving a very useful tool.

ACKNOWLEDGEMENTS

The author is grateful to the Welcome Trust of Great Britain for a Travel Grant to attend the meeting at which this paper was given. It is a pleasure to express my thanks to Dr. G. B. Ansell and Professor P. B. Bradley for their interest in the work described.

REFERENCES

(1) Hydén, H., in The Cell, vol. 4 (Ed. J. Brachet and A. E. Mirsky) p. 215 (1960).
(2) Roots, B. I. and Johnston, P. V., Ultrastruct. Res. 10: 350 (1964).
(3) Korey, S., in Biology of Neuroglia (Ed. W. F. Windle) p. 203 (1958).
(4) Rose, S. P. R., Nature, Lond. 206: 621 (1965).
(5) Rose, S. P. R., Biochem. J. 102: 33 (1967).
(6) Bloomstrand, C. and Hamberger, A., J. Neurochem. 16: 1401 (1969).
(7) Norton, W. T. and Poduslo, S. E., Science 167: 1144 (1970).
(8) Giorgi, P. P., Biochem. J. 122: 50P (1971).
(9) Brody, T. M. and Bain, J. A., J. biol. Chem. 195: 685 (1952).
(10) Hebb, C. O. and Smallman, B. N., J. Physiol. 134: 385 (1956).
(11) Hebb, C. O. and Whittaker, V. P., J. Physiol. 142: 187 (1958).
(12) Whittaker, V. P., Progr. Biophys. Molecular Biol. 15: 39 (1965).

(13) de Duve, C., in Methods of Separation of Subcellular Structural Components, Biochem. Soc. Symp. 23 (Ed. J.K. Grant) p. 1 (1963).
(14) de Duve, C. in Subcellular Particles (Ed. T. Hayashi) p. 128 (1958).
(15) Anderson, N.G., Fractions 1: 2 (1965).
(16) Gray, E.G. and Whittaker, V.P., J. Anat. Lond. 96: 79 (1962).
(17) Autilio, L.A., Norton, W.T. and Terry, R.D., J. Neurochem. 11: 17 (1964).
(18) Kurokawa, M., Sakamoto, T. and Kato, M., Biochem. J. 97: 833 (1965).
(19) Abdel-Latif, A.A., Biochim. biophys. Acta 121: 403 (1966).
(20) Diamond, I. and Kennedy, E.P., J. biol. Chem. 244: 3258 (1969).
(21) Haga, T., J. Neurochem. 18: 781 (1971).
(22) Thompson, E.J., Goodwin, H. and Cumings, J.N., Nature, Lond. 215: 168 (1967).
(23) Kornguth, S.E., Anderson, J.W. and Scott, G., J. Neurochem. 16: 107 (1969).
(24) Whittaker, V.P., Biochem. J. 106: 412 (1968).
(25) Koenig, H., Gaines, D., McDonald, T., Gray, R. and Scott, J., J. Neurochem. 11: 729 (1964).
(26) Anderson, N.G., Science 154: 103 (1966).
(27) Anderson, N.G., The Development of Zonal Centrifuges and Ancilliary Systems for Tissue Fractionation and Analysis, Natl. Cancer Inst. Monogr. 21 (1966).
(28) Reid, E., Preparations with Zonal Rotors (1971).
(29) Spanner, S. and Ansell, G.B. in Separations with Zonal Rotors (Ed. E. Reid) p. V-3.1 (1971).
(30) Dekirmenjian, H. and Brunngraber, E.G., Biochim. biophys. Acta 177: 1 (1969).
(31) Lemkey-Johnston, N. and Dekirmenjian, H., Expt. Brain Res. 11: 392 (1970).
(32) Koenig, H. in Handbook of Neurochemistry, vol. II (Ed. A. Lajtha) p. 255 (1969).
(33) Marchbanks, R.M., Biochem. J. 104: 148 (1967).
(34) Iversen, L.L. and Snyder, S.H., Nature, Lond. 220: 796 (1968).
(35) Kuhar, M.J., Shaskan, E.G. and Snyder, S.H., J. Neurochem. 18: 333 (1971).

(36) Michaelson, I.A. and Whittaker, V.P., Biochem. Pharmacol. 12: 203 (1963).
(37) Rodriguez de Lores Arnaiz, G. and de Robertis, E.D.P., J. Neurochem. 9: 503 (1962).
(38) Ansell, G.B. and Spanner, S., Biochem. J. 110: 201 (1968).
(39) Chakrin, L.W. and Shideman, F.E., Int. J. Neuropharm. 7: 337 (1968).
(40) Sellinger, O.Z. and Borens, R.N., Biochim. biophys. Acta 173: 176 (1969).
(41) Baldessarini, R.J. and Vogt, M., J. Neurochem. 18: 951 (1971).

THE ISOLATION AND CHARACTERIZATION OF SYNAPTOSOMAL PLASMA MEMBRANES (1)

I. G. MORGAN,* M. REITH,* U. MARINARI,◆
W. C. BRECKENRIDGE✿ and G. GOMBOS✱

Centre de Neurochimie du CNRS and Institut de Chimie Biologique de la Faculté de Médecine
67 - Strasbourg, France

The importance of methods for the isolation of fractions enriched in neuronal material, synaptosomal fractions (1, 2), synaptic vesicles (3, 5) and synaptosomal plasma membranes (3, 4, 6) can be seen from the large amount of work which has been published on their properties (for review, see 7-10). Unfortunately, despite the proliferation of papers in this field, these fractions have never been thoroughly characterized enzymatically and chemically. We therefore decided to attempt to isolate one of these fractions, the synaptosomal plasma membrane, in a form suitable for further study of its composition.

---

(1) This work has been, in part, supported by grants from the Institut National de la Santé et de la Recherche Médicale (Contract 71 11698) and the Fondation pour la Recherche Médicale Française.

* Boursier étranger de l'Institut National de la Santé et de la Recherche Médicale.

* Present address: Rudolf Magnus Institute, Utrecht, The Netherlands.

◆ Present address: Istituto di Patologia Generale, Università di Genova, Italy.

✿ Fellow of the Medical Research Council of Canada.

✱ Attaché de Recherche au Centre National de la Recherche Scientifique.

I. ISOLATION OF SYNAPTOSOMAL MEMBRANES

The method used is based on the methods of Gray and Whittaker (1), and De Robertis, Rodriguez De Lores Arnaiz, Pellegrino De Iraldi and Salganicoff (2), but with several modifications which we have shown to be of some importance for the isolation of pure fractions (11). The modifications concern three steps in the preparation (Fig. 1).

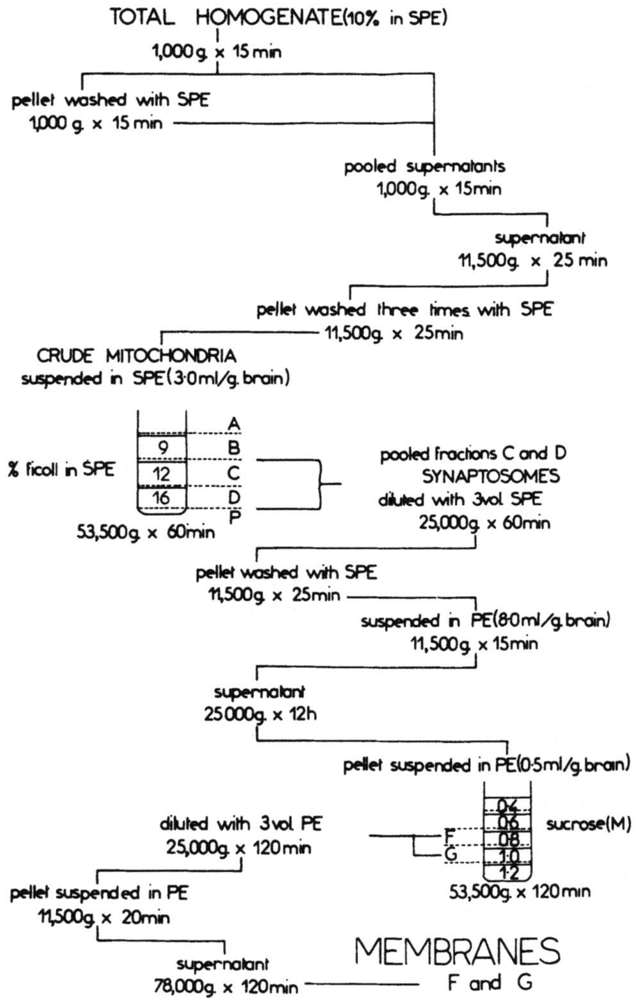

Fig. 1. Flow sheet of the preparation of synaptosomal plasma membrane fractions FL and GL. SPE stands for 0.32 M sucrose, 0.1 mM EDTA in 1 mM phosphate buffer (pH 7.6). PE stands for 0.1 mM EDTA in 1 mM phosphate buffer (pH 7.6).

A) The crude mitochondrial fraction was prepared by centrifugation at 11,500 g for 25 min, since Cotman, Brown, Harrell and Anderson (12) have shown that more vigorous centrifugation conditions sediment significant amounts of endoplasmic reticulum. The crude mitochondrial fraction was then washed three times (and the isolated synaptosomes are washed once more), washes necessary to eliminate otherwise unacceptably high amounts of microsomal contamination.

B) The well-washed mitochondrial fraction was then fractionated on an isotonic Ficoll-sucrose gradient (13, 14) rather than by sucrose density gradient centrifugation. Isotonic Ficoll-sucrose density gradients have many advantages over the more classical hypertonic sucrose gradients. Synaptosomes prepared under isotonic conditions are more sensitive to osmotic shock than synaptosomes prepared on sucrose density gradients, apparently because under hypertonic conditions the synaptosomes shrink, becoming less easily disrupted. Another advantage of isotonic conditions has been demonstrated by Day, McMillan, Mickey and Appel (15). These workers have shown that under hypertonic conditions, myelin tends to form artifactual aggregates with other subcellular particles. Under isotonic conditions, in Ficoll-sucrose, such aggregates can be avoided. Finally, as we suggested from an analysis of the distributions of enzyme markers on Ficoll-sucrose gradients, synaptosomes prepared on Ficoll-sucrose appear to be less contaminated with myelin and membrane fragments of the type described by Whittaker (16). This finding is in accord with the results of Cotman, Herschman and Taylor (17) on fractionated glial cells. When glial cells were fractionated by the technique of Gray and Whittaker (1) a large amount of membrane was found in the crude mitochondrial pellet. Upon sucrose density gradient fractionation, this membrane material was found in the "synaptosome region". Much less of this material was found in the "synaptosome region" on Ficoll-sucrose gradients. On the contrary, these membranes tended to sediment in the "myelin region".

All these data would suggest that synaptosomes prepared on Ficoll-sucrose gradients are much less contaminated with myelin, glial membranes, and other membranes, than are synaptosomes prepared on sucrose gradients. This is in accord with the morphology of the fraction (Fig. 2). As can be seen, there is little evidence of contamination with axonal fragments of the type described

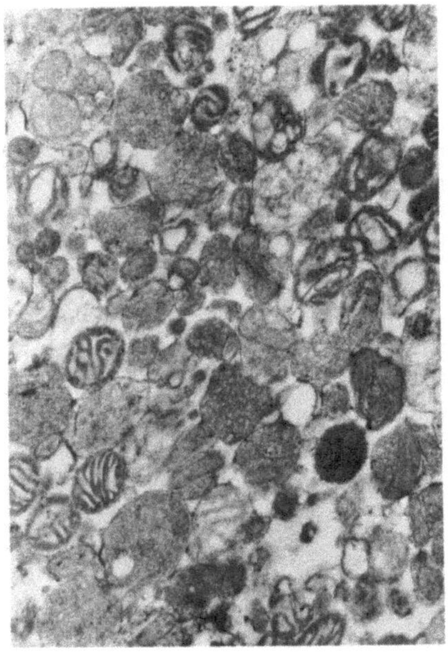

Fig. 2. Synaptosomal fraction prepared from adult rat brain on Ficoll-sucrose density gradients. Note the predominance of membrane-limited particles containing synaptic vesicles and mitochondria. The fractions contain in addition extra--synaptosomal mitochondria and lysosomes, and some membrane-limited particles which do not contain synaptic vesicles. Magnification 46,000 x.

by Lemkey-Johnston and Larramendi (18) and Dekirmenjian and Lemkey-Johnston (19). It should however be noted that these fractions are more contaminated with free mitochondria than synaptosomes prepared on sucrose gradients. Since this method has been developed to isolate the plasma membranes, it is more desirable to eliminate endoplasmic reticulum contamination at the expense of mitochondrial contamination, although, as will be seen later (Table I), this mitochondrial contamination introduces a contamination of the final plasma membrane preparation with external mitochondrial membrane.

C) After separation, the membranes were sedimented approximately under the same conditions as were used to isolate the crude mitochondrial fraction, in order to eliminate material whose sedimentation properties were not significantly changed by the osmotic shock. Although this led to considerable loss of plasma membrane markers (gangliosides and (Na+K)-ATPase), this cen-

Table I - Contamination of Synaptosomal Plasma Membrane Fractions FL and GL.

Contaminating fraction	Marker	% Contamination
Soluble proteins	Lactate dehydrogenase	0
Lysosomes	Acid phosphatase β-galactosidase β-glucosidase	<1
Endoplasmic reticulum	RNA NADPH-cytocrome c reductase	<1
Myelin	Cerebrosides and sulphatides	<1
Inner mitochondrial membranes	Cytochrome c oxidase Succinate dehydrogenase	<1
Outer mitochondrial membranes	Monoamine oxidase	5-10
Golgi apparatus	Galactosyl transferase	<15

The % contamination of the synaptosomal membrane fraction with other subcellular fractions are calculated from the data of ref II, except for the contamination with membranes of the Golgi apparatus (see section III).

trifugation was necessary to eliminate residual myelin, which was easily demonstrable by electron microscopy, gel electrophoresis, and chemical assay in the rejected pellets.

In summary then, when preparing synaptosomal plasma membranes the crude mitochondrial pellet should be thoroughly washed. Synaptosomes should be isolated on isotonic Ficoll-sucrose density gradients, and the material whose properties were not changed by the osmotic shock should be rejected. Several other technical points may be of importance. Cotman and Matthews (20) have shown that synaptosomal plasma membrane markers are better separated from mitochondrial markers if the synaptosomes are osmotically shocked at alkaline pH. Controls of ionic strength and pH appear to be important to avoid adsorption of soluble enzymes such as choline acetylase (21, 22).

II. ENZYMATIC AND CHEMICAL CHARACTERIZATION OF SYNAPTOSOMAL PLASMA MEMBRANES

The two membrane fractions (FL and GL) were then characterized enzymatically and chemically (Table I). The only appreciable subcellular contamination was with the external mitochondrial membrane, which accounted for around 5-10 % of the protein of the fractions. Contamination with myelin, inner mitochondrial membrane, smooth and rough endoplasmic reticulum was insignificant; nor was Golgi apparatus contamination significant (see section III). Using as markers acid phosphatase, β-galactosidase and β-glucosidase, lysosomal contamination was found to be insignificant. Mahler and Cotman (23) have shown that acid phosphatase and β-hexosaminidase activities cosedimented in part with the synaptosomal plasma membrane, but a more detailed analysis has suggested that the activities detected are due either to contamination with partially disrupted lysosomes, or adsorbed soluble lysosomal enzymes (20). Contamination with soluble proteins was tested using lactate dehydrogenase as a marker. Only very low activities were detected, but this in no way excludes that other soluble proteins may be artifactually bound to the membranes during isolation as is the case with choline acetyltransferase (21, 22). From the detectable contaminants, the preparations appeared to be over 80 % plasma membrane.

However there was some contradiction in our initial data (11), in that, while myelin was shown to be absent by the lack of cerebrosides and sulphatides, as well as by the absence (24) of the three protein bands characteristic of myelin (25), there was a significant activity of 2', 3'-cyclic AMP 3'-phosphohydrolase. This enzyme is generally believed to be a marker for myelin (26-28), and we suggested that this enzyme, which is highly concentrated in the central and peripheral nervous systems (29), might also be associated with the glial cell plasma membrane from which myelin is derived (30). This hypothesis was verified since the enzyme was detected in clone C-6 glial cells in culture (31). However it has recently been reported that approximately equal specific activities of the enzyme have been found in isolated neurons, glial cells and in myelin (32). This might suggest that the 2', 3'-cyclic AMP 3'-phosphohydrolase activity detected in the synaptosomal plasma membranes is an intrinsic activity, but if we assume that all the activity detected in the membrane preparations is due to glial plasma membrane, assuming the same enrichement of 2', 3'-cyclic AMP 3'-phosphohydrolase in the glial cell plasma membrane as is found for 5'-nucleotidase in liver plasma membranes, the level of glial cell plasma membrane contamination would be of the order of 10 %.

The morphology of the synaptosomal fractions (Fig. 2) gives further evidence for glial contamination, since in addition to the synaptosomes (identified by the presence of synaptic vesicles and more rarely by the synaptic junctions), there are some membrane-limited bodies with clear diffuse cytoplasm. These bodies may be derived from the neuronal cell body, or from glial cells (33). Some of these bodies contain ribosomes (34-36) and may be responsible for the cycloheximide-sensitive protein synthesis observed in synaptosomal fractions (37, 38). We have attempted to estimate the level of contamination with these bodies by estimating a soluble glial marker, the S-100 protein fraction (39). This protein is pre-

Table II - Partial Fatty Acid Composition of Membrane Sphingomyelins.

	% of sphingomyelin total fatty acids		Reference
	C 24:0	C 24:1	
Synaptosomal plasma membrane	1.3	0.9	
Myelin	8.0	40.0	(41)
Oligodendrocytes	11.6	23.1	(41)
Astrocytes (in vivo)	1.3	2.8	(43)
Astrocytes (in tissue culture)	9.2	13.8	

sent in the soluble fraction derived from the synaptosomal fractions at approximately one fifth of the level in total brain soluble fractions. Assuming that the neuronal: glial ratio in total brain is 1:1, and that the soluble proteins accurately represent the corresponding membrane proteins, approximately 10 % contamination with glial material can be calculated. It should be noted that these calculations, already rather hypothetical, do not take into account the existence of the two main classes of glial cells (oligodendrioglia and astroglia) which may differ chemically and enzymatically, and in their behaviour upon subcellular fractionation.

Another possible means of calculating the contamination with glial material is shown in Table II.

Long-chain fatty acids in the sphingomyelin are characteristic of non-neuronal tissues (40), and isolated myelin and oligodendrocytes (41). C-6 clone astrocytes (42) in tissue culture also contain long-chain fatty acid sphingomyelins (unpublished results) although astrocytes in vivo may contain sphingomyelins devoid of long-chain fatty acids (43, 44). The sphingomyelins of synaptosomal plasma membranes, by contrast, contain very low amounts of these long-chain fatty acids (45, 47). If such fatty acids are found in the sphingomyelin of all non-neuronal cells, this would be a sensitive index of contamination of neuronal membranes with non-neuronal material. Assuming that all the detected long-chain fatty acids are due to glial contamination, then a contamination of the order of 10 % can be calculated on a lipid basis. Neuronal plasma membra-

Table III - Contamination of the Synaptosomal Plasma Membrane Fractions with Glial Membranes.

Marker	% Contamination
S-100 protein fraction	5-10
2', 3'-cyclic AMP 3'-phosphohydrolase	5-10
Sphingomyelin long-chain fatty acids	<10

nes appear to be much less rich in sphingomyelin (45-47) than are other plasma membranes (48-50), and thus on a protein basis the contamination would be much lower. Thus all the methods used for estimating glial contamination (Table III) agree in setting a limit of 10 % to such contamination. It must however be stressed that all the calculations require certain assumptions of contestable validity.

This problem can also be approached by considering the enrichment of putative neuronal plasma membrane markers in the membrane fractions (Table IV). We have previously calculated that a synaptosomal plasma membrane fraction should be enriched in a specific marker of the neuronal plasma membrane around 14-fold (11). Gangliosides have been claimed to be markers for the neuronal plasma membrane (51-53) but the ganglioside GM1 has been reported to be a myelin component (54-56). Gangliosides have also been reported in isolated astrocytes (57) and oligodendrocytes (32) as well as in astrocytes in culture <u>in vivo</u> (58). Moreover it has recently been shown that GM3 is found in the reticulo-endothelial cells of brain (59). These results, together with the low levels of gangliosides in isolated neurons (57-60), makes it unlikely that gangliosides can be used as a specific neuronal marker, but even so the membranes are enriched over 10-fold in ganglioside sialic

Table IV - Enrichment of Putative Plasma Membrane Markers in the Synaptosomal Plasma Membrane Fractions.

Marker	Enrichment (1)
(Na+K)-ATPase	9-14
Gangliosides	8-11
5'-nucleotidase	1-4
Acetyl-cholinesterase	0.5-1.0

(1) Specific activity of fraction/specific activity of homogenate.

acid. This is not due to a preferential enrichment in di- and tri-sialo gangliosides since the ganglioside pattern is not very different from total brain (unpublished results). It is interesting to note that non-neural tissue gangliosides are rich in the long-chain fatty acids (61, 62) found also in non-neural sphingomyelin (40), whereas neural gangliosides are in general very rich in stearic acid and contain little long-chain fatty acid (63).

Another possible marker for the neuronal plasma is ouabain--sensitive (Na+K)-ATPase, an enzyme which is highly enriched in neural tissue (64). Astrocytic tumors and astrocytes in tissue culture contain very little (Na+K)-ATPase activity (65, 66) and indeed it might be expected that neurons would have a greater need for ion transport than would glial cells. However isolated astrocytes have been reported to have (Na+K)-ATPase activities equal to, or higher than, those of neurons (67, 68). Nevertheless this enzyme, previously reported to be concentrated in synaptosomal plasma membranes (69), is also concentrated over 10-fold in our membrane preparations.

It is interesting to note that two other putative plasma membrane markers, 5'-nucleotidase and acetyl-cholinesterase, are only slightly enriched, if at all, in the membrane preparations. However acetyl-cholinesterase is known from histochemical studies to be found in the endoplasmic reticulum and neurotubules as well as on the plasma membrane (70), and thus is most unlikely to be a valid plasma membrane marker. 5'-nucleotidase is known to be associated with the glial cell plasma membrane (71) and, although in other tissues it seems to be the best plasma membrane marker (72, 73), this does not appear to be the case for brain, at least as regards the synaptosomal plasma membrane. Several studies have claimed parallel distributions of gangliosides, acetyl-cholinesterase, (Na+K)-ATPase and 5'-nucleotidase, but these results demonstrate a parallel only between gangliosides and (Na+K)-ATPase. The failure to observe enrichment of acetyl-cholinesterase and 5'-nucleotidase in synaptosomal plasma membranes has recently been confirmed (20).

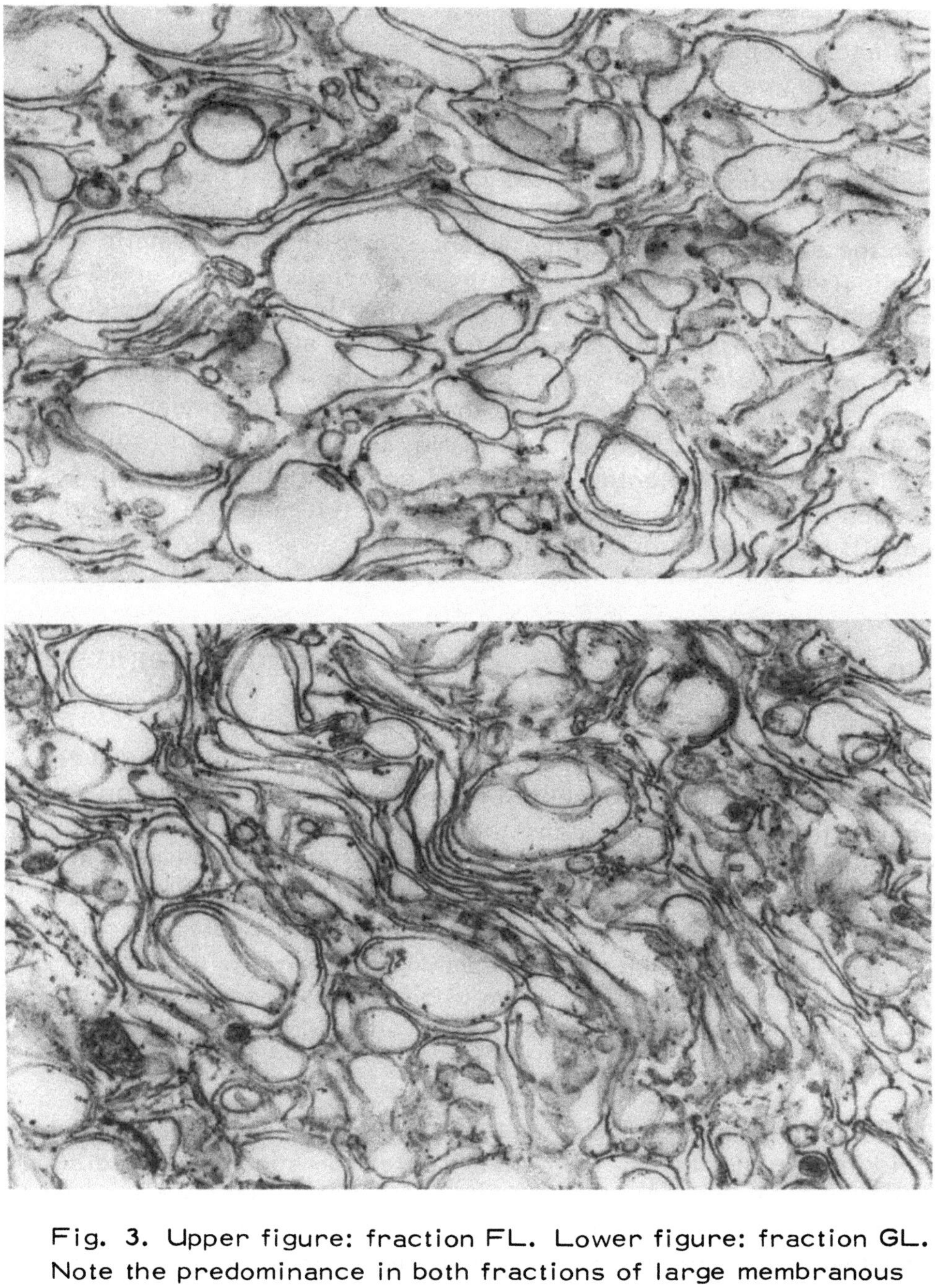

Fig. 3. Upper figure: fraction FL. Lower figure: fraction GL. Note the predominance in both fractions of large membranous vesicles of synaptosomal dimensions. Fraction GL appears to contain more long membranous sheets than fraction FL. Magnification 100,000 x.

The morphology of these membrane fractions under the electron microscope is shown in Fig. 3. Fraction FL consists of large membranous vesicles of synaptosomal dimensions and of a limited number of membrane sheets. Fraction GL, by contrast, consists primarily of membrane sheets with fewer membranous vesicles. These morphological differences do not seem to have enzymatic correlates, and it is probable that the two fractions are derived from the same basic type of membrane. Despite the enzymatic and chemical data which suggest that the preparations are over 80 % neuronal plasma membrane, very few synaptic junctions are observed in the final membrane fractions. This may be due to the rejection of the material which sediments at sucrose densities higher than 1.0 M sucrose, since the cleft material appears to be denser (74). Moreover intact clefts may be concentrated in the heavier fractions rejected at the final centrifugation step. Other methods of staining are more specific for cleft material (75, 76) and will be tried on these preparations.

III. GLYCOPROTEIN BIOSYNTHESIS IN NERVE-ENDINGS

Several lines of evidence have been advanced which suggest that nerve-endings are capable of synthesizing glycoproteins, at least to the extent of modifying the carbohydrate chains of these molecules. The kinetics of incorporation of glucosamine into synaptosomal fractions differs from the kinetics of incorporation of leucine (77, 78) in a manner which has been interpreted as reflecting local incorporation of glucosamine. However neither fucose (79, 80) nor sialic acid (81) seem to be rapidly incorporated into nerve-ending glycoproteins.

Synaptosomal fractions incubated *in vitro* seem to incorporate a wide range of sugars, from their nucleotide sugar intermediates, into glycoproteins (82). Festoff, Appel and Day (83) have reported *in vitro* incorporation of glucosamine by synaptosomes. In addition, synaptosomal fractions contain a significant proportion of the nucleotide sugar: glycoprotein-glycosyl transferase activities in embryonic chicken brain (84, 85). Three glycosyl transferases (mannosyl transferase, galactosyl transferase and N-acetyl-glucosaminyl transferase using endogenous acceptors) have been reported in synaptosomal subfractions (86). However there does not appear to be appreciable fucosyl transferase activity in synaptosomal fractions (80, 87).

It is difficult to allow in these studies for the rapid transport of glycoproteins known to occur in vivo (88-90), or for contamination of the fractions examined in vitro. Contamination of synaptosomal fractions with endoplasmic reticulum and membranes of the Golgi apparatus could explain the apparent capacity of synaptosomes to synthesize glycoproteins. Indeed a contamination of synaptosomal subfractions by Golgi apparatus membranes has been claimed, based on the presence of thiamine pyrophosphatase activities (91).

We have examined this problem by comparing the distribution of two putative Golgi apparatus markers, thiamine pyrophosphatase and UDP-galactose-N-acetyl-glucosamine galactosyl transferase.

Galactosyl transferase, which in liver is a reliable marker for the Golgi apparatus (92-95), does not appear to be enriched in the synaptosomal plasma membrane fractions when compared to the homogenate, but it is found to be enriched in one of the fractions (EH) rejected during the preparation of the membranes (Table V). The distribution of this enzyme does not parallel that of the gangliosides, nor that of (Na+K)-ATPase, and it is thus unlikely that the major part of this cerebral galactosyl transferase is associated with the synaptosomal plasma membrane. We believe it is more likely that the activity detected in the synaptosomal plasma membrane preparations indicates a contamination with Golgi apparatus membranes. Until we have isolated a Golgi apparatus fraction from brain, and determined its galactosyl transferase activity, it will not be possible to calculate a % contamination. However if we compare the fractions to fraction EH (which is unlikely to be pure Golgi apparatus membranes) the contamination is of the order of 10-15 %. In comparison with the enrichments of galactosyl transferase in Golgi apparatus membranes isolated from liver (93, 94), the contamination of the plasma membrane fractions with Golgi ap-

Table V - Enrichment of Putative Golgi Apparatus Markers in the Synaptosomal Plasma Membrane Fractions.

Fraction	Enrichment (1)	
	Thiamine pyrophosphatase	Galactosyl transferase
EL	5.0	0.08
EH	10.4	8.40
FL	5.8	0.64
FH	9.2	1.55
GL	6.2	1.00
GH	10.4	2.04

(1) Specific activity of fraction/specific activity of homogenate.

paratus would be less than 1 %. In brain, thiamine pyrophosphatase is an enzyme with two pH maxima (96), neither of which corresponds to the conditions used for histochemical localization (97) or biochemical studies (91). The same pH curve is observed for all subcellular fractions and for glial cells in tissue culture (unpublished results), and thus the results presented here are given only at pH 7, to facilitate comparison with other work. This enzyme is found to be highly concentrated in the plasma membrane fractions (5-fold) but in contrast to the gangliosides and (Na+K)--ATPase it is even more concentrated in the rejected heavy fractions. Thus the enzyme does not parallel these two markers, nor does it distribute in parallel with the galactosyl transferase. This distribution suggests that thiamine pyrophosphatase cannot be used in biochemical studies as a marker for Golgi apparatus membranes from brain. In fact, despite its use as a histochemical marker (97), this enzyme cannot be used as a biochemical marker in liver (93, 94, 98). In brain thiamine pyrophosphatase may not be a valid histochemical marker for the Golgi apparatus since Lane (99) has histochemically detected thiamine pyrophosphatase activity in the plasma membranes of insect neurons and glial cells.

The results indicate that the activity of one galactosyl transferase, UDP-galactose-N-acetyl-glucosamine galactosyl transfera-

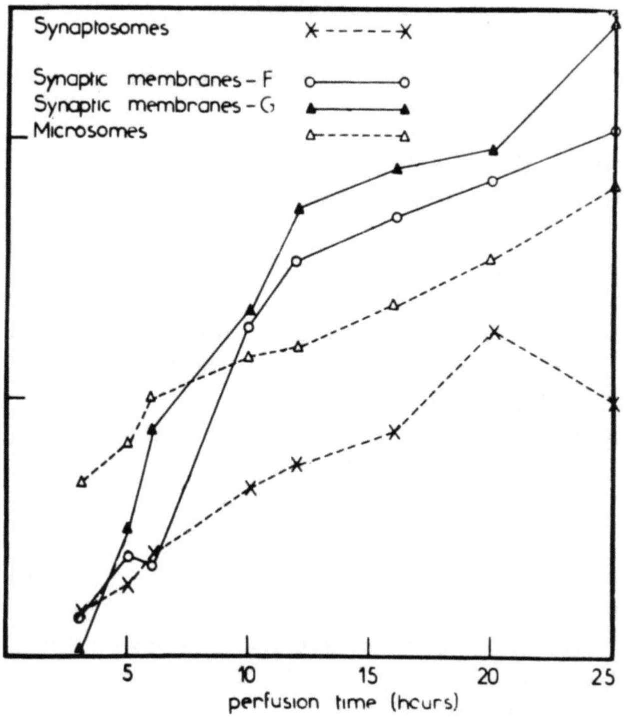

Fig. 4. Time course of incorporation of 14C-glucosamine into the chloroform:methanol insoluble residue of synaptosomal plasma membrane fractions.

se, is low in synaptosomal plasma membranes isolated from rat brain. Since this galactosyl transferase is among the glycosyl transferases found to be concentrated in synaptosome-rich fractions from embryonic chicken brain (84, 85), our results obtained using purer fractions, reopen the question of the synaptosomal localization of most of the glycosyl transferases previously studied. It is obviously necessary to test these preparations for a range of glycosyl transferases with a variety of acceptors (endogenous and exogenous; glycoprotein and glycolipid). But as shown in Fig. 4, the kinetics of incorporation of glucosamine into these synaptosomal plasma membrane fractions can be interpreted in terms of axonal flow (unpublished results), which would suggest that the ability of synapses to synthesize glycoproteins may be very limited, if it exists at all. It seems more likely that glycoprotein synthesis in brain takes place by the same pathways, in the same subcellular structures, as in other tissues.

REFERENCES

(1) Gray, E. G. and Whittaker, V. P., J. Anat. London 96: 79 (1962).

(2) De Robertis, E., Pellegrino De Iraldi, A., Rodriguez De Lores Arnaiz, G. and Salganicoff, L., J. Neurochem. 9: 23 (1962).

(3) Whittaker, V. P., Michaelson, I. A. and Kirkland, R. J. A., Biochem. J. 90: 293 (1964).

(4) De Robertis, E., Rodriguez De Lores Arnaiz, G., Salganicoff, L., Pellegrino De Iraldi, A. and Zieher, L. M., J. Neurochem. 10: 225 (1963).

(5) Lapetina, E. G., Soto, E. F. and De Robertis, E., Biochim. Biophys. Acta 135: 33 (1967).

(6) Rodriguez De Lores Arnaiz, G., Alberici, M. and De Robertis, E., J. Neurochem. 14: 215 (1967).

(7) Whittaker, V. P., Ann. N. Y. Acad. Sci. 137: 982 (1966).

(8) Michaelson, I. A., Ann. N. Y. Acad. Sci. 144: 387 (1967).

(9) De Robertis, E. and Rodriguez De Lores Arnaiz, G., in Handbook of Neurochemistry, Ed. A. Lajtha, Plenum Press, New York, vol. II, p. 365 (1969).

(10) Whittaker, V. P., in Handbook of Neurochemistry, Ed. A. Lajtha, Plenum Press, New York, vol. II, p. 327 (1969).

(11) Morgan, I. G., Wolfe, L. S., Mandel, P. and Gombos, G., Biochim. Biophys. Acta 241: 737 (1971).

(12) Cotman, C., Brown, D. H., Harrell, B. W. and Anderson, N. G., Arch. Biochem. Biophys. 136: 436 (1970).

(13) Kurokawa, M., Sakamoto, T. and Kato, M., Biochem. J. 97: 833 (1965).

(14) Abdel-Latif, A. A., Biochim. Biophys. Acta 121: 403 (1966).

(15) Day, E. D., McMillan, P. N., Mickey, D. D. and Appel, S. H., Anal. Biochem. 39: 29 (1971).

(16) Whittaker, V. P., Biochem. J. 106: 412 (1968).

(17) Cotman, C., Herschman, H. and Taylor, D., J. Neurobiol. 2: 169 (1971).

(18) Lemkey-Johnston, N. and Larramendi, L. M. H., Exp. Brain Res. 5: 326 (1968).

(19) Lemkey-Johnston, N. and Dekirmenjian, H., Exp. Brain Res. 11: 392 (1970).

(20) Cotman, C. W. and Matthews, D. A., Biochim. Biophys. Acta, in press.

(21) Fonnum, F., Biochem. J. 103: 262 (1967).

(22) Fonnum, F., Biochem. J. 109: 389 (1968).
(23) Mahler, H.R. and Cotman, C.W., in Protein Metabolism of the Nervous System, Ed. A. Lajtha, Plenum Press, New York, p. 151 (1970).
(24) Waehneldt, T.V., Morgan, I.G. and Gombos, G., Brain Res. 34: 403 (1971).
(25) Waehneldt, T.V. and Mandel, P., FEBS Letters 9: 209 (1970).
(26) Kurihara, T. and Tsukada, Y., J. Neurochem. 14: 1167 (1967).
(27) Kurihara, T. and Tsukada, Y., J. Neurochem. 15: 827 (1968).
(28) Kurihara, T. and Mandel, P., in Les Mutants Pathologiques chez l'Animal, Ed. M. Sabourdy, Editions CNRS, Paris, p. 57 (1970).
(29) Drummond, G.I., Iyer, N.T. and Kieth, J., J. Biol. Chem. 237: 3535 (1962).
(30) Bunge, M.B., Bunge, R.P. and Pappas, G.D., J. Cell Biol. 12: 448 (1962).
(31) Zanetta, J.P., Benda, P., Gombos, G. and Morgan, I.G., J. Neurochem., in press.
(32) Norton, W.T. and Poduslo, S.E., in Abstracts 3rd Int. Meet. I.S.N., Budapest, Ed. J. Domonkos, A. Fonyó, I. Huszák and J. Szentágothai, Akadémiai Kiadó, p. 30 (1971).
(33) Lemkey-Johnston, N. and Larramendi, L.M.H., J. Cell Biol. 39: 80A (1968).
(34) Morgan, I.G., FEBS Letters 10: 273 (1970).
(35) Gambetti, P.L., Autilio-Gambetti, L. and Gonatas, N.K., J. Cell Biol. 47: 68A (1970).
(36) Cotman, C.W. and Taylor, D.A., Brain Res. 29: 366 (1971).
(37) Morgan, I.G. and Austin, L., J. Neurochem. 15: 41 (1968).
(38) Autilio, L.A., Appel, S.H., Pettis, P. and Gambetti, P.L., Biochemistry 7: 2615 (1968).
(39) Cicero, T.J., Cowan, W.M., Moore, B.W. and Suntzeff, V., Brain Res. 18: 25 (1970).
(40) Svennerholm, E., Ställberg-Stenhagen, S. and Svennerholm, L., Biochim. Biophys. Acta 125: 60 (1966).
(41) Fewster, M.E. and Mead, J.F., J. Neurochem. 15: 1303 (1968).
(42) Benda, P., Lightbody, J., Sato, G.H., Levine, L. and Sweet, W., Science 161: 370 (1968).
(43) Pomazanskaya, L.F., Freysz, L. and Mandel, P., Zh. Evol. Biokhim. Fiziol. 5: 523 (1969).
(44) Nussbaum, J.L., Neskovic, N. and Mandel, P., J. Neurochem. 18: 1529 (1971).
(45) Kishimoto, Y., Agranoff, B.W., Radin, N.S. and Burton, R. M., J. Neurochem. 16: 397 (1969).

(46) Cotman, C., Blank, M. L., Moehl, A. and Snyder, F., Biochemistry 8: 4606 (1969).
(47) Breckenridge, W. C., Gombos, G. and Morgan, I. G., in Abstracts of 3rd Int. Meet. I. S. N., Budapest, Ed. J. Domonkos, A. Fonyó, I. Huszák and J. Szentágothai, Akadémiai Kiadó, p. 245 (1971).
(48) Weinstein, D. B., Marsh, J. B., Glick, M. C. and Warren, L., J. Biol. Chem. 244: 4103 (1969).
(49) Keenan, T. W. and Morre, D. J., Biochemistry 9: 19 (1970).
(50) Ray, T. K., Shipski, V. P., Barclay, M., Essner, E. and Archibald, F. M., J. Biol. Chem. 244: 5528 (1969).
(51) Lowden, J. A. and Wolfe, L. S., Canad. J. Biochem. 42: 1587 (1965).
(52) Spence, M. W. and Wolfe, L. S., Canad. J. Biochem. 45: 671 (1967).
(53) Derry, D. M. and Wolfe, L. S., Science 158: 1450 (1967).
(54) Suzuki, K., Poduslo, S. E. and Norton, W. T., Biochim. Biophys. Acta 144: 375 (1967).
(55) Suzuki, K., Poduslo, J. F. and Poduslo, S. E., Biochim. Biophys. Acta 152: 576 (1968).
(56) Suzuki, K., J. Neurochem. 17: 209 (1970).
(57) Norton, W. T. and Poduslo, S. E., J. Lipid Res. 12: 84 (1971).
(58) Shein, H. M., Britva, A., Hess, H. H. and Selkoe, D. J., Brain Res. 19: 497 (1970).
(59) Eto, Y. and Suzuki, K., J. Neurochem. 18: 503 (1971).
(60) Tamai, Y., Matsukawa, S. and Satake, M., J. Biochem. 69: 235 (1971).
(61) Puro, K. and Keränen, A., Biochim. Biophys. Acta 187: 393 (1969).
(62) Tao, R. V. P. and Sweeley, C. C., Biochim. Biophys. Acta 218: 373 (1970).
(63) Trams, E. G., Giuffrida, L. E. and Karmen, A., Nature 193: 680 (1962).
(64) Bonting, S. L., Caravaggio, L. L. and Hawkins, N. M., Arch. Biochem. Biophys. 98: 413 (1962).
(65) Embree, L. J., Shein, H. M., Benda, P. and Hess, H. H., J. Neuropath. Exptl. Neurol. 28: 155 (1969).
(66) Embree, L. J., Hess, H. H. and Shein, H. M., Brain Res. 27: 422 (1971).
(67) Medzihradsky, F., Nandhasri, P. S., Idoyaga-Vargas, V. and Sellinger, O. Z., J. Neurochem. 18: 1599 (1971).

(68) Henn, F., Blomstrand, C. and Hamberger, A., in Abstracts 3rd Int. Meet. I.S.N., Budapest, Ed. J. Domonkos, A. Fonyó, I. Huszák and J. Szentágothai, Akadémiai Kiadó, p. 243 (1971).
(69) Hosie, R.J.A., Biochem. J. 96: 404 (1965).
(70) Kokko, A., Mautner, H.G. and Barrnett, R.J., J. Histochem. Cytochem. 17: 625 (1969).
(71) Torack, R.M. and Barrnett, R.J., J. Neuropath. Exptl. Neurol. 23: 46 (1964).
(72) Bosmann, H.B., Hagopian, A. and Eylar, E.H., Arch. Biochem. Biophys. 128: 51 (1968).
(73) Emmelot, P., Bos, C.J., Benedetti, E.L. and Rümke, Ph., Biochim. Biophys. Acta 90: 126 (1964).
(74) Davis, G.A. and Bloom, F.E., J. Cell Biol. 47: 46A (1970).
(75) Bloom, F.E. and Aghajanian, G.K., Science 154: 1575 (1966).
(76) Bloom, F.E. and Aghajanian, G.K., J. Ultrastruct. Res. 22: 361 (1968).
(77) Barondes, S.H., J. Neurochem. 15: 699 (1968).
(78) Barondes, S.H. and Dutton, G.R., J. Neurobiol. 1: 99 (1969).
(79) Zatz, M. and Barondes, S.H., J. Neurochem. 17: 157 (1970).
(80) Zatz, M. and Barondes, S.H., J. Neurochem. 18: 1125 (1971).
(81) De Vries, G.H. and Barondes, S.H., J. Neurochem. 18: 101 (1971).
(82) Bosmann, H.B. and Hemsworth, B.A., J. Biol. Chem. 245: 363 (1970).
(83) Festoff, B.W., Appel, S.H. and Day, E., Fed. Proc. 28: 734 (1969).
(84) Den, H. and Kaufman, B., Fed. Proc. 27: 346 (1968).
(85) Den, H., Kaufman, B. and Roseman, S., J. Biol. Chem. 245: 6607 (1970).
(86) Broquet, P. and Louisot, P., Biochimie, in press.
(87) Zatz, M. and Barondes, S.H., J. Neurochem. 18: 1625 (1971).
(88) Elam, J.S., Goldberg, J.M., Radin, N.S. and Agranoff, B.W., Science 170: 458 (1970).
(89) Forman, D.S., McEwen, B.S. and Grafstein, B., Brain Res. 28: 119 (1971).
(90) Bondy, S.C., Exptl. Brain Res. 13: 127 (1971).
(91) Seijo, L. and Rodriguez De Lores Arnaiz, G., Biochim. Biophys. Acta 211: 595 (1970).
(92) Morre, D.J., Merlin, L.M. and Keenan, T.W., Biochem. Biophys. Res. Comm. 37: 813 (1969).
(93) Fleischer, B., Fleischer, S. and Ozawa, H., J. Cell Biol. 43: 59 (1969).

(94) Cheetham, R. D., Morre, D. J. and Yunghans, W. N., J. Cell Biol. 44: 492 (1970).

(95) Schachter, H., Jabbal, I., Hudgin, R. L., Pinteric, L., McGuire, E. J. and Roseman, S., J. Biol. Chem. 245: 1090 (1970).

(96) Kiessling, K. H., Acta Chem. Scand. 14: 1669 (1960).

(97) Novikoff, A. B. and Goldfischer, S., Proc. Natl. Acad. Sci. U. S. 47: 802 (1961).

(98) Cheetham, R. D., Morre, D. J., Pannek, C. and Friend, D. S., J. Cell Biol. 49: 899 (1971).

(99) Lane, N. J., J. Cell Biol. 37: 89 (1968).

SECTION IV

CHEMICAL PATHOLOGY AND DIAGNOSIS OF LIPID AND MUCOPOLYSACCHARIDE STORAGE DISEASES

GLYCOLIPID, MUCOPOLYSACCHARIDE AND CARBOHYDRATE DISTRIBUTION IN TISSUES, PLASMA AND URINE FROM GLYCOLIPIDOSES AND OTHER DISORDERS. Complex Nature of the Accumulated substances.

M. PHILIPPART

Mental Retardation Center

The Neuropsichiatric Institute

Los Angeles, Calif. 90024

The concept of storage disorders has been long a favorite with clinicians and pathologists. It may have become somewhat distorted. For the biochemist the study of these conditions has been rewarding, providing a wealth of data about which meaningful discussions of the pathogenesis are now possible. Indeed the characterization of a substance accumulated in tissues allows us to predict and test precisely the enzymes involved in its metabolism. The development of de Duve's lysosomal theory (1) and its application by Hers (2) to the problem of storage disorders secondary to the deficiency of a specific lysosomal enzyme has now provided a rational explanation to the pathogenesis of the sphingolipidoses. Mucopolysaccharide storage disorders have also been assumed to represent other examples of lysosomal pathology (3). The deficiency of specific proteins, possibly lysosomal enzymes, which have not been characterized yet, have now been demonstrated in several variants (Hurler, Hunter, San Filippo a and b, and Maroteaux-Lamy) (4). The catabolism of mucopolysaccharides in man is still poorly understood and the heterogeneous nature of the substances accumulated or excreted in a single variant has until now hindered the discovery of the single enzyme deficiency which is theoretically expected.

The mixed character of the substances found in most and possibly all storage disorders has been frequently misinterpreted, overemphasized by some and dismissed by others. A rational explanation for this prevalent phenomenon may be offered in most instances. A first underlying mechanism has been suggested by Hers (2). It is based on the relative lack of specificity of the lysosomal acid hydrolases which are capable of acting on a variety of substrates having only a terminal moiety in common. This property, as a matter of fact, has been the key to the progress in the diagnosis of the storage disorders, since it generally allows the use of inexpensive, sensitive, artificial substrates wich are commercially available. However, mixed storage of substances having the same terminal substituent is limited to a few sphingolipidoses such as GM_1 ganglioside and keratan sulfate in GM_1 gangliosidosis (5, 6), fucosyl sphingolipid and polysaccharides in fucosidosis (7), ceramide trihexoside and digalactoside in Fabry's disease (8).

A second and probably the most common mechanism for the accumulation of heterogeneous, chemically unrelated substances is the physical affinity of such substances for each other. The anatomical site of storage disorders is found in "granules" which are generally considered to be engorged lysosomes. Such granules have been isolated by ultracentrifugation and their composition has been determined. As expected, the main substance in the granules is that which has accumulated as a result of the basic enzyme deficiency (5, 9-11). The bulk of the granule, however, is made up of a complex mixture of lipids and proteins. The physical nature of such aggregates has been elegantly demonstrated by Samuels (12) who, by mixing *in vitro* the different constituents in the right proportion, was able to prepare "artificial granules" with an identical electron microscopic appearance. These granules are present indeed with a morphology which is characteristic of a given disorder, emphasizing that we are not dealing with a purely random phenomenon. A given storage substance thus acts as the organizing nucleus of the granules, preferentially trapping the molecules with the best shape to fit them, and then determining the resulting microscopic configuration. Of course this takes place inside the lysosome where the primary storage does occur. Trapped molecules are thus removed from the action of the surrounding active enzymes. The "storage" will then proceed in a snowballing fashion. Pseudomembranous, lamellar structures may arise from these molecular interactions, which are similar to those thought to be operative at the normal assembly of membranes. In the case of cellular membranes, however, specific proteins under precise genetic control are likely to direct the organization.

A third mechanism for generating the accumulation of substances unrelated to the basic storage material has been largely overlooked. In this case, the original impairment interferes progressively with many basic processes at the level of the cell as well as of the organ. Consequently an imbalance may arise between the cellular organelles or the cell types found in a given organ. The cells which represent the primary site of the storage may eventually die. Defense or repair mechanisms such as fibrosis can significantly alter the make-up of an organ. Inasmuch as the chemical composition tends to vary from one subcellular organelle or one cell type to another, the analysis of a pathological organ in toto is likely to reflect the underlying alteration in the distribution of its constituents. This in turn may be falsely interpreted as evidence for a mixed storage.

The contribution of the pathologists to the topographic description of such processes will always remain invaluable. A simple illustration of the effect of profound changes in the make-up of a tissue may be found in the demyelinating diseases in which, independent of the etiology, abnormal ganglioside distribution has frequently been observed. In leukemia, a large excess of cells, quite normal in their chemical constitution may greatly overtax the capability of the reticulo-endothelial system to degrade their glucosyl ceramides with its β-glucosidase (13, 14). This situation actually mimics Gaucher's disease and emphasizes the fact that a relative (or local) deficiency of an enzyme may differ only in the degree of the resulting damage.

It is now apparent that several of the storage disorders described by the pathologists do not fit the definition of lysosomal disorders as proposed by Hers (2). A first group is represented by a variant of Niemann-Pick's disease, Crocker's type C (15). In such cases there is no striking accumulation of a single lipid giving the clue to the metabolic deficiency. Of course there is a moderate increase in sphingomielin justifying its classification among the sphingomyelinoses. The existence of a relationship between types A and C of Niemann-Pick's disease is further strengthened by an equal increase in a host of substances other than sphingomyelin such as lysobisphosphatidic acid, glycolipids, monosialogangliosides and sometimes cholesterol (16-20). The slightly decreased level of sphingomyelinase activity found in some organs cannot explain the pathogenesis of Niemann-Pick type C (20). We have here a complex storage of seemingly unrelated substances and no underlying enzyme deficiency has been discovered.

In another perplexing variety of storage disorder individualized by Leroy and DeMars (21) under the descriptive label of Inclusion-cell or I-cell disease, there is a moderate deficiency of several acid hydrolases (22) but only minimal amounts of identified storage substances (23). A recent report showed that antibodies against soluble lysosomal enzymes in toto elicit the I-cell phenomenon in a culture of normal fibroblasts (24). I-cell disease could thus represent a type of immune disorder. The most puzzling type of storage is exemplified by the group of the so-called juvenile amaurotic idiocies, questionably gathered together under the eponym of Batten's disease (25). Despite the morphological evidence for the accumulation of amorphous (lipofuscin) or pseudomembranous (curvilinear or fingerprint bodies) granules, there is no documented increase in any substance and no known enzyme deficiencies. Most pathologists have been reluctant to dissociate these disorders without actual biochemical storage from the proven lipid storage disorders. The finding of a decreased turnover of most lipids in the brain from a child with lipofuscin storage brought forth the first biochemical argument for an alteration in lipid metabolism (26). A slight increase in most lipid fractions resulting from a decreased turnover will not affect the normal distribution of tissue lipids. One would expect some elevation in the total lipid concentration over a period of years unless there were a concomitant decrease of other cells or cellular organelles having a similar lipid composition. The morphological appearance of storage would essentially result from an imbalance between the normal quantity of cell membranes and the amount of their lipid constituents accumulated inside the lysosomes, following a hypothetical failure to complete their catabolism. The hypothesis of "premature aging" would fit the idea of decreased lipid turnover. The finding of normal hydrolase activities and the lack of demonstrable inhibitors do not favor the notion of decreased lysosomal function. It has been suggested (27) that lipid peroxidation might result in the entrapment of other lipid molecules which would thus become inaccessible to lysosomal digestion. This falls to explain the primary-probably diverse-genetic mechanisms or the lack of demonstrable lipid accumulation.

CHOICE OF THE MATERIAL TO STUDY

There are no simple rules to guide a choice of diagnostic studies for the elucidation of a clinical problem. Thus, it is necessary to use as many clinical and pathological clues as possible. An apparently normal psychomotor development followed by progressive deterioration and loss of prior acquisitions is a feature shared by most storage disorders. Cherry-red spots (Gangliosidoses, Niemann-Pick), pigmentary degeneration of the retina (lipofuscin storage and variants), increased spinal fluid protein (Krabbe, metachromatic leukodystrophy, Refsum), increased nerve conduction time (same diseases) bone lesions (mucopolysaccharidoses) and hepatomegaly (Pompe, Wolman, Niemann-Pick, Gaucher, mucopolysaccharidoses) represent the most frequently encountered signs (Table I). Pathological findings of interest include the demontration of foam cells in the bone marrow and of lysosomes enlarged by lipid or water-soluble substances in the liver. Widespread pheno-

Table I - Useful Clues in the Diagnosis of Storage Disorders.

Clue	Disease
Cherry-Red spot	Gangliosidoses, Niemann-Pick
CSF protein ↗	Krabbe, sulfatidosis, Refsum
Nerve conduction time ↗	Krabbe, sulfatidosis, Refsum
Splenomegaly	Gaucher
Hepatomegaly	Wolman, Pompe, Niemann-Pick, mucopolysaccharidoses
Bone alterations	Mucopolysaccharidoses
Recurrent pain	Fabry
Pigmentary degeneration of the retina	Lipofuscinosis, "Batten", Refsum

typic variations are the rule and few, if any, symptoms are necessary or sufficient to establish a diagnosis. When the site of a biopsy is considered, the tissue should be suitable for pathological studies (including electron microscopy) and biochemical determination of both tissue composition and enzyme activities. Although the liver will be a convenient site in many cases, it will not allow the demonstration of an increased level of substrate in disorders which involve mostly the nervous system (Krabbe, Tay-Sachs).

A rational approach to the diagnosis requires an in-depth knowledge of the normal as well as the pathological distribution of lipids and mucopolysaccharides in all the different tissues. Each storage disorder tends to have a characteristic topography which is a reflection of the physiological site of synthesis of the accumulating substrate. Lysosomal digestion, however, is not necessarily the obligatory fate of an intracellular substance. Part of the excess material can be excreted as in the case of mucopolysaccharides (28) which are then found in increased concentration not only in the urine but also in plasma (29) and saliva (30). Body fluids also carry cell debris of the tissues they bathe to yield an "indirect biopsy" (31) of these tissues. Therefore analysis of body fluids very often helps in the diagnosis of storage disorders (mucopolysaccharidoses, Fabry, metachromatic leukodystrophy, Sandhoff's variant of Tay-Sachs).

METHODS OF ANALYSIS

Ideally, the assay of enzyme activities is the most direct route to determine the cause of a storage disorder. However, with the exception of a few cases where the diagnosis is already partly established on a clinical and pathological basis, it is impractical to analyze vast arrays of enzyme activities without some idea of the nature of the accumulating substrate. Because of this, chromatographic techniques represent a much more powerful tool to investigate conveniently lipid or mucopolysaccharide distribution since complex mixture of these compounds can be analyzed rapidly and quantitatively. The introduction of 2-dimensional systems of thin-layer chromatography (32) has been a major improvement over previous techniques and should be used routinely.

Many of the lipids stored in these diseases are glycolipids and thus it is of interest to see if this storage is reflected in the carbohydrate composition of the whole tissue. This is done by isolation and analysis of total crude glycolipids followed by the estimation of the carbohydrate composition of the residue obtained after lipid extraction (nucleic acids, glycogen, and glycoproteins). Fractionation on silicic acid (33) combined with DEAE-cellulose when sulfatides are present (34) provides good recovery of the glycolipids. Recently we have used Hakomori's technique (35, 36) of acetylation of the total lipid extract followed by Florisil fractionation. This is faster than silicic acid and it gives purer glycolipids. Under these conditions, however, some of the sulfatide sulfate ester may be lost. The tissue residue after lipid extraction may be methanolyzed and the carbohydrates analyzed and quantitated by gas-liquid chromatography as their trimethyl silyl derivatives (37). This technique also provides a check on the efficiency of the prior lipid extraction since residual fatty acids can be extracted with hexane from the methanolysate and analyzed separately. Deoxyribose is destroyed under these conditions but fucose, ribose, xylose, mannose, galactose and glucose are not. Hexosamines and sialic acids can also be determined after reacetylation but recoveries tend to be poor either due to inadequate hydrolysis or lack of sensitivity of the hydrogen-flame detector or both. In selected cases, the lipid-extracted residue is digested with pronase and glycopeptides are separated from mucopolysaccharides (37). Carbohydrate distribution in glycopeptides is studied by gas--liquid chromatography after methanolysis. Mucopolysaccharides can be either methanolyzed or fractionated on CF_{11} cellulose (37).

IMPORTANCE OF A PRECISE DIAGNOSIS

Besides the obvious research interest, the usefulness of early diagnosis is rapidly growing. Genetic counseling is passing from the stage of impersonal calculation of risks to that of identification of the heterozygotes and presymptomatic homozygotes. Amniocentesis has already allowed a few families to obtain normal children or to have legal abortions performed on the basis of the presence or absence of a given lysosomal enzyme in amniotic cells obtained during the first weeks of pregnancy. Enzyme substitution is in an active experimental stage. Kidney transplantation has given very encouraging results in Fabry's disease (38). Methods of decreasing the synthesis of the accumulating substrate are being sought (39-41). The era of rational therapy is thus being entered. One essential condition for success will be of course that therapy be started before irreversible tissue damage has taken place.

THE PROBLEM OF CONTROLS

It is exceedingly difficult to collect meaningful controls for a variety of reasons. The material available in suitable conditions is limited. The definition of normal controls is quite loose in most publications. Changes introduced by growth and development have been better studied, especially at the level of the brain where myelination introduces profound alterations in lipid distribution. Developmental changes in organs are seemingly less pronounced although no definitive or extensive data have been published yet. Except for the newborn period few children die. Most deaths result from severe conditions likely to disturb the metabolism in one way or another, such as severe infections, congenital malformations, tumors, burns and so on. Storage disorders may so distort the normal anatomy of an organ as to preclude any meaningful comparison with the normal. A brain with gliosis and demyelination secondary to birth trauma might be a better control for metabolic demyelination than a normal brain matched for age. The secondary effect of infection, malnutrition or chronic anoxia on lipid or mucopolysaccharide distribution in tissues have barely been explored, and it is unwarranted to assume these are negligible. For example, patients with cardiac malformations may have a significant decrease in the proportion of monounsaturated fatty acid in their brain cerebrosides (42). To make matters worse there is abundant evidence that there is a wide range of variations in a so-called normal population. Blood groups and tissue types have provided meager chemical clues to the genetic heterogeneity of man. The mutation of a protein like hemoglobin has mostly functional consequences. Differences in phenotypic expression should be discussed at the biochemical level. Let us consider simple consequences of variable genetic constitution on the expression of a given enzyme deficiency. Blood group substances of the Lewis type accumulate in fucosidosis (7). Patients with the Lewis B gene who synthesize an antigen with two fucosyl residues (43) might be more severely effected by a fucosidase deficiency than patients with Lewis A type whose antigen contains only one fucosyl residue. A deficiency - still to be discovered - of the mammalian enzyme α-N-acetylgalactosaminidase might only affect individuals with blood type A, sole natural substrate presently known for that enzyme. Patients with blood group B might be more severely affected by Fabry's disease (44) which is secondary to a deficiency in α-galactosidase, the enzyme capable of destroying the antigenic activity of the B substance by removing its terminal galactosyl residue. We have observed such a patient in whom an increase in titratable substance B could be demonstrated in blood and saliva (Arcilla and Philippart, unpublished data). The patient, however, did not seem more incapacited by the disease than patients with different blood groups, although before receiving a kidney transplant he had the highest plasma level of ceramide trihexoside we have ever observed.

RESULTS

Plasma Glycolipids

The study of plasma is most helpful in Gaucher's and Fabry's diseases. Plasma (five ml) is mixed with an equal volume of water and extracted with 50 ml of chloroform:methanol (2:1 v/v). After standing overnight at 4°, the extract is filtered and the aqueous upper phase is discarded. The lower phase is dried and processed as described in the methods.

Glucosyl ceramide is the major glycolipid in plasma (Table II). In Gaucher's disease, its concentration is at least twice the upper normal level in agreement with previous results (45, 46). The largest increase, about 10-fold above the highest normal limit, was observed in a woman recently splenectomized and now developing severe bone lesions. This unfavorable consequence of splenectomy is well known (47). Other plasma glycolipids also tend to be elevated.

A characteristic increase in ceramide trihexoside is found in Fabry's disease (48). Only in case 1 was there a striking increase in ceramide dihexoside. Here we found a high level of ceramide digalactosides which could never be demonstrated in other cases. In case 2, the CTH level barely rose above the upper normal limit suggesting that chronic hemodialysis may help in reducing the plasma level. Immediately following this procedure, the level decreased by 25%. In patient 5 a reduction of 75% was obtained under the same conditions, but the lowest level could not be maintained as in case 2.

CTH concentration is always equal or higher than that of CDH in patients with Fabry's disease, while in controls and in patients with Gaucher's disease, there is consistently more CDH than CTH.

Table II - Plasma Glycolipids.

Donor	Age	Sex	nmoles/ml Plasma					
			CMH		CDH	CTH	CTet	Gangliosides
			Glu	Gal				
Normal ('')	5-42		0.9-4.4	0-0.1	0.5-2.8	0.2-1.3	0.02-0.5	0.1-0.6
Gaucher 1 (')	22	F	41.3	0.4	6.8	3.4	2.5	N.D.
Gaucher 2	34	M	10.1	0.01	2.2	0.4	0.4	N.D.
Gaucher 3	16	F	15.8	0.06	3.4	1.7	1.5	3.7
Fabry 1	49	M	7.3	0.8	10.5 (°)	7.8	2.4	2.6
Fabry 2 (^)	37	M	3.7	0.2	1.0	1.7	0.8	N.D.
Fabry 3	20	M	1.8	1.0	0.9	3.4	0.5	N.D.
Fabry 4	46	M	3.8	0.2	2.8	5.2	1.9	1.5
Fabry 5 (^)	35	M	4.0	0	3.2	11.3	0.7	N.D.
Fabry 6 (=)	39	M	3.0	0.2	2.1	4.2	1.3	2.1
Fabry 7	8		4.3	0.5	2.9	3.8	0.5	0.3

('') 5 males, 5 females; (') Splenectomized; (^) Hemodialysis; (°) including 3.5 nmoles of ceramide digalactoside; (=) Nephrectomized.

Abbreviations: Glu = Glucosyl ceramide; Gal = Galactosyl ceramide; CMH = Ceramide monohexoside; CDH = Ceramide dihexoside; CTH = Ceramide trihexoside; CTet = Ceramide tetrahexoside.

Table III - Urinary Glycolipids.

Donor	Age	Sex (number)	nmoles/100 ml							
			CMH		CDH		CTH	CTet	Ganglioside	Sulfatide
			Glu	Gal	Lac	Dgal (')				
Normal (°)	1-68	M (16)	14.8 ± 10.9	3.0 ± 3.1	14.8 ± 10.9	trace	2.6 ± 2.9	2.5 ± 2.6	N.D.	1.1 ± 2.8
(^)			2.3 - 68.8	0 - 10.5	1.2 - 44.8		0.3 - 12.5	0.2 - 10.4		0.1 - 22.0
Normal (°)	1-68	F (17)	34.1 ± 20.3	3.1 ± 4.7	13.7 ± 12.3	trace	2.5 ± 1.6	2.1 ± 1.6	0.7 ± 0.6	0.7 ± 0.9
(^)			2.3 - 82.0	0 - 15.5	0.7 - 38.6		0.1 - 4.5	0.1 - 4.6		0.04 - 2.9
Gaucher 1	22	F	25.7	5.5	19.2	N.D.	10.8	11.4	22.7	2.0
Gaucher 2	34	M (2)	15.6 - 22.8	0.1 - 3.4	4.6	N.D.	0.7	7.4	N.D.	N.D.
Gaucher 3	14	F	106.8	337.1	22.9	N.D.	1.7	3.2	N.D.	N.D.
	16	F	36.7	2.4	3.9	N.D.	2.3	0.7	0.2	N.D.
Fabry (^)	3-49	M (39)	10.7 - 249.0	0.1 - 37.7	1.7 - 917.7	0 - 70.3	16.7 - 591.0	2.9 - 289.0	0.1 - 30.5	N.D.
Fabry (^)	4-25	F (7)	8.9 - 76.6	0 - 76.6	4.5 - 28.9	0 - 15.1	1.0 - 23.9	1.4 - 11.3	0.5 - 2.3	N.D.
Sulfatidosis (^)	1-22	M (12)	1.5 - 110.8	0 - 24.6	0.9 - 69.9	N.D.	0.1 - 2.2	0.1 - 1.1	0.3 - 2.0	22.3 - 674.0
Sulfatidosis (^)	1-15	F (11)	4.0 - 67.7	0 - 9.9	1.6 - 56.7	N.D.	0.4 - 4.8	0.5 - 4.0	0.1 - 1.6	37.2 - 980.0
Tay-Sachs 1	1	F	27.2	0.5	33.1	N.D.	2.7	3.7	0.2	N.D.
Tay-Sachs 2	1	M	55.1	7.3	17.3	N.D.	4.5	4.2	0.1	3.8
GM_1-Gangliosidosis 1	1	F	18.4	1.8	2.8	N.D.	0.9	0.8	0.1	N.D.
GM_1-Gangliosidosis 2	1	M	27.8	94.5	11.2	N.D.	3.2	2.1	1.3	8.6

(°) mean ± SD; (^) mean range; (') ceramide digalactoside.

This ratio may thus help to identify patients with CTH level at the upper normal limit.

Urinary Glycolipids

A characteristic lipid excretion is found in Fabry's disease and sulfatidosis (metachromatic leukodystrophy) (34, 48, 49) (Table III). Normal variations are wide in the controls as well as in the patients. This reflects to a large extent the variation expected in the amount of cells and debris comprising the urinary sediment. Thus the proportion of a given glycolipid in relation to the others has more meaning than its absolute concentration. In Gaucher's disease urinary glycolipids may be normal (case 2). The expected increase in glucosyl ceramide is found only occasionally, even in the same patient (cases 1 and 3). Higher levels may be observed in other lipid storage disorders. Various patients sporadically excreted high amounts of galactosyl ceramide, normally a minor urinary lipid. This was never consistent. We have not been able to confirm the galactosyl ceramide increase observed in Krabbe's disease (49).

With the exception of a few cases of Fabry's disease with a generalized increase in glycolipid excretion, the concentration of ceramide tetrahexoside was always low. An increase in that fraction would thus appear diagnostic for Sandhoff's disease, a Tay-Sachs variant (49).

Hematoside (GM_3 ganglioside) is the major ganglioside present in plasma and urine. Its level was elevated in a few cases with increased excretion of all glycolipid fractions. No GM_1 or GM_2 ganglioside was ever encountered in ganglioside storage disorders. Since gangliosides are both lipid- and water-soluble, it was of interest to find normal amounts of sialic acid in dialyzed urine from patients with Tay-Sachs' disease.

Kidney Glycolipids

There is little resemblance between kidney and urine glycolipids. This indicated that the latter fraction either originates from a separate pool within the kidney or is degraded to a large extent by urinary enzymes. The idea that urinary sediment represents an "indirect biopsy of the kidney" (31) thus should be received with

Table IV - Kidney Glycolipids.

	Sex	Age	CMH (nmoles/100 mg of dry weight)		CDH	Sulfatides	CTH	CTet	Gangliosides
			Glu	Gal					
Controls (°)			26.0 - 41.0		45.0 - 63.0	56.0 - 72.0	96.0 - 124.0	127.0 - 183.0	20.0
Encephalopathy (^)	M	25	13.9	27.5	65.1	72.7	182.0	55.5	18.4
Cockayne	M	9	18.0	13.1	23.2	32.0	99.0	73.2	3.8
GM_1-Gangliosidosis (formalin)	F	1	10.2	6.1	90.1	30.0	37.8	19.2	162.8
Krabbe #1	F	$\frac{3}{12}$	30.0	22.8	21.1	107.7	25.4	77.4	16.2
Krabbe #2	M	1	7.4	7.1	21.7	54.9	85.9	99.4	10.1

(°) Calculated from E. Martensson (57); (^) (52).

caution. Although it would be of obvious interest, we have rarely been able to compare glycolipid distribution in the kidney and urine from the same patient. An increased concentration of gangliosides in the kidney from GM_1 gangliosidosis (Table IV) was not observed in the urine (Table III) from the same patient. The normal level of kidney galactosyl ceramide in this case suggests that the increased excretion of this glycolipid in the urine of GM_1 gangliosidosis case 2 is not characteristic. Surprisingly urinary galactosyl ceramide was not increased in Krabbe's disease despite a profound deficiency of galactocerebroside galactosidase (51). The largest level of galactosyl ceramide was observed in a case of undefined encephalopathy, where the clinical and pathological data formally precluded the diagnosis of Krabbe's disease (52). Despite the sulfotransferase deficiency which has also been incriminated in Krabbe's disease (53), sulfatide levels in the two studied kidneys were normal or high. Our data on kidney and urine glycolipids do not confirm the possibility of diagnosing Krabbe's disease by analysis of the urine sediment (49).

Liver Glycolipids

Glycolipid distribution in the liver (Table V) is less complex than it is in the kidney. There are smaller amounts of CTH and CTet in the liver and no demonstrable mono- and dihexosyl-sulfatide. Analysis of pathological specimens revealed that only the case of undefined encephalopathy had a definitely increased level of galactosyl ceramide. As in the kidney, the glycolipid distribution, including galactosyl ceramide, was normal in Krabbe's disease. Thin--layer chromatography of gangliosides isolated from Tay-Sachs li-

Table V - Liver Glycolipids.

	Sex	Age	nmoles/100 mg of dry weight						
			CMH		CDH	CTH	CTet	GM_3 (')	Gangliosides.
			Glu	Gal					
Normal (°) mean ± SD Range	M+F	4-65	30.0±12.0 6.0-48.0	<1.5 0.3-2.4	48.0± 30.0 18.0-120.0	24.0±18.0 14.0-70.0	18.0±12.0 6.0-34.0	12.0± 5.0 6.0-24.0	
Encephalopathy (^)	M	25	35.8	40.7	37.9	14.2	8.6	7.9	5.8
Cockayne	M	9	8.5	2.3	12.6	6.7	3.0	5.6	
GM_1-Gangliosidosis 3	F	1	32.9	3.4	30.2	29.6	18.0	10.9	43.0 (GM_1) 7.5
Hurler	M	1/2	8.3	1.5	8.6	1.8	4.6	8.0	5.3
GM_2-Gangliosidosis	F	2	8.0	3.8	1.5	4.3	1.5	2.5	1.2 (GM_2) 0.1
Krabbe 2	M	1	9.4	1.0	18.5	11.4	5.8	11.4	2.3
Krabbe 3	F	1/2	37.6	5.7	24.2	1.2	0.5	0.5	
Niemann-Pick A 1	F	1	138.0	5.1	119.1	33.5	8.3	4.3	3.3
Niemann-Pick A 2	F		107.8	2.9	143.7	37.3	8.7	3.1	
Niemann-Pick C 1	M	5	74.2	0	68.5	23.5	8.0	3.9	
Niemann-Pick C 2	M	2	101.9	11.5	128.4	45.9	16.4	4.3	
Fucosidosis	F	5	82.4	7.7	88.8	24.1	27.8	(229.80 Fucosyl Sphingolipids)	

(°) Calculated from Kwiterovich et al. (54); (') GM_3 = Hematoside; (^) (52).

ver demonstrated the characteristic stored lipid. On hydrolysis, this material had the carbohydrate composition expected for GM_2 ganglioside. A significant increase of GM_1 ganglioside was also detected in GM_1 gangliosidosis. Analysis of liver from cases of Niemann-Pick's disease, Crocker's group A and C (15), showed similar alterations in both variants. There was a 2.5 to 5-fold increase in glucosyl ceramide. Although a much larger elevation would be expected in Gaucher's disease, the differential diagnosis with Niemann-Pick's disease type C is not always easy since the enormous accumulation of sphingomyelin characteristic of type A is lacking. CDH levels are also augmented by a factor of 1.5 to 3, while the more complex glycolipids, including hematoside, remain within normal limits. This increase in liver glycolipids in Niemann-Pick's disease is by no means specific since it is also observed in fucosidosis. Here, however, the bulk of the glycolipid fraction is made up of fucosyl-sphingolipids which are not detectable in normal liver.

Spleen Glycolipids

Although we have not yet established the normal range for the distribution of spleen glycolipids (Table VI), our pathological data confirm the alterations noted in liver and kidney. There is no increase of galactosyl ceramide in Krabbe's disease. Gaucher's disease is characterized by a striking accumulation in glucosyl ce-

Table VI - Spleen Glycolipids.

	Sex	Age	nmoles/100 mg of dry weight							
			CMH		CDH	CTH	CTet	GM3	Gangliosides
			Glu	Gal					
Control	M	25	36.2	19.3	62.2	19.9	22.9	23.0	N.D.
Krabbe 1	F	3-1/2	26.5	16.1	5.7	3.9	2.3	2.6	2.2
Krabbe 2	M	1	13.4	2.4	53.0	11.5	9.9	9.4	3.9
Gaucher 4	M	2	24,464.0	16.4	891.6	294.1	320.0	610.5	
Gaucher 5	F	13	1,245.4	17.9	54.5	56.4	32.0	17.3	
Niemann-Pick C 2	M	2	127.8	2.9	75.0	19.2	6.8	2.2	7.0
Niemann-Pick C 3	F	5	471.6	37.9	241.3	63.3	26.4	6.7	
Cockayne	M	9	35.3	7.0	58.2	23.8	9.7; 7.0	34.4	11.6
Mycobacteria Infection	M	41	153.9	0.8	74.1	11.4	6.9	10.1	

ramide. Data on two cases, both belonging to the prevalent chronic type of the disorder, were chosen as representing the extremes in the degree of storage encountered. It is apparent that the accumulation does not simply reflect the duration of the disease, since the youngest case has by far the highest glycolipid concentration. The concomitant increase in other glycolipid fractions in Gaucher case 4 is similar to that observed in those patients with Fabry's disease who excrete the largest amount of ceramide trihexoside in their urine.

In Niemann-Pick type C case 3, the glucosyl ceramide level reached 38% of the value found in Gaucher case 5. This emphasizes the possible confusion between the two disorders, as already mentioned above for liver. The case of mycobacteria infection is listed here to give an example of glucosyl ceramide increase unlikely to result from a primary disorder in lipid metabolism.

Peripheral Nerve

Peripheral nerve biopsies have been seldom used for biochemical study. This approach is of peculiar interest in disorders such as Krabbe's disease and metachromatic leukodystrophy which involve the metabolism of glycolipids largely restricted to the nervous tissue, namely galactocerebroside and sulfatide. Although differences between peripheral and central myelin have been documented (55), peripheral nerve biopsy provides a more convenient way of investigating nervous tissue than does brain biopsy. The few reports available (55, 56) suggest that the ratio of cerebroside to sulfatide which approaches 4 in normal brain, may reach 6 or 8

Table VII - Peripheral Nerve Glycolipids.

Donor		Sex	Age	nmoles/100 mg dry weight				
				CMH (')	CDH	Sulfatide	Total Glycolipids	C/S (^) Ratio
Controls (7)	A			1,240-2,113(2)	trace	310-690(2)	(1,781)(711-2,844)	3.0-4.0(2)
Krabbe 1	B	F	3/12	1,232	46	131	1,409	9.4
Krabbe 2	A	M	1	1,069	23	292	1,384	3.7
Krabbe 3	B	F	1/2	1,810	153	259	2,222	7.0
	A			1,703	N.D.	N.D.	N.D.	N.D.
Krabbe 4	A	F	1-1/2	1,618	40	101	1,759	16.0
MLD 1	B	F	1-1/2	+ ('')		+++ ('')	1,188	N.D.
MLD 2	B	F	12	+ ('')		+++ ('')	1,037	N.D.
MLD 3	B	F	8	+ ('')		+++ ('')	1,473	N.D.
MLD 4	B	F	3	211		976	1,187	0.21
MLD 5	B	M	13	11		37	48	0.30
MLD 6	B	M	22	284		144	428	2.0
MLD 7	B	F	2	79		692	771	0.11
Hurler	B	M	2	457	23	110	590	4.2

(') Galactocerebroside only; (^) Cerebroside/sulfatide; ('') As estimated by thin-layer chromatography; A = Autopsy; B = Biopsy; N.D. = Not determined; MLD = Metachromatic leucodystrophy.

in peripheral nerve. We were unable to establish the normal ratio for peripheral nerve in the present work since we realized only recently that the faster preparative method (36) discussed in the introduction entailed significant degradation of sulfatide.

For this analysis, a few milligrams of sural nerve are easily obtained and shared with the pathologist. Whenever possible, connective tissue is carefully dissected away to minimize any dilution that its presence will cause on the glycolipid content when expressed in relation to the dry weight of the tissue. The persistance of an excess of connective tissue is probably responsible for some of the low values reported here (Table VII). The lipids are extracted and processed as usual. The aliquot of the total lipid extract is examined by thin-layer chromatography to estimate the ratio of cerebroside to sulfatide. Only the total lipid galactose could be accurately determined in the first samples studied.

Cerebroside levels seemed within normal limits in Krabbe's disease, while they were markedly reduced in metachromatic leukodystrophy. In Krabbe case 3, sciatic nerve had a composition similar to sural nerve; data on autopsy and biopsy material were identical. A consistent increase in CDH was also found, while sulfatides were definitely decreased in Krabbe cases 1 and 4.

Sulfatides were significantly increased in 5 out of 7 cases of metachromatic leukodystrophy. Total glycolipid levels remained within normal limits in these patients. Interestingly, normal glycolipid level with a reversed cerebroside/sulfatide ratio has been observed in the central myelin isolated in such a case (58). Low values found in MLD case 5 may result from heavy contamination with connective tissue, or severe demyelination or both. The abnormal cerebroside/sulfatide ratio was still demonstrable. In MLD case 6, the oldest of the series, a low glycolipid content correlated well with the marked demyelination. This was the only case where sulfatide was less concentrated than cerebroside and proportionately less increased than in the other cases.

Brain Glycolipids

Galactolipids are specifically localized in myelin. A reduction in white matter cerebrosides may be the earliest sympton of demyelination as seen in the Niemann-Pick biopsy (Table VIII). Results on the autopsy material from the same case illustrated the rapid pace of the demyelination which was obviously a terminal phenomenon secondary to the loss of neurons and glial cells. Since the sulfatide concentration tends to fall less rapidly, the cerebroside/sulfatide ratio is often decreased. There is however no direct correlation between the loss of glycolipids and the reduced affinity of the myelin for specific stains.

In the case of combined Schilder's and Addison's disease, a sex-linked demyelinating disease, the glycolipid content remained normal in the grey matter. As a matter of fact, the sulfatide concentration was higher in this case than in controls. High levels of grey matter sulfatides were also observed in the Niemann-Pick biopsy and in Krabbe case 2. The youngest case (No. 3) of Krabbe's disease had a lesser reduction in glycolipids than the two others. Case 2 actually had more glycolipids in grey matter than in white matter. Case 4, the oldest, had most of the grey matter glycolipids. The cerebroside/sulfatide ratio was highly variable in the white matter which contained relatively high amounts of CDH. As previously discussed for the visceral organs, there is no obvious storage of galactocerebroside in the brain in Krabbe's disease.

In the case of Lowe's syndrome, despite a moderate loss of glycolipids no actual demyelination could be demonstrated. The di-

Table VIII - Brain Glycolipids.

Donor	Sex	Age	nmoles/100 mg of dry weight							
			Grey Matter				White Matter			
			CMH	Sulfatide	C/S (')	CDH	CMH	Sulfatide	C/S (')	CDH
Normal	3 M; 4 F	2-18	1,200-2,300	445-665	1.8-5.2	< 60	10,200-15,400	3,000-4,340	2.8-5.4	< 20
Niemann-Pick C 1										
Biopsy	M	4	1,930	1,333	1.4		7,000	2,780	2.5	
Autopsy	M	5	1,080	334	3.2		3,020	1,560	1.9	
Schilder + Addison	M	12	2,050	890	2.3		4,340	1,000	4.3	
Krabbe 2	M	1	1,640	910	1.8		1,530	640	2.4	83
Krabbe 3	F	1/2					2,420	460	5.3	130
Krabbe 4	F	1-1/2	193	110	1.8		1,930	210	9.2	520
Lowe Biopsy	M	9	872	163	5.3	6	6,615	2,215	3.0	83
Autopsy	M	10	1,510	400	3.8	126	6,337	1,888	4.4	81
Hurler Biopsy	M	4					7,725	858	9.0	447
San Filippo Frontal	M		657	132	5.0	292	1,445	613	2.4	122
Cerebellum			1,757	275	6.4	113	9,512	1,315	7.2	401
Fucosidosis 1	M	4	480	222	2.2		3,860	445	8.7	
2	F	5	79	17	4.6	(9) ('')	1,473	565	2.6	(20) ('')

(') Ratio of cerebroside to sulfatide; ('') Fucosyl-sphingolipid.

screpancy between biopsy and autopsy data in the grey matter may reflect differences in sampling. It is sometimes difficult to separate completely grey from white matter, especially when dealing with frozen specimens.

The cases of Hurler's and San Filippo's disease were remarkable by the high content of CDH in the white matter. In the frontal lobe the neuronal population was severely depleted and the white matter demyelinated. In fucosidosis, the extensive demyelination correlated well with the low glycolipid concentration. Fucosyl--sphingolipids which accounted for the bulk of the glycolipids in the liver were barely detectable in the brain.

In all cases of primary or secondary demyelination, ganglioside content tended to be slightly decreased in the grey matter and slightly increased in the white matter. Ganglioside distribution was essentially within normal limits. However, some of the normally minor fractions (GD_3, GM_2 and GM_3) were frequently elevated (Table IX).

Monosialogangliosides (GM_2 and GM_3) were generally more increased in grey than in white matter while the reverse was true for disialoganglioside GD_3 which was increased in all cases of severe demyelination.

Table IX - Minor Gangliosides in Brain.

Donor	Sex	Age	Percent of Total N-Acetylneuraminic Acid					
			Grey Matter			White Matter		
			GD_3	GM_2	GM_3	GD_3	GM_2	GM_3
Normal	3 M; 4 F	2-18	0.8-4.8	0.5-2.6	trace	1.2-4.5	0.3-1.5	trace
Krabbe 1	F	3/12	--	10.6	trace	19.6	7.8	3.9
Krabbe 2 Biopsy	M	9/12	--	12.0	trace	8.8	5.4	trace
Autopsy	M	15/12	3.2	trace	trace	9.1	3.2	3.5
Krabbe 3	F	6/12	8.2	3.4	3.7	6.1	1.1	trace
Krabbe 4	F	18/12	8.8	7.0	5.0	8.0	2.1	trace
Schilder + Addison 2	M	12	--	--	--	8.0	4.9	trace
Schilder + Addison 3	M	9	2.9	2.3	trace	12.5	2.4	4.9
Fucosidosis	F	5	6.4	5.6	trace	10.0	4.0	trace
Alpers	M	4	14.5	1.0	0.4			
Niemann-Pick C 1								
Biopsy	M	4		15.8	15.7			
Autopsy	M	5		13.7	8.0; 4.3 (1)		12.2	4.0; 2.9 (1)

(1) Unidentified component with a higher Rf than GM_3.

Non-Lipid Carbohydrate in Brain

Since substances other than lipid may be increased in some storage disorders, we routinely processed the residue left after lipid extraction. Only a few milligrams of dry residue are needed to determine the carbohydrate concentration (using sorbitol as an internal standard) and distribution by gas-liquid chromatography. A similar study has recently been published (59). The low recovery of hexosamine and the difficulty of checking the identity of unknown peaks on limited amounts of material are serious drawbacks to this method. Carbohydrates in the lipid-extracted tissue derives from nucleic acids, glycoproteins, glycogen and mucopolysaccharides. Optimal conditions for hydrolysis differ for these various substances. A relatively mild procedure (IN HCl, 65°C 18 hours) is routinely used to avoid destruction of pentoses and hexoses. No corrections for losses of hexosamines have been introduced in tables X and XI, since they vary depending on the hydrolysed material which is complex and not well characterized. An adequate series of controls has not been completed at this time. Aside from the case of Fucosidosis only small increases in total carbohydrate content were found in the grey matter from most pathological samples (Table X). In Fucosidosis, there was a 10-fold increase in total carbohydrate. The overall carbohydrate distri-

Table X - Brain Grey Matter. Carbohydrate Distribution in Lipid Extracted Residue.

Donor	Sex	Age	Content (')	Percent of total carbohydrate							
				Ribose	Fucose	Mannose	Galactose	Glucose	$GalNH_2$ ('')	$GlcNH_2$ (°)	Inositol
Normal	2 M; 2 F	2-12	1.2-7.0	16.1	4.7	17.1	21.1	17.7	2.0	8.8	2.8
Krabbe 2	M	15/12	9.2	2.5	7.8	20.7	10.4	4.9	1.2		51.2
Krabbe 4	F	18/12	6.9	2.8	3.9	14.6	14.1	1.7	3.2		45.9
Schilder + Addison 2	M	12	5.1	6.4	5.5	15.3	14.9	7.6	1.2	9.5	34.3
Tay-Sachs 2	M	1	11.8	3.3	4.8	19.5	21.6	7.9	2.2	4.3	28.9
GM_1-Gangliosidosis 3	F	15/12	7.3	4.5	6.6	28.2	49.4	7.4	1.0	28.2	
GM_1-Gangliosidosis 4	M	1	10.9	7.3	4.1	21.1	22.9	6.6		21.1	
GM_1-Gangliosidosis 5 (liver)	F	18/12	33.8 (^)	1.4	1.3	17.5	28.0	5.4	10.3	32.0	
Cockayne	M	9	16.7	5.4	8.5	13.1	18.0	27.4		7.6	
Neuroaxonal Dystrophy	M	2	9.6	3.1	1.6	24.8	24.7	33.0	4.7		3.5
Fucosidosis 2	F	5	57.5	0.1	7.3	19.3	37.5	8.8		7.0	6.5
I-Cell	M	2	7.8	2.3	3.4	19.2	20.0	23.3	9.8	15.7	
Lowe	M	10	16.1	2.1	5.9	23.1	10.4	21.5		22.9	2.5

(') Microgram of carbohydrate/mg of dry weight (after lipid extraction); ('') $GalNH_2$ N-acetylgalactosamine; (°) $GlcNH_2$ N-acetylglucosamine; (^) Normal 15.1-40.9.

Table XI - Brain White Matter. Carbohydrate Distribution in Lipid Extracted Residue.

Donor	Sex	Age	Content (')	Percent of total carbohydrate							
				Ribose	Fucose	Mannose	Galactose	Glucose	$GalNH_2$ ('')	$GlcNH_2$ (°)	Inositol
Normal	2 M; 2 F	2-12	1.9-8.5	10.5	5.0	26.2	22.3	14.1	0.3	8.4	8.6
Krabbe 2	M	15/12	13.1	2.9	2.8	15.6	19.5	5.8	3.2	7.1	33.4
Krabbe 4	F	18/12	7.6	4.2	3.8	15.2	15.9	3.6	4.7	5.9	34.5
Schilder + Addison 2	M	12	15.1	6.1	2.1	6.2	9.0	8.4	13.5	9.3	17.3
Tay-Sachs 2	M	1	6.9	2.4	4.6	17.0	16.8	7.6	2.7	1.5	42.9
Cockayne	M	9	17.8	3.3	2.5	20.7	17.0	47.3	4.9	6.3	
Neuroaxonal Dystrophy	M	2	9.6	3.3	1.6	24.8	24.7	33.0	4.7		3.5
Fucosidosis 2	F	5	53.8	2.6	21.5	25.1	34.5	2.7	7.8	2.3	2.2
I-Cell	M	2	20.4	3.9	7.6	17.7	36.1	12.2		11.8	
Lowe	M	10	7.1	8.3	6.0	26.5	36.2	2.9		19.2	

(') Microgram of carbohydrate/mg of dry weight (after lipid-extraction); ('') $GalNH_2$ N-acetylgalactosamine; (°) $GlcNH_2$ N-acetylglucosamine.

bution including fucose, remained close to normal. Galactose was increased at the expense of ribose and glucose. In the white matter (Table XI) the findings were similar, but fucose and galactosamine were significantly increased. High amounts of inositol have been detected in normal brain (59). Although our own controls did not confirm it, this observation limits the significance of the high percentage of inositol present in both grey and white matter from our cases of Krabbe's, Schilder's associated with Addison's, and Tay-Sachs' disease. Non-lipid hexosamine was not increased in Tay-Sachs' disease where GM_2-ganglioside thus remains the only substrate known to accumulate. GM_1-gangliosidosis was characterized by a significant increase in glucosamine and to a lesser extent in galactose. This indicates an accumulation in the brain of a keratan-sulfate like glycoprotein which has been previously described in the liver from these patients (5, 6). The only available white matter specimen (not listed in the table) had been preserved in formaldehyde and did not show any increase in galactose or glucosamine content. The stored glycoprotein may have been lost in the formaldehyde solution. In the brain from I-Cell disease, there was only a small relative increase in grey matter glucosamine and white matter galactose. In Lowe's syndrome, glucosamine was elevated in both grey and white matter, probably reflecting an elevated glycogen content.

Mucopolysaccharides

Only small amounts of mucopolysaccharides are obtained from normal lipid-extracted tissues after pronase digestion (37). Beside the "keratan sulfate" which was quite heterogeneous according to colorimetric analyses, hyaluronic acid and chondroitin-4-sulfate were the major fractions. The few pathological samples analyzed were selected on the basis of elevated total carbohydrate or hexosamine (Table XII). It is apparent that demyelination - be it primary as in Schilder's combined with Addison's disease or secondary as in Fucosidosis - is not necessarily accompained by any quantitative or qualitative alteration in mucopolysaccharides. The liver from San Filippo's disease did not contain the expected increase in heparan-sulfate which had probably been dissolved into the formaldehyde solution. Little can be expected from the analysis of specimens preserved under such conditions. Interestingly, the total uronic acid content (protein-bound) was still much higher than in the control. Analysis of the liver from I-Cell disease failed to reveal any obvious

Table XII - Mucopolysaccharides.

Tissue	Donor	Sex	Age	Uronic Acid (')	Per cent of total uronic acid: "Keratan Sulfate" (^)	Hyaluronic Acid	Heparan Sulfate	Chondroitin 4-Sulfate	Chondroitin 6-Sulfate ('')	Dermatan Sulfate	Heparin
Brain	Control (White Matter)	M	5	1.8	27.8	27.4	2.5	35.6	N.D.	3.7	3.0
	Schilder + Addison 2 Grey Matter	M	12	2.4	29.8	12.8	10.8	30.0	N.D.	4.0	12.3
	White Matter			2.5	20.0	41.5	5.6	18.3	6.0	5.0	3.2
	Fucosidosis 2 Grey Matter	F	5	2.3	21.8	12.0	7.6	38.0	N.D.	11.7	9.0
	White Matter			1.6	20.0	13.9	0	29.6	N.D.	11.9	24.6
Liver	Control	F	15	2.1	22.1	45.2	2.1	24.5	N.D.	0.5	5.6
	Tay-Sachs 2	M	1	1.7	29.6	38.2	10.4	14.4	0	3.3	4.1
	I-Cell	M	2	1.7	34.1	14.7	4.7	7.0	16.3	7.7	15.5
	San Filippo (°)	M		10.0	42.7	10.2	3.0	33.0	6.3	1.7	3.0
	Encephalopathy	M	19	9.5	38.4	36.8	5.0	5.2	5.6	5.4	3.7
Spleen	Mycobacteria Infection (°)	M	41	0.8	60.8	5.3	9.0	6.8	18.0	0.5	0
	Gaucher 5	F	13	0.8	64.3	4.6	4.1	14.8	10.1	0.8	1.5

(') Microgram per mg of dry weight (after lipid-extraction); (^) Heterogeneous fraction with a low carbazole/orcinol ratio; ('') This fraction frequently contained pigments interferring with the carbazole reaction; (°) Formaldehyde-fixed tissue.

disturbance in mucopolysaccharide metabolism. A high uronic acid content with a relative increase in "keratan sulfate" and hyaluronic acid was observed in a case of undefined encephalopathy.

ACKNOWLEDGEMENTS

We are most grateful to Dr. John Callahan for his very kind assistance in preparing this manuscript, and to Miss Colette Meurant and Mr. Seiji Nakatani for their excellent technical work. We thank Dr. P. Durand, Dr. E. Farkas, Dr. A. Franceschetti, Dr. G. Guazzi, Dr. H. Hirt, Dr. G. Lyon, Dr. C. Marchal, Dr. J.J. Martin, Dr. H. Roels, Dr. E. Schneegans, Dr. G. Sugarman, Dr. F. Van Hoof, and Dr. M. Vidailhet for providing pathological specimens.

This study was supported in part by funds from the State of California, Department of Mental Hygiene, the Mental Retardation Program, NPI, UCLA and research grants NB 06938, HD 04612, MCH 927, HD 00345, and HD 05615 from the National Institutes of Health.

REFERENCES

(1) De Duve, C. and Wattiaux, R., Annual Review of Phys. 28: 435 (1966).

(2) Hers, H.G., Progress in Gastroenterology 48: 625 (1965).
(3) Van Hoof, F. and Hers, H.G., C.R. Acad. Sci. Paris 259: 1283 (1964).
(4) Barton, R.W. and Neufeld, E.F., J. Pediatrics, in press.
(5) Suzuki, K., Suzuki, K. and Kamoshita, S., J. Neuropath. & Exptl. Neurol. 28: 25 (1969).
(6) Callahan, J.W. and Wolfe, L.S., Biochim. Biophys. Acta 215: 527 (1970).
(7) Philippart, M., Neurology 19: 304 (1969).
(8) Sweeley, C.C. and Klionsky, B., J. Biol. Chem. 238: 3148 (1963).
(9) Samuels, S., Korey, S.R., Gonatas, J., Terry, R.D. and Weiss, M., J. Neuropathology 22: 81 (1963).
(10) Kamoshita, S., Aron, A.M., Suzuki, K. and Suzuki, K., Am. J. Dis. Children 117: 379 (1969).
(11) Suzuki, K., Suzuki, K. and Chen, G.C., J. Neuropath. & Exptl. Neurol. 26: 537 (1967).
(12) Samuels, S. and Aleu, F., Inborn Disorders of Sphingolipid Metabolism, Pergamon Press, p. 317 (1967).
(13) Albrecht, M., Proc. 10th Congress of European Soc. of Hematology, Strasbourg (1965).
(14) Kampine, J.P., Kanfer, J.N., Gal, A.E., Bradley, R.M. and Brady, R.O., Biochim. Biophys. Acta 137: 135 (1967).
(15) Crocker, A.C., J. Neurochem. 7: 69 (1961).
(16) Rouser, G., Kritchevsky, G., Yamamoto, A., Knudson, A. and Simon, G., Lipids 3: 287 (1968).
(17) Philippart, M., Martin, L., Martin, J.J. and Menkes, J.H., Arch. Neurol. 20: 227 (1969).
(18) Fredrickson, D.E., The Metabolic Basis in Inherited Disease, Ed. 2, McGraw-Hill Book Co., New York (1966).
(19) Seng, P.N., Hoppe-Seyler's Z. Physiol. Chem. 352: 280 (1971).
(20) Callahan, J.W. and Philippart, M., Neurology 21: 442 (1971).
(21) Leroy, J.G. and Demars, R.I., Science 157: 804 (1967).
(22) Leroy, J.G. and Spranger, J.W., New England J. Medicine 284: 110 (1971).
(23) Dawson, G., Matalon, R. and Dorfman, A., Third Int'l. Meeting of the Int'l. Soc. for Neurochem., Budapest, p. 344 (1971).
(24) Tulkens, P., Trouet, A. and Van Hoof, F., Nature 228: 1282 (1970).
(25) Donahue, S., Zeman, W. and Watanabe, I., Inborn Disorders of Sphingolipid Metabolism, Pergamon Press, p. 3 (1967).
(26) Philippart, M., Third Int'l. Meeting of the Int'l. Soc. for Neurochem., Budapest, p. 343 (1971).

(27) Hartroft, W. S. and Porta, E. A., American J. Med. Science 250: 324 (1965).
(28) Fratantoni, J. C., Hall, C. W. and Neufeld, E. F., Proc. Nat'l. Acad. Science 60: 699 (1968).
(29) Calatroni, A., Donnelly, P. V. and Di Ferrante, N., J. Clinical Investigation 48: 332 (1969).
(30) Philippart, M. and Sugarman, G. I., Lancet 2: 854 (1969).
(31) Desnick, R. J., Sweeley, C. C. and Krivit, W., J. Lipid Res. 11: 31 (1970).
(32) Rouser, G., Kritchevsky, G., Galli, C. and Heller, D., J. Am. Oil Chem. Soc. 42: 215 (1965).
(33) Vance, D. E. and Sweeley, C. C., J. Lipid Res. 8: 621 (1967).
(34) Philippart, M., Sarlieve, L., Meurant, C. and Mechler, L., J. Lipid Res. 12: 434 (1971).
(35) Hakomori, S. and Andrews, H. D., Biochim. Biophys. Acta 202: 225 (1970).
(36) Saito, T. and Hakomori, S., J. Lipid Res. 12: 257 (1971).
(37) Svejcar, J. and Robertson, W. V., Analytical Biochem. 18: 333 (1967). Sweeley, C. C., Bentley, R., Makita, M., and Wells, W. W., J. Am. Chem. Soc. 85: 2497 (1963).
(38) Philippart, M., Advances in Experimental Med. Biology, to be published, Plenum Publishing Co.
(39) DeJong, B. P., Robertson, W. V. and Schafer, I. A., Pediatrics 42: 889 (1968).
(40) Madsen, J. A. and Linker, A., J. Pediatrics 75: 843 (1969).
(41) Philippart, M., Soc. for Pediatric Research, 81st Annual Meeting, p. 61 (1971).
(42) Svennerholm, L. and Stallberg-Stenhagen, S., J. Lipid Res. 9: 215 (1968).
(43) Watkins, W. M., Science 152: 172 (1966).
(44) Kint, J. A. and Dacremont, G., New England J. Medicine 285: 121 (1971).
(45) Hillborg, P. O. and Svennerholm, L., Acta Paed. 49: 707 (1960).
(46) Vance, D. E., Krivit, W. and Sweeley, C. C., J. Lipid Res. 10: 188 (1969).
(47) Herrlin, K. M. and Hillborg, P. O., Acta Paed. Scand. 51: 137 (1962).
(48) Philippart, M., Sarlieve, L. and Manacorda, A., Pediatrics 43: 201 (1969).
(49) Desnick, R. J., Dawson, G., Desnick, S. J., Sweeley, C. C. and Krivit, W., New England J. Medicine 284: 739 (1971).

(50) Martensson, E., Akademisk Avhandling Goteborg (1966).
(51) Suzuki, K. and Yoshiyuki, S., Proc. Nat'l. Acad. Sciences 66: 302 (1970).
(52) Klein, H. and Dichgans, J., Arch. Psychiat. Nervenkr. 212: 400 (1969).
(53) Austin, J., Suzuki, K., Armstrong, D., Bachhawat, B., Schlenker, J. and Stumpf, D., Arch. Neurol. 23: 502 (1970).
(54) Kwiterovich, P.O., Sloan, H.R. and Fredrickson, D.S., J. Lipid Res. 11: 322 (1970).
(55) O'Brien, J.S., Sampson, E.L. and Stern, M.B., J. Neurochem. 14: 357 (1967).
(56) Kishimoto, Y., J. Neurochem. 18: 1365 (1971).
(57) Martensson, E., Biochim. Biophys. Acta 116-117: 296 (1966).
(58) O'Brien, J.S. and Sampson, E.L., Science 150: 1613 (1965).
(59) Hultberg, B., Öckerman, P.A. and Dahlqvist, A., J. Clin. Invest. 49: 216 (1970).

NEUROPATHOLOGY OF GLYCOPEPTIDES DERIVED FROM BRAIN GLYCOPROTEINS (1)

E. G. BRUNNGRABER

Illinois State Psychiatric Institute

1601 West Taylor Street - Chicago, Illinois 60612

In another chapter in this volume we described the current status of research on the heteropolysaccharide chains of neural glycoproteins. We have described a procedure (1) in which it is possible to isolate and determine gangliosides, glycoproteins, and glycosaminoglycans in brain tissue samples. Gangliosides are extracted with chloroform-methanol (2:1 and 1:2, v/v). The defatted tissue is treated with papain or pronase to liberate glycopeptides (derived from brain glycoproteins) and glycosaminoglycans. The solubilized glycopeptides are separated into nondialyzable and dialyzable glycopeptide fractions. Glycosaminoglycans are separated from the nondialyzable glycopeptides by precipitation with cetylpyridinium chloride. Application of the method to normal samples provided a clearcut separation of glycopeptides, glycosaminoglycans, and gangliosides.

The studies with rat brain material have suggested that the broad outlines of the structure of the predominant glycopeptides in the nondialyzable glycopeptide preparation are similar to that of several glycopeptides obtained from purified glycoproteins, the structures of which have been studied by other workers (2, 3, 4).

---

(1) An account of this work was presented at the IVth International Congress of Human Genetics, Paris, France, on the 9th of Sept. 1971 at the round-table on Mucopolysaccharidoses and Mucolipidoses organized by Dr. P. Durand, Gaslini Institute, Genoa, Italy.

These glycopeptides have in common the feature that mannose occupies an internal position. Galactose is located in an external position as the penultimate sugar to which NANA or fucose is attached. Glucosamine serves to attach the entire polysaccharide unit to the protein in the most internal position, and also appears in a more external position by linking the external galactose to the more internally located mannose residues. A hypothetical structure for these glycopeptides is taken from the one proposed by Spiro (3) for fetuin. Six genetically linked enzyme deficiency diseases are

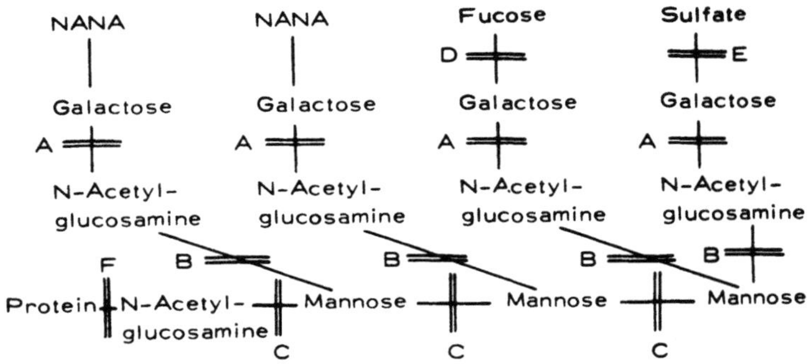

known which might affect the catabolic breakdown of the carbohydrate chains of the nondialyzable glycopeptides. Failure to cleave the galactose residues (GM_1 gangliosidosis) would cause accumulation of all of the chains which contain galactose (block at A). Glycopeptides containing galactose, glucosamine, and mannose would accumulate. In Tay-Sachs' disease, the deficiency of hexosaminidase A might cause a block at position B. Glycopeptides containing mannose and glucosamine would accumulate. In mannosidosis (block at point C) glycopeptides containing glucosamine and mannose would accumulate, although one would expect that these glycopeptides would contain only one molecule of glucosamine. As we shall see, Öckermann has reported the isolation of oligosaccharides containing only one glucosamine residue but which contain several mannose residues. In fucosidosis, (block at D) those chains which contain fucose in a terminal position would not be amenable to attack and abnormal glycopeptide material rich in fucose would accumulate. In sulfatase deficiency, a block at point E would cause those polysaccharide chains which contain sulfate groups to accumulate. An elevation of protein-bound hexosamine has been reported in metachromatic leukodystrophy (see below). Accumulation of aspartylglucosamine in aspartylglycosaminuria occurs because of the lack of

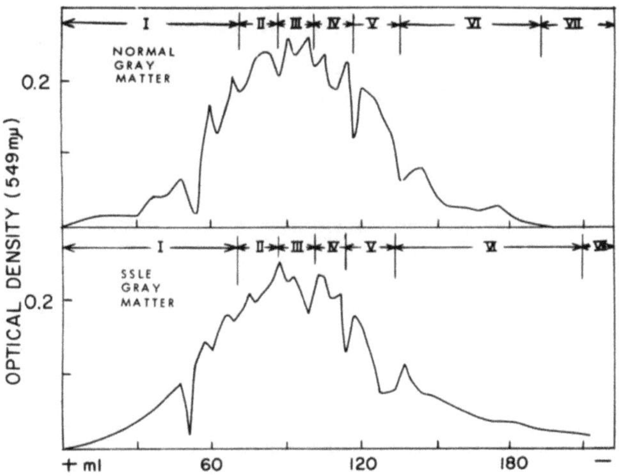

Fig. 1. The nondialyzable glycopeptides from normal 8--year old gray matter and gray matter from a 8-year old child with subacute sclerosing leukoencephalitis were isolated and fractionated by column electrophoresis utilizing the LKB column apparatus No. 3340. Electrophoresis was conducted at 3° in 0.1 M glycine-NaOH buffer, pH 10.3 at 200 V and 25-27 mA for 55-60 hours. The column was eluted with the same buffer and 3 ml fractions were collected. The fractions were analyzed for NANA. Contents of test tubes (see top of graph) containing materials corresponding to fractions I through VI were combined, dialyzed, and concentrated. The fractions were analyzed for carbohydrate composition. (See Table I). The Anode is at the left.

the enzyme which splits glucosamine from asparagine at the protein-polysaccharide linkage region (block at F).

The nondialyzable glycopeptide preparation from human brain was subjected to column electrophoresis (Fig. 1) and fractions corresponding to fractions I, II, III, IV, V, and VI (see top of graph) were analyzed for their sugar contents (Table I). The data obtained were remarkably similar to that obtained by similar treatment of rat brain nondialyzable glycopeptide preparations (see Chapter, "Chemistry and metabolism of glycopeptides from brain Glycoproteins" in this volume). As was the case with rat brain, fractions I and II contain more galactose and hexosamine than can be accounted for by assuming that the bulk of the nondialyzable glycopeptide preparation consists of a basic structure con-

Table I - Carbohydrate Composition of Nondialyzable Glycopeptides from Normal Human Gray Matter (8-year old) Obtained by Column Electrophoresis.

Fraction	Mannose	Galactose	N-Acetyl-glucosamine	NANA	Fucose
I	1	2.6	2.5	1.8	0.35
II	1	1.7	2.4	1.6	0.41
III	1	1.0	1.7	0.9	0.36
IV	1	0.9	1.5	0.6	0.43
V	1	0.9	1.6	0.5	0.67
VI	1	0.7	1.2	0.2	0.43
Totals	1	1.1	1.6	0.7	0.43

taining 3 molecules of mannose, 4 molecules of galactose, 5 molecules of N-acetylglucosamine and variable amounts of NANA and fucose attached to this structure. In the case of the rat brain, it was postulated that the nondialyzable glycopeptide preparation contains, as a minor component, a glycopeptide or glycopeptide population of high electrophoretic mobility with a high ratio of NANA/hexose, and which does not contain mannose or fucose as a constituent sugar. This appears to be true also for human brain.

The dialyzable glycopeptide preparation consists of three glycopeptide types. The predominant glycopeptide or glycopeptides contain a heteropolysaccharide chain consisting almost solely of mannose and N-acetylglucosamine. Approximately 70% of the total hexose and hexosamine of the dialyzable glycopeptide preparation is associated with these glycopeptides. They do not contain sulfate-ester groups or NANA and they contain very little (if any) galactose and fucose. These glycopeptides are easily separated from the others since they have a low electrophoretic mobility. A second class of glycopeptides correspond to the glycopeptide fractions that were found in the nondialyzable glycopeptide preparation. Only small amounts of the larger molecular weight glycopeptides were found. Appreciable amounts of the NANA-containing, fucose-rich glycopeptides of lower molecular weight corresponding to the gly-

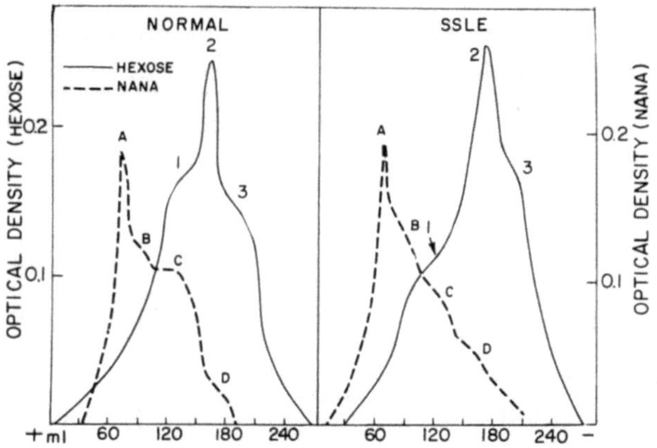

Fig. 2. Electropherogram obtained upon subjection of the dialyzable glycopeptides from normal and SSLE cerebral gray matter on formalated cellulose using the LKB electrophoresis apparatus. The anode is at the left. Letters A, B, C, and D denote four peaks corresponding the NANA-containing glycopeptides as determined by the analysis of eluate fractions for NANA. Numbers 1, 2, and 3 denote three peaks corresponding to three glycopeptide fractions as determined by the analysis of eluate fractions for hexose. Peaks 2 and 3 are known to contain glycopeptides which do not contain NANA as a constituent sugar. Peaks 2 and 3 are elevated 2-3 fold in a case of Tay-Sachs' disease.

copeptides of fractions V and VI of the nondialyzable glycopeptide preparation appear in the dialyzable glycopeptide preparation. The third class of glycopeptides in the dialyzable glycopeptide preparation are minor components. These include glycopeptides which contain galactosamine as the predominant hexosamine, and which may contain glucose. These glycopeptides appear to have chemical properties that differ markedly from those which we have been discussing. Many of these glycopeptides contain large amounts of NANA and show a high electrophoretic mobility (Fig. 2).

Data on the carbohydrate composition of nondialyzable and dialyzable glycopeptides obtained from pathological brain tissues from various cases analyzed in this laboratory are presented in Tables II and III.

Table II - Carbohydrate Composition of Nondialyzable Glycopeptides from Normal and Pathological Human Brain Tissue.

	NANA	Fucose	Hexosamine	Hexose
GRAY MATTER				
Normal 8-year old	0.72	0.55	1.94	2.17
SSLE, 8-year old	0.63	0.40	1.59	1.79
Neurolipidosis, with failure of myelination, 4 years	0.31	--	1.33	--
Generalized lymphosarcoma, 3 years	0.70	0.47	2.32	1.77
Galactosemia, 25-year old	0.60	0.29	1.20	1.85
Late infantile amaurotic idiocy (Bielschowsky), 11-year old	0.44	--	0.83	1.44
Tay-Sachs, 20-month old	0.77	0.49	1.81	2.18
Infantile Neuroaxonal Dystrophy, 9½-year old	0.46	0.33	1.32	1.66
WHITE MATTER				
Normal 8-year old	0.46	0.24	1.10	1.30
SSLE, 8-year old	0.50	0.26	1.17	1.38
Neurolipidosis with failure of myelination, 4 years	0.32	--	1.35	--
Galactosemia, 25-year old	0.59	0.21	1.20	1.67
Late infantile amaurotic idiocy (Bielschowsky), 11-year old	0.46	--	1.14	1.17
Tay-Sachs, 20-month old	0.65	0.38	2.36	1.79
Infantile Neuroaxonal Dystrophy, 9½-year old	0.48	0.26	1.15	1.60

All values are expressed in terms of μmoles/g fresh tissue.

Table III – Carbohydrate Composition of Dialyzable Glycopeptides from Normal and Pathological Human Brain Tissue.

	NANA	Fucose	Hexosamine	Hexose
GRAY MATTER				
Normal 8-year old	0.21	0.22	1.03	1.05
SSLE, 8-year old	0.24	0.25	1.49	1.48
Neurolipidosis, with failure of myelination, 4 years	0.57	--	0.91	--
Generalized lymphosarcoma, 3 years	0.24	0.27	1.65	1.33
Galactosemia, 25-year old	0.27	--	0.40	--
Late infantile amaurotic idiocy (Bielschowsky), 11 years	0.25	--	0.28	--
Tay-Sachs, 20-month old	0.24	0.37	1.93	3.57
Infantile Neuroaxonal Dystrophy, 9½-year old	0.12	0.15	0.65	0.84
WHITE MATTER				
Normal 8-year old	0.18	0.17	1.08	0.81
SSLE, 8-year old	0.13	0.20	1.17	1.02
Neurolipidosis, with failure of myelination, 4 years	0.37	--	0.75	--
Galactosemia, 25-year old	0.27	--	0.35	--
Late infantile amaurotic idiocy (Bielschowsky), 11 years	0.14	--	0.25	--
Tay-Sachs, 20-month old	0.18	0.34	1.72	1.51
Infantile Neuroaxonal Dystrophy, 9½-year old	0.16	0.20	0.80	0.97

All values are expressed in terms of μmoles/g fresh tissue.

Tay-Sachs' Disease (5)

Autopsy material was derived from a 20-month old girl of Ashkenazy Jewish parents. The clinical history, ocular and neural signs were typical of Tay-Sachs' disease. Hexosaminidase A was shown to be absent in the retina, spleen, jejunum, and liver. A four-fold elevation of gangliosidic NANA was noted in both gray and white matter and 80% of the ganglioside was shown to be GM_2. Although the gangliosidic NANA increased four-fold due to the accumulation of the neuraminidase-resistant ganglioside GM_2, the concentration of NANA in the nondialyzable and dialyzable glycopeptides remained normal (Table II and III). It is known that Tay--Sachs ganglioside becomes elevated in this disease due to the terminal location of N-acetylgalactosamine which cannot be cleaved due to the deficiency of hexosaminidase A. Gangliosidic NANA becomes elevated since the cleavage of NANA is dependent upon the prior removal of the terminal N-acetylgalactosamine. This structural feature appears to be absent in the glycopeptides. This agrees with the finding that all of the NANA in the glycopeptides is removed by neuraminidase, but 30-40% of the NANA in the gangliosides is resistant to the action of this enzyme.

A three-fold elevation in the concentration of hexose associated with the dialyzable glycopeptides was found (Table III). The increased level of hexose was due to an elevation of mannose; galactose levels remained normal. The hexosamine concentration in the dialyzable glycopeptides was twice that found in normal tissue. A possible interpretation of these findings is that the enzyme hexosaminidase A, absent in Tay-Sachs' disease, participates in the metabolic breakdown of the heteropolysaccharide chains of the brain glycoproteins. The hydrolytic enzymes which catabolize such chains are exo-glycosidases, and the heteropolysaccharide chains are degraded sequentially from the non-reducing end of the chain. For example, the hydrolytic degradation of the nondialyzable glycopeptides in Tay-Sachs' disease would cease after the galactose residues had been removed. The block is at point B in the hypothetical formula shown above. The resultant mannose-rich glycopeptides accumulate and appear in the dialyzable glycopeptide fraction.

Column electrophoresis of the nondialyzable glycopeptide preparation and analysis of the six fractions obtained (Fig. 1, Table I) showed no abnormality in the composition of the six fractions. The elevation of those glycopeptides which do not contain NANA (peaks 2 and 3, Fig. 2) was demonstrated by column electrophoresis of the dialyzable glycopeptide preparation (5).

Mannosidosis

Kjellman et al. (6) and Öckerman (7) reported on a 4 and 1/3 year old patient afflicted with mannosidosis. The clinical aspects of this case resembled that of Hurler's syndrome. A periodic acid--Schiff positive storage material was found in the cytoplasm of neurones. The enzyme, α-mannosidase, was found to be deficient in the liver (7). Oligosaccharides containing 3, 5, or 8 molecules of mannose were isolated from gray matter (7, 8). Each of these oligosaccharides contained only one glucosamine molecule. Of special interest was the finding that all of the glucosamine was in a reducing, presumably terminal, position.

In cases of mannosidosis, catabolism of the heteropolysaccharides chains of the glycoproteins would proceed so that neuraminidase, α-fucosidase, β-galactosidase, and β-hexosaminidase would remove the externally located sugars leaving the mannose-rich, glucosamine-containing internal portion of the glycopeptide intact. Degradation then ceases since the enzyme, α-mannosidase, is absent (block at point C in formula above). The water soluble mannose-rich fragment obtained by Öckerman is presumably derived by the action of an enzyme which splits the bond between N-acetylglucosamine and asparagine. The oligosaccharide then accumulates since it cannot be further degraded (9). Öckerman's oligosaccharide may be derived from the catabolism of the nondialyzable glycopeptides. However, the nondialyzable glycopeptides from rat brain contain only two or three mannose residues per molecule of glycopeptide. The analytical values obtained for human brain suggest that this is probably the case in this tissue also. Consequently, it appears more likely that Öckerman's oligosaccharides are largely derived from the mannose-rich glycopeptides of the dialyzable glycopeptide preparation. These may contain as many as 8 mannose residues. The oligosaccharide containing only three mannose residues may be derived from the heteropolysaccharide chains in the nondialyzable glycopeptide preparation.

Fucosidosis

Van Hoof and Hers (10) and Durand et al. (11) described cases in which the tissues lacked the enzyme α-fucosidase. Fucosidosis is a neurovisceral disease with severe and progressive cerebral degeneration, mental retardation, eventual respiratory difficulties and death at approximately 4 to 5 years of age. Nerve cell bodies are enlarged and distended. The cells appeared optically empty, and periodic acid-Schiff-positive material was seen in only some cells. The disease is characterized by an increase in the fucose content of the "mucopolysaccharide" fraction. Glycoproteins are the only brain constituent that are currently known to contain fucose. Consequently, the absence of α-fucosidase would affect only this constituent; the catabolism of the glycosaminoglycans should remain unaffected. As far as is known, fucose residues are present solely in terminal positions of the heteropolysaccharide chains. Hence, in fucosidosis, the block in biological degradation (at point D in formula above) occurs at an initial stage of the catabolic sequence. Chains which contain fucose in the terminal position would not be amenable to subsequent attack by β-galactosidase, β-hexosaminidase, and α-mannosidase. These enzymes are believed to act in sequence by attacking the non-reducing ends of the chains. In this disease, one would expect an increased level of polysaccharide chains which contain fucose as a terminal group. The fucose-galactose-N-acetylglucosamine-portion of the heteropolysaccharides will accumulate; the NANA-galactose-N-acetylglucosamine- chains are degraded (9).

Fucose and mannose are normally absent in gangliosides and glycosaminoglycans of neural origin, but they are characteristic components of glycoproteins. Consequently, mannosidosis and fucosidosis are diseases which primarily affect the metabolism of neural glycoproteins.

GM_1 Gangliosidosis

An accumulation of glycopeptides in visceral organs occurs in GM_1 gangliosidosis. The major compound stored contains heteropolysaccharide chains consisting largely of glucosamine and galactose (12) and was thought to resemble keratan sulfate in chemical properties. However, there was no increase in either keratan sulfate or glycopeptides in brain tissue. Wolfe et al. (13) found a decrease in the NANA-containing nondialyzable glycopeptides in brain tissue and this loss was not considered specific for the disease. It appears that the deficiency of β-galactosidase which normally cleaves the terminal galactose of GM_1 ganglioside is not involved in the catabolism of the heteropolysaccharide chains of the glycoproteins.

Metachromatic Leukodystrophy (MLD)

In metachromatic leukodystrophy, demyelination in cerebral and cerebellar white matter is accompanied by gliosis which leads to a progressive sclerosis. The axons are well preserved. The primary disturbance is an accumulation of sulfatides due to the deficiency of sulfatase A (14-17). Cases with elevated amounts of protein-bound hexosamine or with normal levels of protein-bound hexosamine were studied by Austin (14). Increase in protein-bound hexosamine were reported by several laboratories (18-23). It is not known whether the increase in protein-bound hexosamine, especially noteworthy in white matter, is due to an elevation of the heteropolysaccharide chains of glycoproteins, glycosaminoglycans, or both.

Late Infantile Amaurotic Idiocy (Jansky-Bielschowsky type) (24)

The brain tissue studied was that from a 10 year 9 month old boy whose deterioration started at two years of age. There was no storage of ganglioside material, although neuronal storage of unidentified substances appeared to be ubiquitous with modest to severe ballooning of the nerve cells. Gangliosidic NANA was 33% of the normal concentration. Although white matter contained normal amounts of nondialyzable glycopeptides, there was a significant decrease (40%) in gray matter. The composition of the nondialyzable glycopeptides from white and gray matter appeared to be normal. A marked abnormality was noted in the dialyzable glycopeptides. Although the NANA content of this fraction was normal, the hexosamine content was reduced by 75% in both white and gray matter. The data indicated that the mannose-rich, NANA-free glycopeptides were markedly reduced in amount or absent. Similar cases were reported by Landolt (25), Tingey et al. (26) and Borri et al. (27).

Early Infantile Neurolipidosis with Failure of Myelination (28)

Brain tissue is a three year 11 month old Jewish girl afflicted with amaurotic idiocy and neuronal storage of gangliosides was studied. Onset of disease commenced at 6 weeks of age. The cytomorphological changes were typical of Tay-Sachs' disease. The case was marked by a virtual absence of stainable myelin in cerebellar and cerebral hemispheres; cerebroside content of white matter was 15% of the normal value. The lack of myelin was not due to demyelination as a consequence of a primary destruction of neuronal cell bodies. Rather, the process of myelination was never initiated in most fiber tracts. Cerebral gray matter showed a marked reduction in nondialyzable glycopeptides and abnormalities in their composition was indicated by abnormal values for the ratio hexosamine/NANA in both gray and white matter. NANA values in dialyzable glycopeptide preparations were elevated in both white and gray matter while the dialyzable glycopeptide-hexosamine was normal. There was no storage of glycoprotein material. The nature of these changes in the glycopeptides was quite different from that noted in the typical Tay-Sachs case discussed above; the failure of myelination was especially distinctive.

Infantile Neuroaxonal Dystrophy (29)

The brain of a 9 and a half year-old girl with infantile neuroaxonal dystrophy was studied (29). This disease is characterized pathologically by the presence of eosinophilic axonal spheroids. The concentration of gangliosidic NANA and hexosamine of white and gray matter fell within the normal range with only minor or nonspecific changes in the thin-layer chromatographic patterns of the gangliosides. Carbohydrate analysis of the nondialyzable glycopeptides revealed little or no change in their composition, however there was a 40% reduction of the amount of nondialyzable glycopeptide material recovered from gray matter. Analysis of fractions obtained by column electrophoresis of the nondialyzable glycopeptides from gray matter revealed that the composition of the nondialyzable glycopeptide preparation was identical to that obtained from the normal brain although the concentration had been reduced by 40%. The concentration of the nondialyzable glycopeptide material in white matter was normal, with the possible exception of a small elevation in hexose content. The dialyzable glycopeptides from the gray matter also showed a marked decrease in concentration; small changes in the ratios of carbohydrate residues suggested possible changes in the composition of this preparation. The concentration of dialyzable glycopeptides was not different from the normal value in white matter, but marked changes in composition were indicated by changes in the values for the ratios of hexose/hexosamine, fucose/hexosamine, NANA/hexose and NANA/fucose.

The nature of the stored material in this disease remains unidentified, since the chemical analysis revealed no storage of gangliosides, glycoproteins, or glycosaminoglycans.

Galactosemia (30)

Galactosemia is a genetically-linked disease due to congenital deficiency of the enzyme galactose-1-phosphate uridyl transferase. We have studied (30) the brain tissue of a 25-year old male diagnosed galactosemic at 13 years of age. The encephalopathy noted in this case was characterized by mental deficiency, extrapyramidal motor disorder, and epilepsy. The gangliosidic NANA content fell within the lower limit of the normal range in gray matter and slightly above the normal range in white matter. There was a small decrease in the amount of nondialyzable glycopeptides in gray matter, and an increase in the concentration of these glycopeptides in white matter. This resulted in the equalization of nondialyzable glycopeptide material in white and gray areas of the galactosemic brain. Normally, the amount of nondialyzable glycopeptides in gray matter is twice that found in white matter. The nondialyzable glycopeptides from galactosemic brain gray matter contain considerably less hexosamine than normal, resulting in a higher hexose/hexosamine ratio for galactosemic gray matter.

The dialyzable glycopeptides from white and gray matter of the galactosemic brain contain only 50% of the hexosamine found in normal white and gray matter, but showed a 30% increase in NANA content. This suggests that the mannose-rich, NANA-free glycopeptides of the dialyzable glycopeptide preparation are markedly reduced in this disease. This effect had been noted in the case of Bielschowsky disease referred to above. One can speculate that the lesion in galactosemia may result in an abnormal balance of nucleotide sugars, which are the immediate precursors of the heteropolysaccharide chains of the glycoproteins. It is conceivable that this would result in a failure to synthesize the appropriate glycoproteins. In view of the advanced age of the patient, however, the difference noted in the glycoproteins in this case may be secondary to the primary lesion.

Subacute Sclerosing Leukoencephalitis

All of the diseases discussed thus far were genetically linked. The etiology of subacute sclerosing leukoencephalitis (SSLE) remains obscure, although there is evidence for a viral origin for this disease. Ledeen et al. (31) had found that the concentration of four gangliosides corresponding to minor components in normal tissue was elevated. The gangliosidic NANA was slightly decreased in gray matter but normal in white matter. We have studied the nature of the protein-bound carbohydrates in the defatted tissue residue obtained by Ledeen (31). In the cerebral gray matter (32), the concentration of nondialyzable glycopeptides was slightly decreased in value. The sugar composition of the total nondialyzable glycopeptide preparation, as well as various fractions obtained there from by column electrophoresis (Fig. 1) was similar in normal and SSLE cerebral gray matter. However, the data suggested that in SSLE the number and type of terminal sugars (NANA and fucose) attached to hexose-hexosamine portions of the various fractions differed. Whether the relatively small differences observed were due to the disease, individual variations, sex, or in variations due to the limitations of the technique used is presently unknown. In the cerebral white matter, the amount and sugar composition of the nondialyzable glycopeptides isolated, per gram of tissue, was not affected by the disease. This was surprising since the white matter in SSLE is affected by marked gliosis, inflammation, and demyelination.

The hexosamine (but not the NANA) associated with the dialyzable glycopeptides was elevated by about 50%. This elevation of hexosamine and hexose associated with the dialyzable glycopeptide preparation was accounted for by an increase in a mannose-rich glycopeptide of relatively low electrophoretic mobility and which does not contain NANA and fucose (Fig. 2). The other glycopeptides in this preparation appear to be relatively unaffected. Except for a decrease in NANA content, the other sugars of the dialyzable glycopeptide preparation from white matter were unaffected by the disease.

The amount of total protein-bound hexosamine of cerebral cortex in SSLE was identical to that found in normal tissue. This conflict with the findings of Cumings (33, 34), who found an elevation, may be more apparent than real. The elevation of the mannose-rich glycopeptide of low molecular weight and electrophoretic mobility provides evidence of an abnormality in one of the macromolecular components the hexosamine content of which contributes to Cumings' higher values for protein-bound hexosamine in SSLE. A compensating loss in the amount of hexosamine associated with the nondialyzable glycopeptides and glycosaminoglycans in the present case accounts for the failure to find an elevation in total protein-bound hexosamine.

Generalized Lymphosarcoma

Dr. Ledeen kindly provided defatted gray matter obtained from a 3-year old child who had died of generalized lymphosarcoma. Since the child showed no neurological symptoms, the material was used to isolate nondialyzable and dialyzable glycopeptides in order to provide control values for 3-year old brain tissue. Nondialyzable glycopeptides from this brain were characterized by a lower value for the ratio of hexose/hexosamine than that of any of the other brain samples of various ages, both normal and pathological, we have studied. The increase in hexosamine content of these glycopeptides was compensated for by a correspondingly equal loss in hexose content. Column electrophoresis of the nondialyzable glycopeptides indicated significantly lower values for the ratio hexose/hexosamine in fractions I, II, III, and IV. Lower values for this ratio were also found in fractions V and VI. At present, it is not clear whether these differences can be ascribed to age differences. For example, no difference in the carbohydrate composition of the six fractions obtained from 20-month old Tay--Sachs gray matter and normal 8-year old gray matter was found (see above).

The hexosamine content of the dialyzable glycopeptides was 60% greater than that of the 8-year old normal control. The 25% elevation in hexose and small elevations of NANA and fucose were of more doubtful significance.

It appears that the differences observed may not be due to age differences, but rather, an effect due to the disease itself. The possibility of neurological involvement in cases of lymphosarcoma is indicated by a study of a 48-year old patient with lymphosarcoma who developed progressive multifocal leukoencephalopathy which was diagnosed before death (35).

REFERENCES

(1) Brunngraber, E. G., Brown, B. D. and Hof, H., Clin. Chim. Acta 32: 159 (1971).
(2) Wagh, P. V., Bornstein, I. and Winzler, R. J., J. Biol. Chem. 244: 658 (1969).
(3) Spiro, R. G., Ann. Rev. Biochem. 39: 599 (1970).
(4) Dunn, J. T. and Spiro, R. G., J. Biol. Chem. 242: 5556 (1967).
(5) Brunngraber, E. G., Witting, L. A., Haberland, C. and Brown, B. D., Brain Res., in press.
(6) Kjellman, B., Gamstorp, I., Brun, A., Öckerman, P. A. and Palmgren, B., J. Pediatrics 75: 366 (1969).
(7) Öckerman, P. A., J. Pediatrics 75: 360 (1969).
(8) Öckerman, P. A., J. Pediatrics 77: 168 (1970).
(9) Brunngraber, E. G., J. Pediatrics 77: 166 (1970).
(10) Van Hoof, F. and Hers, H. G., Eur. J. Biochem. 7: 34 (1968).
(11) Durand, P., Borrone, C. and Della Cella, G., J. Pediatrics 75: 665 (1969).
(12) Suzuki, K., Suzuki, K. and Kamoshita, S., J. Neuropath. & Exptl. Neurol. 28: 25 (1969).
(13) Wolfe, L. S., Callahan, J., Fawcett, J. S., Andermann, F. and Scriver, C. R., Neurology 20: 23 (1970).
(14) Austin, J. H., in Medical Aspects of Mental Retardation, Ed. C. H. Carter, C. C. Thomas, Publisher, Springfield, Illinois, p. 768 (1965).
(15) Austin, J. H., McAfee, D., Armstrong, D., O'Rourke, M., Shearer, L. and Bachhawat, B., Biochem. J. 93: 156 (1964).
(16) Austin, J. H., Trans. Amer. Neurol. Ass. 83: 149 (1958).
(17) Mehl, F. and Jatzkewitz, H., Biochem. Biophys. Res. Commun. 19: 407 (1965).
(18) Hagberg, B., Sourander, P., Svennerholm, L. and Voss, H., Acta Paediat. 49: 135 (1960).
(19) Cumings, J. N., Metab. Clin. & Exptl. 9: 219 (1960).
(20) Edgar, E. W. F., in Cerebral Lipidosis, A Symposium, Ed. L. Van Bogaert, J. N. Cumings and A. Lowenthal, C. C. Thomas, Publisher, Springfield, Illinois, p. 186 (1957).
(21) Edgar, G. W. F., Psychiat. Neurol. Neurochir. 64: 28 (1961).
(22) Bargeton, in Brain Lipids and Lipoproteins and the Leukodystrophies, Ed. J. Folch-Pi and H. Bauer, Elsevier Publishing Co., Amsterdam, p. 90 (1963).

(23) Blackwood, N. and Cumings, J.N., Proc. 5th Int. Congr. Neuropathol., Zürich, Excerpta Medica, Amsterdam, p. 364 (1966).

(24) Haberland, C., Brunngraber, E.G., Witting, L.A. and Hof, H., Neurology, in press.

(25) Landolt, R., in Cerebral Lipidosis II, Ed. A. Nunes Vicente, P. Dustin and A. Lowenthal, Presses Académiques Européennes, Brussels, p. 326 (1968).

(26) Tingey, A.H., Norman, R.M., Urich, H. and Beasley, W.H., J. Ment. Sci. 104: 91 (1958).

(27) Borri, P.F., Hooghwinkel, G.J.M. and Edgar, G.W.F., J. Neurochem. 13: 1249 (1966).

(28) Haberland, C. and Brunngraber, E.G., Arch. Neurol. 23: 481 (1970).

(29) Haberland, C., Brunngraber, E.G. and Witting, L.A., Arch. Neurol., in press.

(30) Haberland, C., Perou, M., Brunngraber, E.G. and Hof, H., J. Neuropath. & Exptl. Neurol. 30: 431 (1971).

(31) Ledeen, R., Salsman, K. and Cabrera, M., J. Lipid Res. 9: 129 (1968).

(32) Brunngraber, E.G., Brown, B.D. and Chang, I., J. Neuropath. & Exptl. Neurol. 30: 525 (1971).

(33) Cumings, J.N., in Encephalitides, Ed. L.V. Van Bogaert, J. Radermaker and A. Lowenthal, Elsevier Publishing Co., Amsterdam, p. 638 (1961).

(34) Cumings, J.N., in Mechanisms of Demyelination, Ed. A.S. Rose and C.M. Pearson, McGraw-Hill, Inc., New York, p. 58 (1963).

(35) Ellison, G.W., J. Neuropath. & Exptl. Neurol. 28: 501 (1969).

EFFECT OF BACTERIAL NEURAMINIDASE ON THE ISO-ENZYMES OF ACID HYDROLASES OF HUMAN BRAIN AND LIVER

J.A. KINT and A. HUYS

Department of Pediatrics Akademisch Ziekenhuis

Ghent 9000, Belgium

A number of lysosomal enzymes are known to occur in multiple forms. Physical and kinetic properties such as thermolability, pH-optimum or electrophoretic mobility are used in differentiating or separating such isoenzymes. As these hydrolases, which have a low specifity, are responsible for the degradation of large macromolecules such as glycolipids, mucopolysaccharides and glycoproteins, the question may arise on the physiological meaning of multiple forms. In this study more details on the occurrence and on the molecular differences of these isoenzymes in human brain and liver are presented.

METHODS

(A) Pulsed power electrofocusing

All separations were carried out with the ORTEC equipment (ORTEC Inc. 100 Midland Road, Oak Ridge, Tennessee 37830, U.S.A.). The polyacrylamide gel was formed in the following conditions: acrylamide (B.D.H. - Analytical grade) 6.5 g %; N,N'-methylenebisacrylamide (B.D.H. - Analytical grade) 660 mg %; Ampholine pH = 3-10 (L.K.B.) 3 %; ammoniumpersulphate (B.D.H. - Analytical grade) 0.5 mg/ml. Total gel volume was 30 ml. The anode buffer consisted of 0.01 M sulphuric acid while at the cathode a 0.05 M diamino-ethane solution was used. Samples (10-70 μl) were applied at the top of the gel (cathodic end). The samples were mixed with concentrated sucrose and the sample slots were closed with some polyacrylamide. Each experiment was started with a voltage of 225 Volts and a pulse frequency of 100 pulses per second. After about one hour the voltage was increased to 450 Volts and the frequency to 250 pulses per second. The initial current of 0.05 Ampères dropped to about 0.01 ampère at the end of the run. Electrofocusing is terminated after about two hours when all proteins were separated. This was controlled at the resolution of some hemoglobin into 4 or 5 very sharp lines.

(B) Isoenzyme detection

The gel is removed from the electrophoresis cell and incubated with 1 mM of the appropriate derivative of 4-methylumbelliferone. The following substrates were used (Koch-Light, Colnbrook, England): 4-methylumbelliferyl-α-D-mannopyranoside, 4-methylumbelliferyl-β-D-galactopyranoside, 4-methylumbelliferyl-α-D-galactopyranoside and 4-methylumbelliferyl-2-acetamido-2-deoxy-β-D-glucopyranoside. After incubation at 37° C for periods up to two hours, the reaction product 4-methylumbelliferone was made fluorescent by concentrated ammonia. For revealing the isoenzymes of acid phosphatases Bergmeyer's method (1) was used.

(C) Preparation of tissue homogenates

The tissues examined consisted of autopsy material of human brain and liver. Small pieces of tissue were homogenized in 10 volumes of distilled water, frozen and thawed four times or treated for 30'' in an ultrason desintegrator. The supernatant of a 10.000 x g centrifugation was analyzed.

(D) Neuraminidase treatment

In a total volume of 200 microliter, tissue homogenate corresponding to 10 mg fresh weight, acetate buffer 0.1 M pH = 4.5 and 5 milliunits neuraminidase from Clostridium perfringens (Sigma) was incubated at room temperature for one hour. As a control the same volume of homogenate was incubated in identical conditions but without the neuraminidase added.

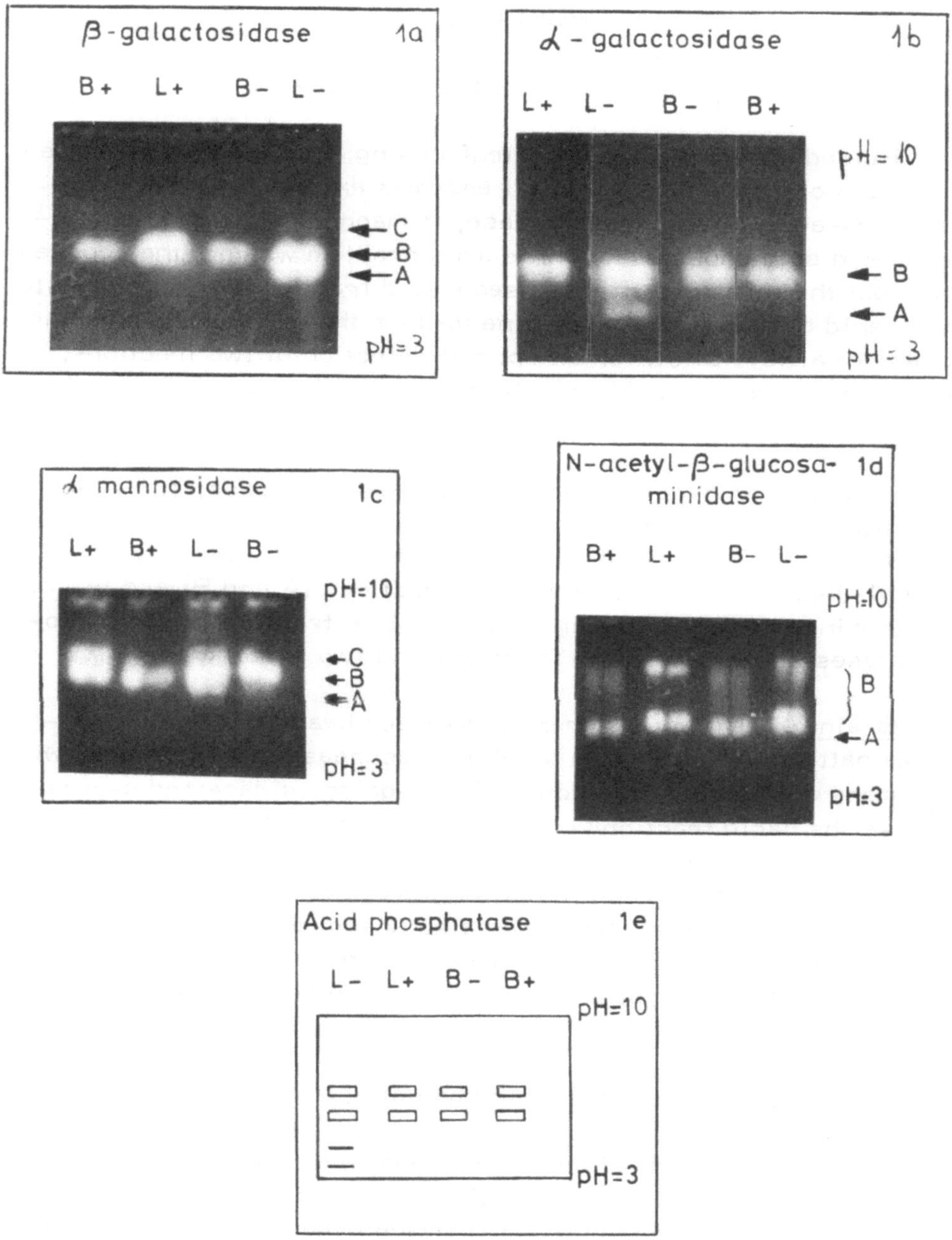

Fig. 1. Isoenzyme patterns of 5 acid hydrolases. L = untreated liver; L+ = liver incubated with neuraminidase; B- = untreated brain; B+ = brain incubated with neuraminidase. 1a = β-galactosidase; 1b = α-galactosidase; 1c = α-mannosidase; 1d = N-acetyl-β-glucosaminidase; 1e = acid phosphatase.

RESULTS

(A) Separation of the isoenzymes

Pulsed constant power electrofocusing resulted in a clear separation of multiple forms of all enzymes examined: β-galactosidase, N-acetyl-β-glucosaminidase, α-mannosidase, α-galactosidase and acid phosphatase. The main problem was not the separation but the localisation of the separated fractions. Indeed, due to the rapid diffusion in the alcaline medium the methylumbelliferone does not always allow for an optimal detection of two fractions, close to one another. All results are shown in Fig. 1a - 1b - 1c - 1d - 1e.

Liver. For α-galactosidase and β-galactosidase, two main fractions (A and B) were seen. α-mannosidase consisted of three fractions (A, B and C) of about the same intensity while for N-acetyl-β-glucosaminidase two main fractions (A and B) and two or three minor components were seen close to fraction B. Acid phosphatases showed two main fractions and two minor fractions.

Brain. The main differences between brain and liver isoenzyme patterns was found in a relative decrease of the fraction with the more acid isoelectric point. For α-galactosidase and α-mannosidase the "acid fraction" could only be detected with a concentrated tissue homogenate (e. g. 70 μl of a 20 % extract). But for acid phosphatase such an "acid fraction" was never seen. For β-galactosidase this "acid fraction" was also absent but a new one at a more alcaline isoelectric point appeared (fraction C). N-acetyl-β-glucosaminidase did not show any difference between liver and brain.

(B) Neuraminidase effect

Liver. Treatment of the organ homogenates with bacterial neuraminidase resulted in dramatic changes of the isoenzyme distribution. For α-galactosidase, α-mannosidase, acid phosphatase and α-galactosidase this change consisted always in a disappearance of the most acid fraction. At the same time an increase of one of the other fractions was observed for α-galactosidase, α-mannosidase and acid phosphatase while for β-galactosidase a fraction at the site of fraction C in brain appeared. For N-acetyl-

-β-glucosaminidase there was no change in the main acid fraction, but the heterogeneity at the neutral fraction disappeared and only one main fraction remained. Total activity was not affected by the neuraminidase treatment.

Brain. In the brain the same changes as in the liver were noted except for β-galactosidase and acid phosphatase. Both these enzymes did not show any difference after the neuraminidase treatment, but as it has already been noted, the same two enzymes were missing in the untreated brain the acid fraction present in the liver. So, one is inclined to identify the normal brain pattern of β-galactosidase and acid phosphatase with the liver pattern after neuraminidase treatment.

Other experiments. In order to make sure that the change in isoenzyme distribution was really due to an enzymatic process, the neuraminidase was heated for 10 minutes at 100° C. After such a heat denaturation no conversion of the isoenzymes could be accomplished. An ultrafiltrate of the neuraminidase also lost this capacity while the concentrated liquid above the membrane still could transform the isoenzymes (Fig. 2).

In Figure 3 an electrofocusing of the N-acetyl-β-glucosaminidase isoenzymes is shown in tissues of a control of a Tay-Sachs patient. It appears - as it has already been demonstrated by O'Brien et al. (2) and by Sandhoff et al. (3) - that one isoenzyme

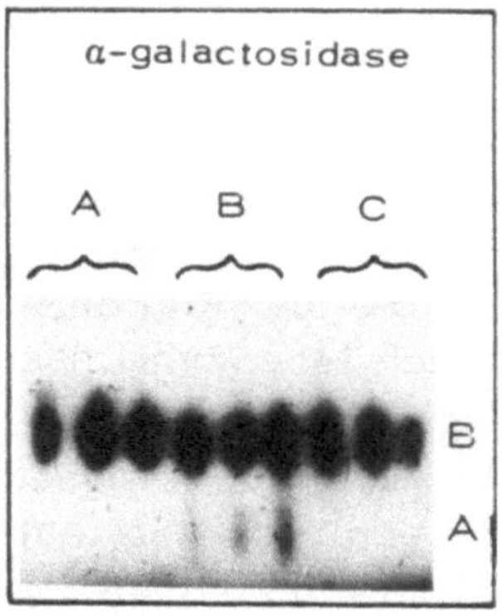

Fig. 2. Effect of neuraminidase on α-galactosidase isoenzymes, before and after ultrafiltration. A = untreated neuraminidase; B = ultrafiltrate of neuraminidase; C = concentrated neuraminidase above the membrane.

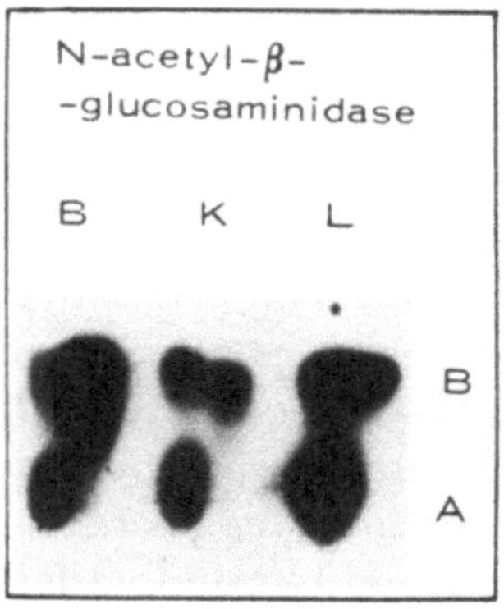

Fig. 3. Isoenzymes of N-acetyl-β-glucosaminidase in Tay-Sach's disease and in a control. B = brain; K = kidney; L = liver.

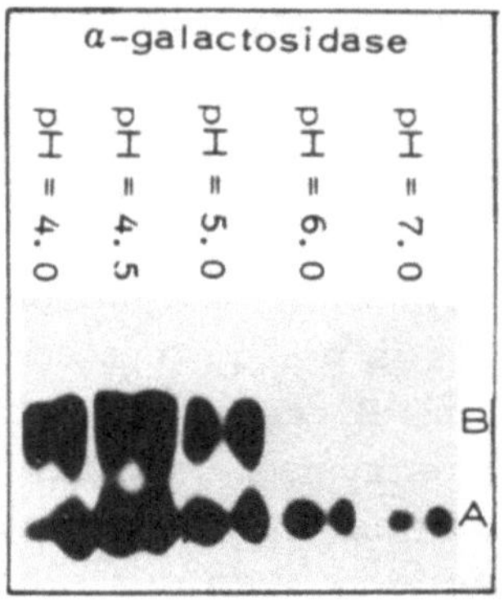

Fig. 4. Isoenzymes of α-galactosidase at different pH-values.

is deficient. From our experiments it can be concluded that the deficient isoenzyme is the one with the more acid isoelectric point. In another experiment (Fig. 4) the differentiation of the α-galactosidase isoenzymes was possible. Indeed at pH = 7.0 the more acid fraction was still present while the other was totally inactive. This therefore presents a mean for measuring both isoenzymes without the need for separating them.

DISCUSSION

The technique of isoelectric focusing in polyacrylamide gel gives a high resolution with a minimum of amount of enzyme in a minimum of time. Using the ORTEC equipment offers a double advantage. First the vertical position of the electrophoretic cell gives a very efficient cooling because the whole cell is immersed in two liters of the buffer liquid. When the electrophoretic tank is put in a cold room or in a refrigerator overnight, the large volume of cold buffer provides a rather efficient way of removing the Joule heat. And there is a second aspect of the equipment which makes this system rather unique, namely the power supply. Due to a special electronic circuit, the output power does not change during electrophoresis, regardless of the variations in cell resistance during the run. Such constant power is achieved in an output wave form with a high voltage pulse with a rapid rise and an exponential decay. These pulses have a low duty cycle: they are off much more than they are on. The voltage pulse when present, can therefore be as high as 1000 Volts, without apparently causing any harm to the enzymes.

The presence of isoenzymes of α-galactosidase and α-mannosidase in men or animals has not been described until now. Our data seem to indicate that the existence of multiple forms is a general feature of the lysosomal enzymes. The physiological meaning of this is not quite clear at the moment: probably each isoenzyme has a more restricted substrate specificity, different from the other isoenzymes. Another possibility may be that not each isoenzyme is located in the same subcellular fraction. It has been reported that there is, besides a lysosomal α-mannosidase, also a non-lysosomal fraction of this enzyme which exists in rat liver (4).

According to the evidence presented in this paper, the molecular differences between the isoenzymes may, at least partly, be ascribed to the presence of neuraminic acid. For α-galactosidase, α-mannosidase and acid phosphatase one fraction is converted into another existing fraction by releasing neuraminic acid. For β-galactosidase this does not seem to be so simple since a new fraction appears after the neuraminidase treatment. And for N-acetyl-β-hexosaminidase only a minor change was seen. This may however not be regarded as an evidence for the absence of neuraminic acid: depending on the conformational position of the bound sialic acid, neuraminidase can not always reach the bond to be hydrolysed. In this respect the source of the neuraminidase may probably also be of significance. The acid hydrolases are thus to be considered as sialic acid containing glycoproteins, a conclusion also found by Goldstone and Koenig after chemical analysis of purified lysosomal proteins (5). The presence of multiple forms for α-galactosidase and α-mannosidase makes it very likely that variants of mannosidosis and of Fabry's disease will be found, just as this was the case in Tay-Sachs' disease and it is recommendable that in patients suspected of having one of these 2 diseases isoenzyme analysis should be carried out, e. g. in a homogenate of isolated white blood cells.

The coupling of the neuraminic acid onto the enzyme glycoprotein proceeds through the action of sialyltransferases. Therefore a genetic defect of such sialyltransferase may lead to the presence of hydrolases which may not have the required specificity. When such an enzyme disturbance will be found, this has to be called a "secondary enzyme defect".

Finally it must be kept in mind that in normal liver and brain lysosomal neuraminidase is also present. If this neuraminidase is capable of cleaving sialic acid from the other hydrolases, and possibly thereby changing the substrate specificity, this neuraminidase will turn out to be a regulating enzyme in lysosomal metabolism. After terminating our experiments a recent paper of Goldstone et al. (6) came to our attention in which these authors reported an effect of neuraminidase on lysosomal enzymes, similar to our results. However no data on α-galactosidase and α-mannosidase were presented nor did these authors report any examination on the difference between brain and liver isoenzymes.

REFERENCES

(1) Bergmeyer, H. U., Methods of Enzymatic Analysis, Academic Press - New York, p. 945 (1965).

(2) Okada, S. and O'Brien, J. S., Science 165: 698 (1969).

(3) Sandhoff, K., FEBS Letters 4: 351 (1969).

(4) Marsh, C. A. and Gourlay, G. C., Biochim. Biophys. Acta 235: 142 (1971).

(5) Goldstone, A. and Koenig, H., Life Sciences 9: 1341, II (1970).

(6) Goldstone, A., Konecny, P. and Koenig, H., FEBS Letters 13: 68 (1971).

PROBLEMS IN THE CHEMICAL DIAGNOSIS OF GLYCOPROTEIN STORAGE DISEASES

P.A. ÖCKERMAN, N.E. NORDÉN

Dept. of Clinical Chemistry, University Hospital
Lund, Sweden

and L. SZABÓ

Dept. of Pediatrics, University Hospital
Szeged, Hungary

In the present paper we shall discuss the only two diseases of glycoprotein storage that have so far been well delineated, mannosidosis and aspartylglucosaminuria.

In order to diagnose chemically the glycoprotein storage diseases you can measure and characterize substances that are stored in the tissues and excreted in the urine and you can measure the activity of the deficient enzymes. Although we shall discuss only chemical problems we want to stress the fact that a very thorough clinical, histological, and perhaps also electron microscopical study is of utmost importance for a correct diagnosis.

The material stored in mannosidosis consists of oligosaccharides with eight, five, and three mannose residues and one glucosamine residue, but no amino acids. The details of the structure of these oligosaccharides are presently unknown, mainly because so far not enough material has been available for a correct characterization. It has not been excluded that also other larger mannose--rich molecules or charged molecules are also stored, but the small uncharged mannose-rich oligosaccharides represent the main storage material. There does not exist any increase of fat soluble mannose-rich substances nor of glycosaminoglycans or glycolipids.

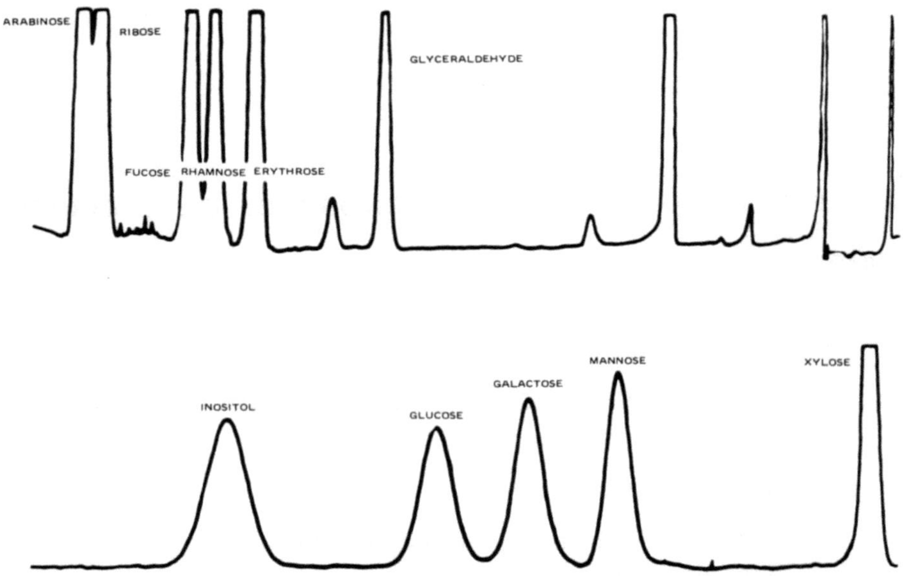

Fig. 1. Gas-liquid chromatography of standard sugars.

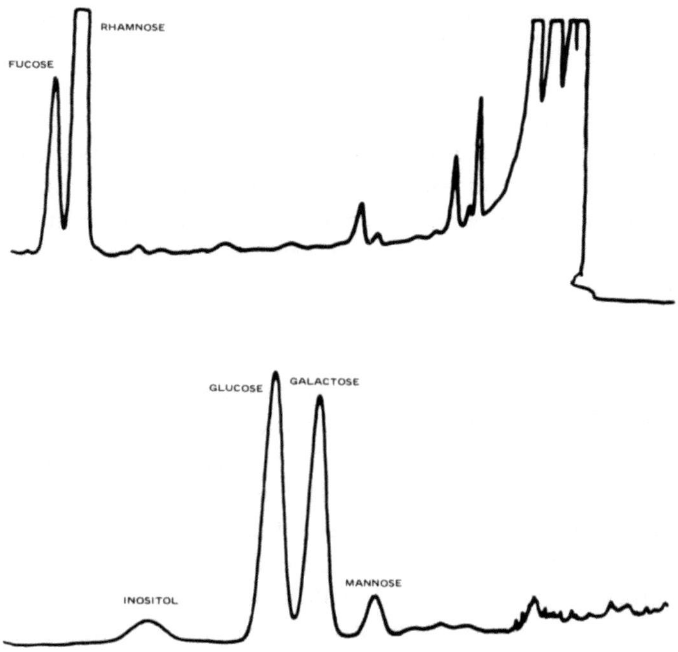

Fig. 2. Gas-liquid chromatography of hydrolysate of urinary oligosaccharides isolated by gel filtration of Sephadex G-25. Internal standard: rhamnose.

Various techniques can be used to isolate and quantitate the stored mannosides.

For assay of hydrolysable hexoses in tissues a lipid-free extract was prepared and hydrolyzed in normal sulphuric acid for three hours. The hydrolysate was neutralized and then used for direct assay of the hexoses using an ion-exchange chromatography system or for assay by gas-liquid chromatography of the alditol acetate derivatives of the sugars. For gas-liquid chromatography a column containing 1.5% ethyleneglycolsuccinate and 1.5% silicone oil (XF-1150) on 100-120 mesh Gas-chrome P was used. Fig. 1-2 show results with the gas-liquid chromatography method, when temperature programming is used.

It can be seen that there is a very good separation of many sugars with this technique. In fact the separation is better than that obtained with the ion-exchange method. The gas-liquid chromatography technique is also much more sensitive and has a higher capacity to analyse many samples in a short time. Preparing alditol acetates takes a little more time than making trimethylsilyl (TMS) derivatives of the sugars, which is also possible for gas--liquid chromatography. Using the TMS derivatives, however, gives much greater difficulties in the interpretation of the curves, since more than one derivative is obtained for each sugar. Using the alditol acetates, you must be sure that there is no fructose in your mixture, since fructose when reduced gives mannitol and sorbitol in equivalent amounts and, therefore, falsely high values for mannose and glucose can be obtained. If the tissues contain preformed alditols these would of course be measured at the same time. This does also mean that the method can be used for quantitative assay of preformed alditols, if the reduction step with borohydride is not performed.

Fig. 3 shows the very markedly increased levels of hydrolysable mannose in liver and brain in a patient with mannosidosis.

The oligosaccharides stored in the tissues are extremely water soluble. Therefore, it is reasonable to believe that the oligosaccharides are excreted in increased amounts in the urine. To assay them there quantitatively might therefore be of great importance. Mannose can be correctly quantitated by concentrating urine, hydrolysing it, neutralising the hydrolysate and then assaying hexoses by ion-

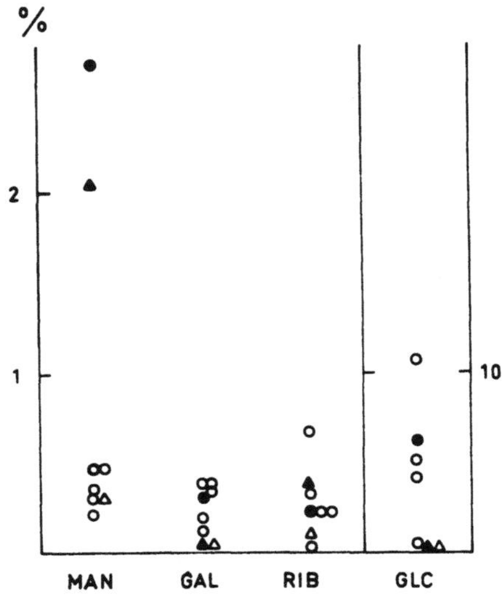

Fig. 3. Sugar levels in brain and liver. Fat-free tissue extracts were hydrolysed and assayed by ion-exchange chromatography. Values in per cent of fat-free dry weight. ● liver, mannosidosis (case from Lund); O liver, controls; ▲ brain, mannosidosis (case from Lund); △ brain, control. Man = mannose; gal = galactose; rib = ribose; glc = glucose.

-exchange chromatography. We have assayed by this technique the amount of free and oligosaccharide-bound mannose in various fractions from normal urine obtained by gel filtration on Sephadex G-25. It seems probable that an increase of mannose-rich oligosaccharides should really appear by this technique, even without prior gel filtration, and not be drowned in the normal variation of free mannose. Of course, the assay can be made both with and without hydrolysis and thereby more certain quantitation of oligosaccharide-bound mannose be obtained. However, the ion-exchange technique is not very simple. The commercially available instruments are easily operated, but fairly expensive and only one assay can be made each day. The same technique can be used without high pressure equipment, but in practice this technique is extremely difficult.

The technique we are working with presently involves concentration of the urine and gel filtration on Sephadex G-25 on a fairly large column. All fractions that contain disaccharides and mono-

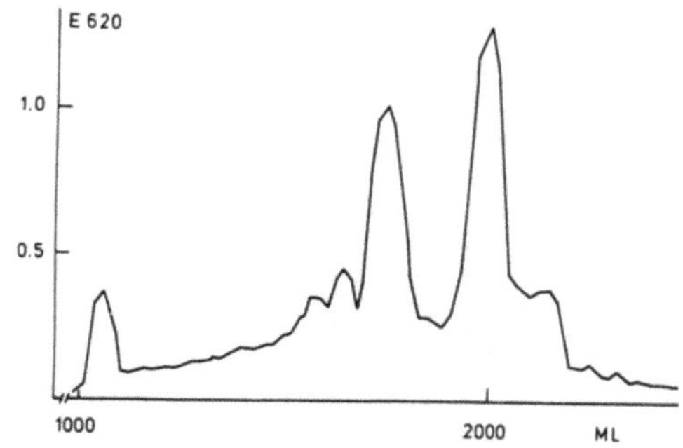

Fig. 4. Gel filtration of 100 ml of normal urine on Sephadex G-25, 5 x 125 cm. Assay of hexoses by the anthrone method.

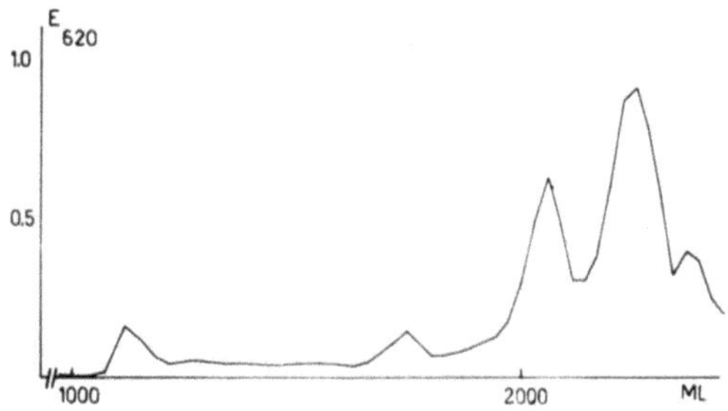

Fig. 5. Gel filtration of urine in mannosidosis (case from Szeged). Technique as in Fig. 4.

saccharides are discarded and the fractions containing trisaccharides and larger molecules are concentrated, hydrolysed, neutralised, reduced by borohydride and acetylated and thereafter analysed by gas-liquid chromatography. Fig. 4 shows the normal pattern of hexoses in the fractions after gel filtration.

In the only patient with mannosidosis so far studied in this respect a pathological excretion of oligosaccharides was visible after gel filtration (Fig. 5).

Gas-liquid chromatography of the hydrolysed oligosaccharide fraction demonstrated clearly a large increase of mannose (Fig. 6).

In summary the mannose-rich oligosaccharides in mannosidosis are stored in the tissues and can there be assayed on surgical biopsy or autopsy material. The same assay can be made on urine. Unfortunately, the technique is in neither case simple and should, therefore, be used only in cases that have been well studied clinically and where mannosidosis is a probable or at least very possible diagnosis. Very probably assay of the mannose-rich oligosaccharides is necessary to obtain a definite diagnosis of mannosidosis.

Diagnosing chemically aspartylglucosaminuria is technically much simpler than diagnosing mannosidosis. It is possible today to state how this disease should be diagnosed by relatively simple means due to the work that has been made in Finland. Aspartyl-

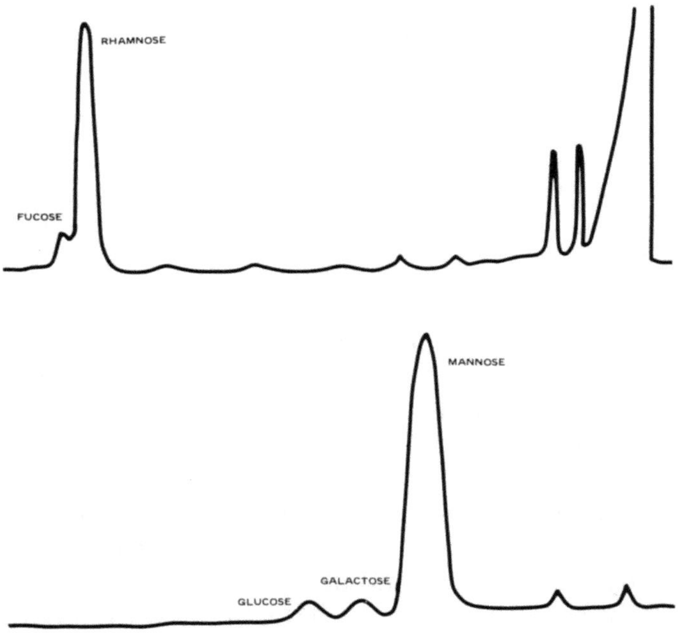

Fig. 6. Gas-liquid chromatography of the hydrolysed oligosaccharide fraction from the experiment shown in Fig. 5. The pattern should be compared to that in Fig. 2. Internal standard: rhamnose.

glucosamine is reacting with ninhydrine and can therefore be assayed easily by the ordinary techniques for amino acids in the urine. We are presently working with high-voltage electrophoresis at pH 2. This technique is very simple and has a large capacity. Fig. 7 shows the results with high-voltage electrophoresis.

Aspartylglucosamine has characteristic staining properties. It rapidly turns grey-brown and after heating at 110° changes in colour and turns more grey-blue. Unfortunately, the interpretation of the chromatogram is not always very simple due to the fact that metabolites of semi-synthetic penicillines, the ampicillins, give a grey-brown band with a mobility very similar to that of aspartylglucosamine. In fact, the ampicillins do not have exactly the same mobility, but in practice cannot always be separated from aspartylglucosamine. Sometimes, the ampicillins give more than one band. However, the metabolites of ampicillin turn brown more slowly than aspartylglucosamine, not until after about 24 hours,

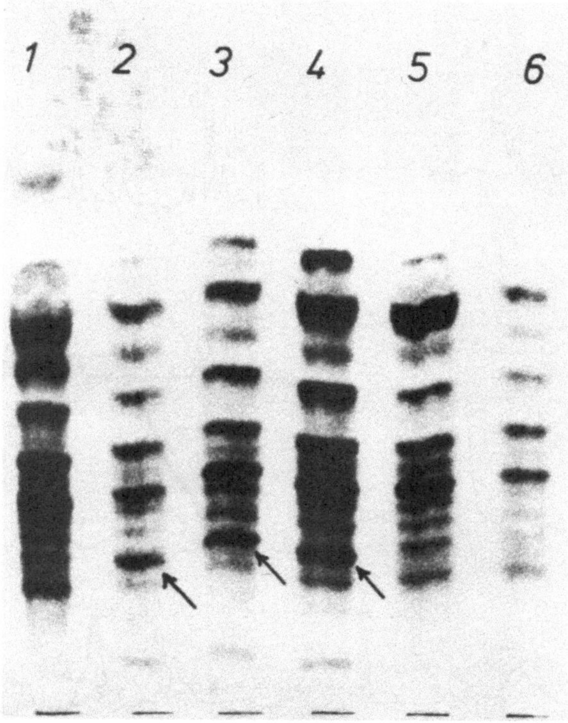

Fig. 7. High-voltage electroporesis of urinary amino acids. No heat treatment. 1 = control; 2-4 = aspartylglucosaminuria (three different cases); 5-6 = controls. Aspartylglucosamine bands indicated by arrows. The colour of the aspartylglucosamine bands was grey-brown one hour after ninhydrin treatment. The urines were kindly provided by prof. J. K. Visakorpi, Helsinki.

and during the first hours the colour is violet. On heating, the ampicillin derivatives do not change colour as clearly as aspartylglucosamine. If there is any doubt as to the identity of the brown band found, it is possible to get a very clear identification by making ion-exchange chromatography or by making high-voltage electrophoresis on a new sample, when the patient is no longer on treatment with ampicillins. In fact, we suggest that a new sample should always be analysed because of the possibility that a patient with aspartylglucosaminuria may be treated with ampicillins and then the interpretation may be difficult. Of course, also in that case ion-exchange chromatography will solve the problems since ampicillin comes out in a very different position to that of aspartylglucosamine (Fig. 8-9).

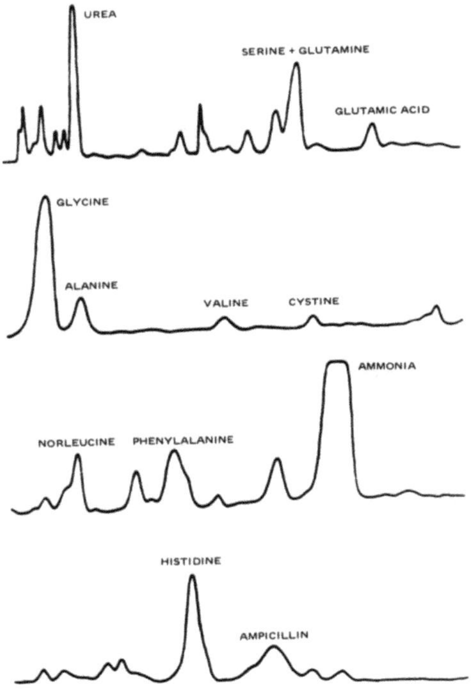

Fig. 8. Ion-exchange chromatography or urinary amino acids (Technicon Amino-Acid Analyzer). Aspartylglucosamine (AGU) appears early in the chromatogram as a small peak immediately after urea.

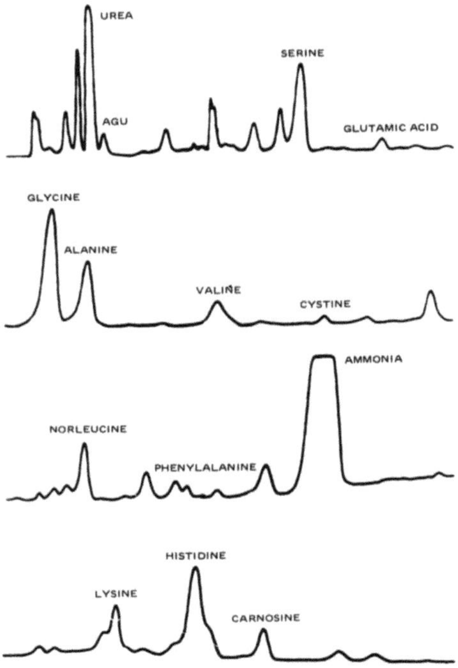

Fig. 9. Urinary amino acids. Ampicillin appears as a broad peak late in the chromatogram after histidine.

There has been found about 40 patients with aspartylglucosaminuria in Finland, and two cases in Great Britain. I have made a screening in Sweden and assayed more than 500 urines with high-voltage electrophoresis. These urines were from patients that had been referred for assay of glycosaminoglycans because they had clinical symptoms and signs that might fit in with one of mucopolysaccharidoses. Many of these patients have symptoms and signs well in accordance with those seen in aspartylglucosaminuria. If aspartylglucosaminuria is as frequent in Sweden as it is in Finland we should have found several cases in our very selected material, but in fact we have not yet seen any definite case of aspartylglucosaminuria. It, therefore, seems probable that the disease is much less common in Sweden than in Finland.

We shall now finally discuss the diagnosis of glycoprotein storage diseases by enzyme assays, and we shall only discuss mannosidosis since we have no personal experience with enzyme assays in aspartylglucosaminuria, and because the technique has been well described in the literature. We have used three different substrates for assay of α-mannosidase, namely para-nitrophenyl-mannoside, methylumbelliferyl-mannoside, and an oligosaccharide containing 5 mannose molecules and one glucosamine molecule, isolated from the tissues of a mannosidosis patient. With all these three substrates a decreased activity of α-mannosidase was found in liver, brain, and spleen from the patient from Lund with mannosidosis as compared to controls (Fig. 10).

Howewer, a certain rest activity was found. A few other lysosomal enzymes had increased, sometimes markedly increased, activities. The decrease of the α-mannosidase activity, therefore, becomes more marked if its activity is related to that of other glycosidases and I believe that in future cases of mannosidosis the activity of α-mannosidase should always be put in relation to that of other glycosidases. In a new case from Szeged, α-mannosidase activity was zero in the liver biopsy, while other lysosomal enzymes were increased.

For assay in future cases liver biopsy should be used. α-mannosidase can be assayed with the extremely sensitive methylumbelliferyl method. In practice fine-needle aspiration biopsy using a 0.7 mm outer diameter needle is sufficient for assay of α-mannosidase, β-galactosidase, β-glucosidase, hexosaminidase, β-glucuronidase, and other lysosomal enzymes, as well as protein on as little as 1-2 mg of tissue. Using such a fine needle the risk to the patient is very small and there is very little reason to hesitate to make this biopsy. However, a surgical biopsy of larger needle biopsy is better for the chemist, since then he can save a bit of the biopsy for future assays and for controls. Also oligosaccharide analyses can be made, if a surgical biopsy has been taken. It would, of course, be preferable to assay the enzymes on some material that is more easy to obtain, for instance plasma, blood cells, and urine, α-mannosidase assay on these materials is very simple, at least concerning blood cells and

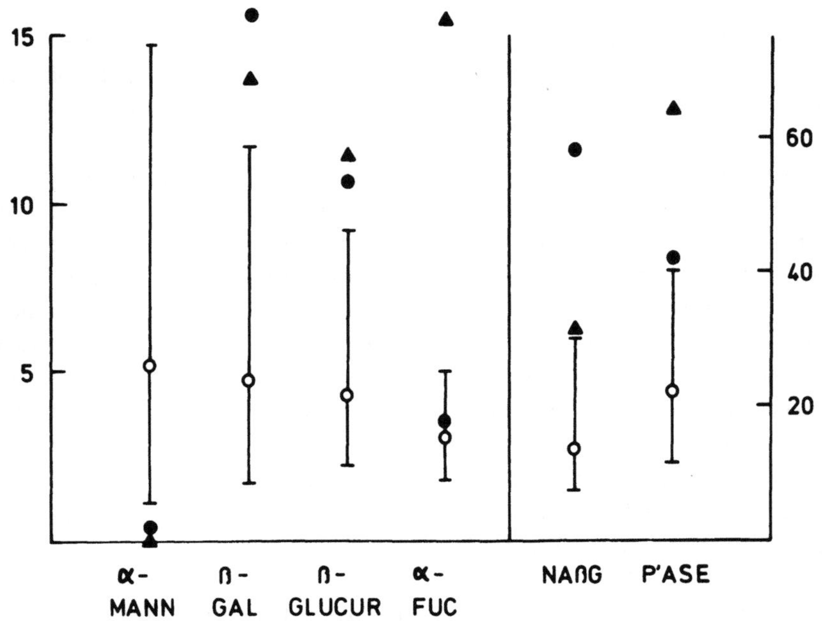

Fig. 10. Activities of acid hydrolases in human liver. All values in IU/g protein. Note different scale for N-acetyl--β-glucosaminidase and acid phosphatase (indicated to the right). α-MANN = α-mannosidase; β-GAL = β-galactosidase; β-GLUCUR = β-glucuronidase; α-FUC = α-fucosidase; NAβG = N-acetyl-β-glucosaminidase; P'ASE = acid phosphatase; O = controls; ● = mannosidosis. Case from Lund; ▲ = mannosidosis. Case from Szeged.

plasma, which can be directly used for the assay. Concerning urine, assay should be made on urine that has been diluted with two volumes of water rather than on urine that has been concentrated by dialysis and lyophilization or by batch treatment with Sephadex or similar techniques, because assay on diluted urine gives higher activity. There is, however, a rather large variation normally of the activity found in urine, plasma, and blood cells. For these reasons, and because of the fact that in the liver there may be a rest activity, it is quite possible that a decrease of the activity in these materials will not be possible to separate definitely from the normal range. Also, in Hurler's disease, where there is a decreased activity of β-galactosidase in the tissues, there is only a very slight decrease of the activity in urine, overlapping the values found in controls. In plasma β-galactosidase activity is increased in Hurler's disease and in white blood cells there is no definite decrease. Therefore, in analogy to Hurler's disease it is quite possible that in mannosidosis there is no measurable decrease of α-mannosidase in urine, plasma or white blood cells and no conclusions can be drawn from such assays, until the value has been tested in definite cases of mannosidosis.

In conclusion mannosidosis should be diagnosed chemically by enzyme assays on liver biopsy and by assay of hydrolysable mannose in tissues and urine. Aspartylglucosaminuria can most easily be diagnosed by amino acid assay in the urine, for instance by high-voltage electrophoresis.

PROGRESS AND PROBLEMS ON FUCOSIDOSIS AND MUCOLIPIDOSIS

P. DURAND

Third Department of Pediatrics, "G. Gaslini" Institute

Genova-Quarto, Italy

Various descriptive names have been given for some years to a group of storage disorders with numerous clinical and pathological features, common to both, the glycolipidoses and mucopolysaccharidoses.

After the discovery of the GM_1 gangliosidoses type I and II (1, 2, 3, 4) and the variant of methachromatic leucodystrophy with mucopolysaccharidúria (5, 6), inborn errors of fucose (7, 8, 9, 10) and mannose metabolism (11, 12) have been described. Other disorders, lipomucopolysaccharidosis or mucolipidosis type I (13, 14, 15, 16) and I-Cell-Disease or mucolipidosis type II (17), were studied. However the biochemical and enzymatic findings of these pathological conditions are, as yet, incomplete and further studies are necessary.

I will assess the clinical, histopathological and biochemical data of fucosidosis and mucolipidosis.

FUCOSIDOSIS

Fucosidosis is a fatal disease with increased accumulation of fucose containing substances, especially fucolipids and fucopolysaccharides, in neural and visceral tissues, due to a congenital defect in specific lysosomal acid hydrolase, α-L-fucosidase. It is trasmitted as an autosomal recessive trait. After our two first cases, four others have been observed (one, Loeb et al., 15, 16; one, Voelz et al., 18; two Donnell, 19).

A family investigation, conducted by Della Cella et al. (20) shows that two other patients in the family studied by us are affected by a similar disease.

Onset of clinical symptoms is at age of 10-16 months.

Clinical features are frequent respiratory tract infections, including bronchopneumonia, followed by a mental and motor deterioration, progressing rapidly after one year of age; hypotonia followed by spastic quadriplegia and a final state of decerebrate rigidity; emaciation; cardiomegaly with signs of myocarditis; thick skin with abundant sweating. The chloride and sodium content of sweat increased in our cases, is normal in other patients; there is also a progressive impairement of gallbladder function. Macrocephaly may develop with frontal bossing; coarsening of facial features or mild gargoylelike characteristic may be present. Dorsolumbar kyphoscoliosis occurs and roentgenographic examination shows hypoplasia and beaking of lumbar vertebrae. Other bony deformities, resembling those seen in mucopolysaccharidoses, may be present.

Late in the disease, tube feeding is usually necessary and the patient is unresponsive to stimuli.

Death occurs within the first decade, usually at about 4-5 years.

Pathologic findings. The essential pathologic feature of fucosidosis is represented by cellular vacuolization which was found in peripheral lymphocytes, in viscera and in neural tissues.

Hepatocytes and also Kupfer cells are distended by abundant stored substance, producing a clear, swollen appearance, weakly PAS positive and, at a later stage, Sudan positive.

The neurophatological process can be summarized as a combination of a neuro-glial dystrophy, in the form of ballooned cells and a dysmyelination and a pallidal lipophanerosis (21).

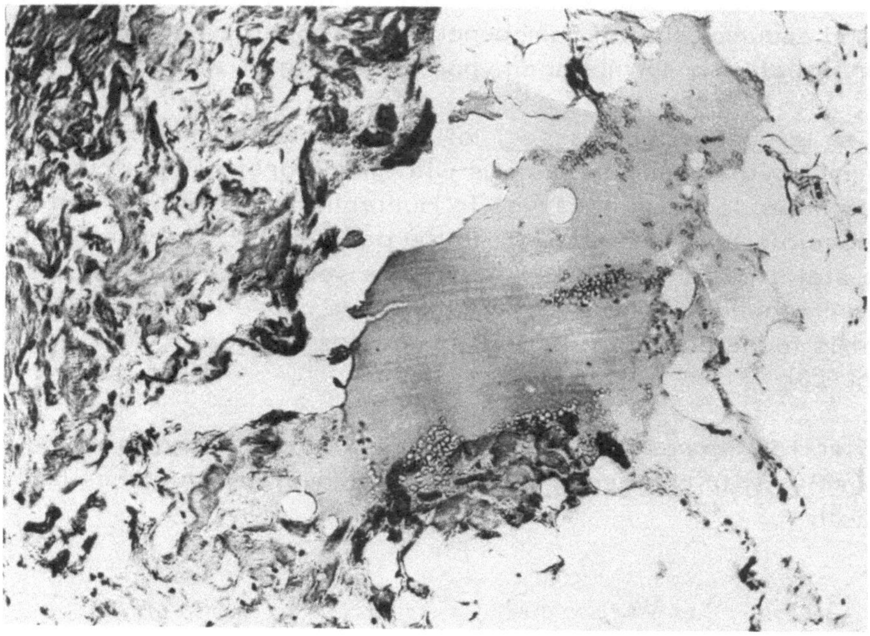

Fig. 1. Skin biopsy in fucosidosis. Note a diffuse connective tissue abnormality and anomalous accumulation of deposits of homogeneous, eosinophilic material between the dermis and hypodermis.

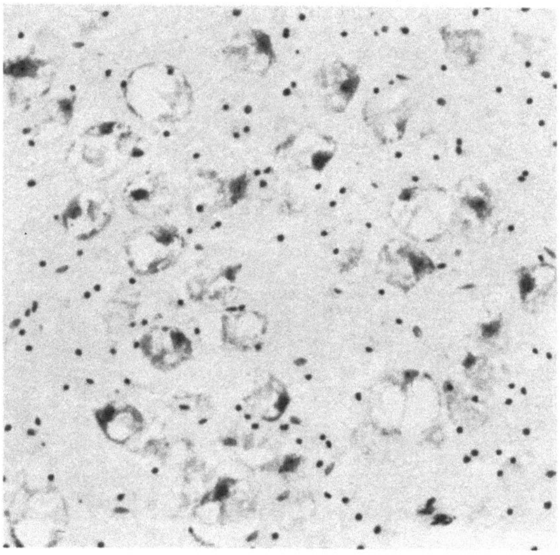

Fig. 2. Cerebral cortex showing cytoplasm ballooning and pallor, granularity and vacuolation.

Skin showed in our case a diffuse connective tissue abnormality and accumulation of homogenous, eosinophilic, picrophilic material between dermis and hypodermis (Fig. 1 and 2).

The ultrastructural aspect of the liver is characterized by abnormal cytoplasmic inclusions with granuloreticular structures, similar to those observed in Hurler's syndrome (but which are possible to differentiate with the periodic-acid-argentic-nitrate stain, 22) and lamellar structures some of which seem as smaller spherules with clear centre and a dense limit. This material is found either in the large vacuoles or free in the cytoplasm (23).

Electron microscopy studies of the brain have shown round corpuscles with enclosed reticular and lamellar structures (16) (Fig. 3).

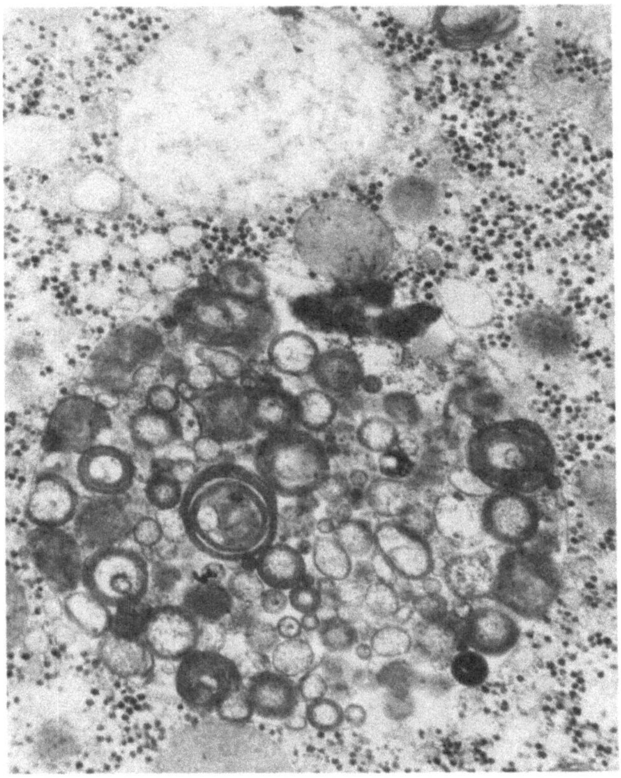

Fig. 3. Ultrastructural photograph of hepatic parenchimal cells in a case of fucosidosis. A vacuole containing numerous lamellar inclusions and above it a clear vacuole. In the upper right corner is a free cytoplasmic lamellar body.

Chemistry of neurovisceral storage substances. Increased amounts of fucose-containing ceramide tetra- and penta-hexosides and of fucose containing polysaccharides were detected in liver and brain. These polysaccharides, with low molecular weight and only small amounts of aminoacids, probably arise in the brain from glycoproteins, that may be, according to Brunngraber (24), the only constituents of neural tissue which contain fucose.

The concentration of the glycolipids in the liver, found in one of our cases, was of 8 μmoles/gram wet weight and in the Spranger's case was of 2,8 μmoles/gram wet weight (25). The concentration of glycolipids in grey matter was of 22 micrograms/100 mg wet weight and in white matter of 44 micrograms/100 mg wet weight.

The concentration of polysaccharide fraction found in our case was 250 micrograms/100 mg wet weight of liver and of 2 micrograms/100 mg wet weight in grey and in white matter (26). These amounts are markedly more elevated than those found in normal tissues.

Dawson and Spranger (25) studying the glycolipid material accumulated in the liver of a Spranger's case of fucosidosis, confirmed our findings of accumulation in the liver of fucose-containing ceramide tetra- and penta-hexosides, and anticipated that other fucolipids will accumulate in fucosidosis. After methanolysis and gas-chromatographic analysis of the trimethylsilyl methylglycosides, the glycosphingolipid was found to contain fucose, galactose, glucose, N-acetylglucosamine in the molar ratio 1:2:1:1:1.

The mucopolysaccharide concentration in the viscera does not appear significantly increased.

Philippart (27) has shown that the storage substances exibit H and Lewis antigenic activity. He was able to fully characterize one of the stored glycosphingolipids which is similar but antigenically different from the known Lewis antigen.

The enzymic defect is specific for fucosidosis. The complete absence of α-fucosidase occurs in all tissues examined, liver, brain, lung, kidney and pancreas (23) and also in urine, serum, leukocytes and fibroblasts (10, 28, 29).

Activity of this enzyme is detectable in cultured amniotic fluid cells and renders this disorder diagnosable also before birth.

The use of urine, serum, leukocytes and fibroblasts assays should make it possible to diagnose fucosidosis simply during life and to differentiate it from mucopolysaccharidoses, mucolipidoses, mannosidosis, generalized gangliosidosis and juvenile GM1 gangliosidosis.

Urinary fucosidase varies in normal subjects between 0.5-1 nanomoles of cleaved substrate per mg protein per minute at 37°C, and, if there is an abnormal proteinuria, the activity may be expressed in relationship to the urinary creatinine.

The urinary α-L-fucosidase was absent in Loeb's and Voelz's patients (15, 18).

Fucosidase in the leukocytes varies between 0.1-0.45 nanomoles of cleaved substrate per mg protein per minute at 37°C and pH 5.5.

According to Zielke et al. (29) serum α-L-fucosidase in normal subjects averaged 488.6 ± 153.2 nanomoles of cleaved substrate per ml of serum per hour at 37°C. Three cases of fucosidosis (studied by Spranger and by Donnell) had no α-L-fucosidase activity in serum (Zielke et al., 29). Activities of other lysosomal acid hydrolases were increased in all tissues examined (acid α-galactosidase and β-xylosidase, 7-8 fold normal; acid α- and β-glucosidase, 3-4 fold normal) or within the range of control values (Van Hoof and Hers, 10).

<u>The presumed developmental pathology of the disease</u> is: absence of α-L-fucosidase, hence, inability to cleave the fucose from glycolipids. Since many glycoproteins, especially of the brain, contain α-fucoside linkages, it is possible that their degradation is affected in this disease. It can be assumed that the fucose containing polysaccharide storage may be due to a block in the degradation of glycoproteins, which have their protein moiety digested by lysosomal cathepsines (Fig. 4). All subsequent reactions could not take place unless fucose is first removed.

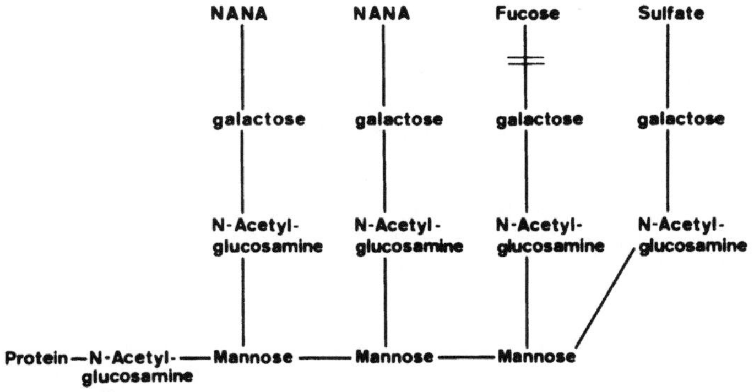

Fig. 4. Schematic representation of the structure of the polysaccharide moiety of a cerebral glycoprotein and metabolic block in fucosidosis.

MUCOLIPIDOSES

Mucolipidoses are diseases with progressive psychomotor retardation, progressive changes of facial and body configuration and skeletal abnormalities resembling Hurler's syndrome but without excessive excretion of urinary mucopolysaccharides.

In the last three years some authors (10, 18, 25, 30, 31, 32) have tried to determine the clinical course and the cytological, ultrastructural, biochemical and enzymatic features of mucolipidoses. This term, however, could create confusion and obviously will have to be modified in the future.

The initial description of lipomucopolysaccharidosis or mucolipidosis type I was that of Spranger et al. (13) and Loeb et al. (15, 16). At least nine patients with a similar disorder have been reported. The patients were mentally retarded and presented gargoyle-like features, mild skeletal abnormalities, vacuolized peripheral lymphocytes and other cytological features similar to the Hurler's syndrome but without hypermucopolysacchariduria. The hepatic ultrastructure shows, in some cases, beyond a finely granuloreticular material, a few osmiophilic spherules and rare myelin-like structures. The existence of mild and more severe forms of the disease may indicate a genetic heterogenicity but a different defect cannot be excluded (Fig. 5).

The activity of 9-13 lysosomal enzymes was increased and especially that of acid β-galactosidase, so that Van Hoof and Hers

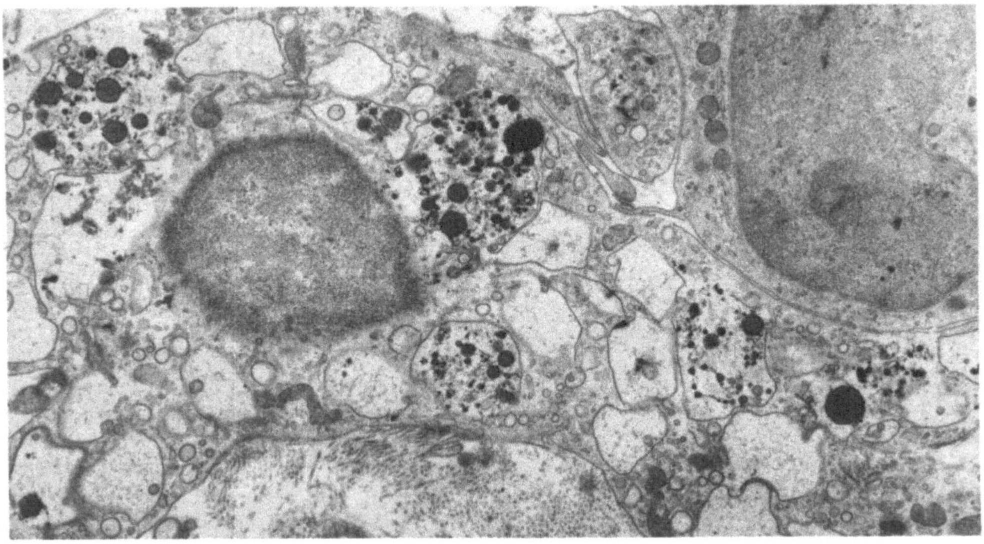

Fig. 5. Ultrastructural fotograph of fibroblasts in a patient with a disorder similar to mucolipidosis I. Vacuoles containing finely granuloreticular material and a few osmiophilic spherules.

designated these cases as mucopolysaccharidosis Gal+. In other patients however a partial defect of mannosidase was reported (31).

Comparison with the moderate deficiency of β-galactosidase activity in a considerable number of mucopolysaccharidosis, even if the acceptance of this enzymic defect, as the basic expression of the critical error,is, as yet, incomplete, should differentiate these disorders. The heterogeneity of clinical features and enzymic data, and the still incomplete informations on pathologic and chemical findings do not permit any definitive conclusion.

It is possible, on the basis of personal experience, that we are in the presence of a few different types of entities.

<u>I-cell disease or mucolipidosis type II</u> was detected by Leroy and De Mars (17). At least sixteen patients are known to suffer from this disorder with autosomal recessive inheritance and characterized by early psychomotor retardation, severe shortness of stature and slowly progressive changes of facial and body configu-

ration. The patients have a gargoyle-like appearance, gingival hyperplasia with hypertrophy of alveolar ridges, no hepatosplenomegaly, skeletal deformities, present soon after birth and inguinal herniae.

In the second year of life and subsequently, the clinical findings and especially the mental situation becomes worse. Ophtalmologic examination is usually negative. Death occurs at about the age of 8 years or before from cardiac failure.

There is no excessive excretion of urinary mucopolysaccharides. Lymphocytes in the peripheral blood and cells in smears of bone marrow showed vacuolization, without metachromatic granules.

The disease has received the name of I-cell because cultured fibroblasts, derived from the patient skin, contain coarse and dark, non metachromatic cytoplasmic granulations under microscopic phase-contrast observation.

Electron microscopy studies have shown that the inclusions are altered lysosomes. Light and electron microscopy have demonstrated that the inclusions were PAS-positive, oil red O positive, metachromatic and osmiophilic (33).

Chemical analysis of the storage material in fibroblasts reveals a marked elevation (three times higher than the one in control fibroblasts) of lipids, part of which have chromatographic characteristic of glucosylceramide, trihexosylceramide, globoside, hematoside and GD_3 ganglioside (34).

The mucopolysaccharide content of I-cell type is elevated or normal. Multiple lysosomal enzyme deficiency is detected in cultured fibroblasts especially β-glucuronidase, β-galactosidase, α-mannosidase, α-fucosidase, arylsulphatase and β-N-acetylhexosaminidase (28, 32). The β-D-glucosidase activity is normal or increased and the acid phosphatase activity is markedly elevated (35). However there is a discrepancy between the findings in I-cells and the ones obtained in autopsy and biopsy specimens of visceral organs; only β-D-galactosidase is partially deficient in liver and brain.

Wiesmann et al. (32) found normal cytoplasmic and mitochondrial enzyme activities (malic dehydrogenase and lactic dehydrogenase) and significantly elevated enzyme activities in the medium in which the fibroblasts of a patient were cultured for 3 days, as compared to the medium in which normal cells were cultured.

Multiple lysosomal enzyme deficiency can be explained, by an increase leakage (or exocytosis) of lysosomal enzymes from the patient's fibroblasts into the medium. This phenomenon, not generalized, supports the working hypothesis that I-cell disease is due to a structural defect in lysosomal membrane.

REFERENCES

(1) Landing, B.H., Silverman, F.N., Craig, J.M., Jacoby, M. D., Lahey, M.E. and Chadwick, D.L., Am. J. Dis. Child. 108: 503 (1964).
(2) O'Brien, J.S., Stern, M.B., Landing, B.H., O'Brien, J.K. and Donnell, G.N., Am. J. Dis. Child. 109: 338 (1965).
(3) Derry, D.M., Fowcett, J.S., Andermann, F. and Wolfe, L. S., Neurology 18: 340 (1968).
(4) O'Brien, J.S., Okada, S., Ho, M.W., Fillerup, D.L., Weath, M.L. and Adams, K., Fed. Proc. 30: 956 (1971).
(5) Austin, J.H., in "Medical Aspects of Mental Retardation", Ed. C.C. Carter, Springfield Ill., C.C. Thomas, p. 768 (1965).
(6) Austin, J.H., Armstrong, D. and Shearer, L., Arch. Neurol. 13: 593 (1965).
(7) Durand, P., Borrone, C. and Della Cella, G., Lancet II: 1313 (1966).
(8) Durand, P., Borrone, C. and Della Cella, G., J. Pediat. 75: 665 (1969).
(9) Durand, P., Borrone, C., Della Cella, G. and Philippart, M., Lancet I: 1198 (1968).
(10) Van Hoof, F. and Hers, H., Europ. J. Biochem. 7: 34 (1968).
(11) Öckerman, P.A., Lancet II: 239 (1967).
(12) Öckerman, P.A., J. Pediat. 75: 361 (1969).
(13) Spranger, J., Wiedemann, H.R., Tolksdorf, M., Graucob, E. and Caesar, R., Z. Kinderheilk. 103: 285 (1968).
(14) Spranger, J.W. and Wiedemann, H.R., Lancet II: 270 (1969).
(15) Loeb, H., Tondeur, M., Jonniaux, G., Mockel-Pohl, S. and Vamos-Hurwitz, E., Helv. Paediat. Acta 24: 519 (1969).
(16) Loeb, H., Tondeur, M., Teppel, M. and Cremer, N., Acta Paediat. Scand. 58: 220 (1969).

(17) Leroy, J. G. and De Mars, R. J., Science 157: 804 (1967).
(18) Voelz, C., Tolksdorf, M., Freitag, F. and Spranger, J., Mschr. Kinderheilk. 119: 352 (1971).
(19) Donnell, G. et al., in preparation.
(20) Della Cella, G., Borrone, C. and Durand, P., unpublished data.
(21) Bugiani, O., Palladini, C. and Durand, P., in preparation.
(22) Tondeur, M., personal communication.
(23) Van Hoof, F. and Hers, H., Rev. Int. Hépat. 17: 815 (1967).
(24) Brunngraber, E. G., in "Protein Metabolism in the Central Nervous System", Ed. A. Lajtha, Plenum Press, New York, p. 383 (1970).
(25) Dawson, G. and Spranger J. W., New Engl. J. Med. 285: 122 (1971).
(26) Philippart, M., personal communication.
(27) Philippart, M., Neurology 19: 304 (1969).
(28) Durand, P. and Borrone, C., Helv. Paediat. Acta 26: 19 (1971).
(29) Zielke, K., Okada, S. and O'Brien, J. S., in press.
(30) Leroy, J. G. and Spranger, J. W., New Engl. J. Med. 283: 598 (1970).
(31) Loeb, H., Tondeur, M. and Vamos-Hurwitz, personal communication.
(32) Wiesmann, U. N., Lightbody, J., Vasella, F. and Herschkowitz, N. N., New Engl. J. Med. 284: 109 (1971).
(33) Hanai, J., Leroy, J. G. and O'Brien, J. S., Am. J. Dis. Child. 122: 34 (1971).
(34) Dawson, G., Matalon, R. and Dorfman, A., in "Abstracts 3th Int. Meeting Int. Society of Neurochem.", Ed. J. Domonkos et al., Budapest, Akademiai Kiadó, p. 344 (1971).
(35) Leroy, J. G., Ho, M. W. and O'Brien, J. S., in "Abstracts 4th Int. Congress of Human Genetics", Ed. De Grouchy et al., Amsterdam, Excerpta Medica, p. 111 (1971).

EARLY DIAGNOSIS OF GLYCOLIPIDOSIS

I. BENSAUDE

Centre d'Etudes sur les Maladies du Métabolisme

chez l'Enfant, Hôpital des Enfantes Malades

Paris, France

Rapid advances of the in vitro techniques, plus rapid advances in the knowledge of metabolic diseases, mostly inborn errors of metabolism, made possible the diagnosis of an increasing number of diseases. The last in vitro technique to be used is the culture of amniotic cells obtained by amniocentesis; we would like to present a brief resumé of what has been done until now and investigate some of the major problems inherent to that method.

Amniotic Fluid

The origin of amniotic fluid is still very unclear, it seems to come at the beginning of pregnancy from both mother and fetus (1). At least, it is known that a term fetus swallows around 450 ml per day and urinates about the same amount. It has also been shown by Gluck et al. (2), that the fetus spits a specific lecithin coming from the fetal lungs, with a specific β-palmitic acid, and found in the amniotic fluid.

The cells found after centrifugation in the amniotic fluid are epitheloid or fibroblast-like. They all come from the fetus (3) and derive mostly from the desquamation of the skin, but also from the bronchi, the gastro-intestinal tract and the genitourinary-tract.

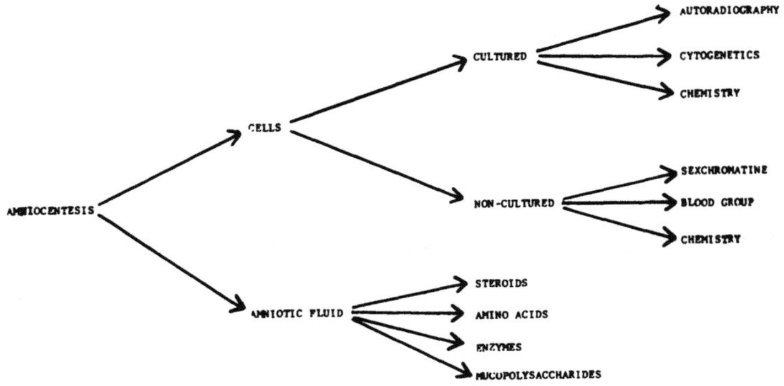

Fig. 1. From Emery et al. (23).

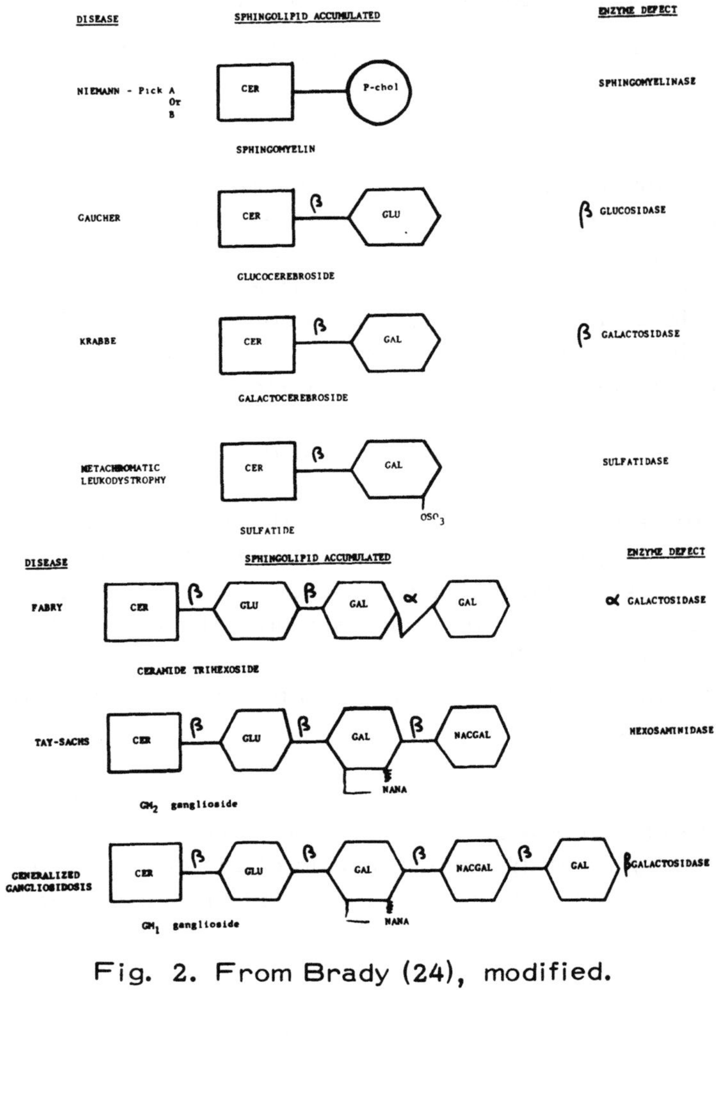

Fig. 2. From Brady (24), modified.

Amniocentesis

Amniocentesis is usually performed, for reasons of security and lack of knowledge, at 14 weeks of pregnancy; at that period 100 to 120 ml of fluid are present (4); 2 weeks later there are around 200 ml of liquid.

Valenti and Kehaty (5) begin by placental localization, Jacobson and Barter (6) introduce a needle directly into the transabdominal wall and puncture around 10 to 20 cc of fluid.

The fluid is centrifuged and the chemical analysis can be done on cultured cells, uncultured cells or amniotic fluid (Fig. 1).

This procedure is quite untraumatic (7) and only 1% accidents occurred on 200 cases.

Among the glycolipidoses we are concerned with, 6 are autosomal recessive (Fig. 2).

Tay-Sachs' disease, where the A component of hexosaminidase is missing (8); Sandhoff's disease (9, 10) where both hexosaminidase A and B are missing; generalized gangliosidosis, where β-galactosidase is involved (11); metachromatic leukodystrophy (12), where arylsulfatase A is lacking; Gaucher's disease (13), where α-glucosidase is absent; and Krabbe's disease, where a galactocerebrosidase is deficient. Only one is X-linked, Fabry's disease, where the enzyme missing is an α-galactosidase (14, 15).

Diagnosis of Tay-Sachs' disease (16) has been already made several times in utero on cultured cells, but also (and this is very important) directly on amniotic fluid (17). Also, prenatal diagnosis of metachromatic leukodystrophy has been published (18, 19).

But very little concrete data has been published until now on diagnosis in utero of glycolipidoses.

The first problem we are concerned with is: what is the evolution of the enzyme from beginning of pregnancy till birth. Some evolution is already known (Fig. 3). A^F, B^F and C^F represent 3 different fetal enzymes, A^F is at the same level during the whole pregnancy (this is the case of β-galactosidase), B^F increases and a wrong diagnosis

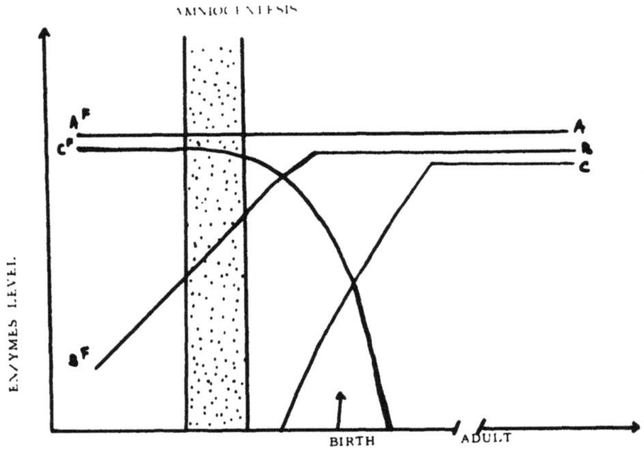

Fig. 3. From Kaback et al. (20).

could be made of heterozygote if this is not known, C^F is hemoglobin, and no comparable changes of a lysosomal enzyme have been found until now (20).

The second problem, we are aware of, is: could we shorten the period between the amniocentesis and the result of the chemical analysis ? The first procedure would be to do the chemical analysis on amniotic fluid or better on uncultured cells.

Table I - Enzyme activities in uncultured and cultured amniotic cells.

Enzyme	Enzyme activity (mμM/h/mg protein)	
	Uncultured cells	Cultured cells
β-galactosidase	10- 30	415-1020
Arylsulfatase A	5- 34	172- 275
β-NAc-glucosaminidase	50-620	2900-6200

Data from M. Kaback et al. (21).

Unfortunately, the enzyme activity on non cultured cells is so low that it is very difficult to differentiate heterozygotes from normals. It is also difficult to explain the widespreading of the enzyme activity on uncultured cells after 4 weeks in flasks, besides O'Brien's explanation that an increasing activity is found in aging tissue-cultures.

To use amniocentesis as a diagnosis, we must be sure of the reliability of the method; the analysis, the tissue-culture technique must be standardized as well as the expression of the results. It seems to be better to express the result as millimicromoles/hour/ /mg of protein that per 10^6 cells.

Lipidoses are classified by Littlefield (22) as severe disorders of higher genetic risk; this shows how promising can be amniocentesis but, at the point where we are now, it must not be used routinely. We must not be too optimistic, we need a great deal more experience on amniotic fluid, uncultivated cells and in all cases the chemical analysis of the fetus or the newborn is an absolute necessity to establish the validity of these techniques and to study the development of the disease in utero.

REFERENCES

(1) Plentl, A.A., Clin. Obstet. Gynec. 9: 427 (1966).

(2) Gluck, L., Landowne, R. and Kulovich, M., Pediat. Res. 4: 352 (1970).

(3) Van Leeuwen, L., Jacoby, H. and Charles, D., Acta Cytol. 9: 442 (1965).

(4) Fuchs, F., Clin. Obstet. Gynec. 9: 449 (1966).

(5) Valenti, C. and Kehaty, N., J. Lab. and Clin. Med. 76: 355 (1969).

(6) Jacobson, C.B. and Barter, R.H., Am. J. Obstet. Gynec. 99: 796 (1967).

(7) Creasman, W.T., Lawrence, R.A. and Thiede, H.A., J. Am. Med. Ass. 204: 949 (1968).

(8) Okada, S. and O'Brien, J.S., Science 165: 698 (1969).

(9) Suzuki, Y., Jacob, J.C. and Suzuki, K., Neurology 20: 388 (1970).

(10) Sandhoff, K., Andreae, U. and Jatzkewitz, H., Life Sci. 7: 283 (1968).

(11) Sloan, H.R., Uhlendorf, B.W. and Jacobson, C.B., Pediat. Res. 3: 368 (1969).

(12) Austin, J., Armstrong, D. and Shearer, L., Arch. Neurol. (Chicago) 13: 593 (1965).
(13) Brady, R.O., Kanfer, J.N., Bradley R.M. et al., J. Clin. Invest. 45: 1112 (1966).
(14) Kint, J.A., Science 167: 1268 (1970).
(15) Bensaude, I., Callahan, J. and Philippart, M., Biochem. Biophys. Res. Comm. 43: 913 (1971).
(16) Schneck, L., Friedland, J., Valenti, C., Adachi, M., Amsterdam, D. and Volk, B.W., Lancet 1: 582 (1970).
(17) Friedland, J., Perle, G., Saifer, A., Schneck, L. and Volk, B.W., Proc. Soc. Exptl. Biol. Med. 136: 1297 (1971).
(18) Nadler, H.L. and Gerbie, A.B., New Engl. J. Med. 282: 596 (1970).
(19) Kaback, M.M. and Howell, R.R., New. Engl. J. Med. 282: 1336 (1970).
(20) Kaback, M.M. and Leonard, C.O., in Proceedings of the Conference on Antenatal Diagnosis, Univ. of Chicago (1970), in press.
(21) Kaback, M.M., Leonard, C.O. and Parmley, T.H., Pediat. Res. (in press).
(22) Littlefield, J.W., New Engl. J. Med. 282: 627 (1970).
(23) Emery, A.E.M., Nelson, M.M., Gordon, G. and Burt, D. Scot. Med. J. 14: 277 (1969).
(24) Brady, R.O., Clin. Chem. 16: 811 (1970).

PATTERNS OF BRAIN GANGLIOSIDE FATTY ACIDS IN SPHINGOLIPIDOSES

B. BERRA and V. ZAMBOTTI

Department of Biochemistry, Medical School

University of Milano (Italy)

In the last few years the development of new methods of analyses resulted in an improvement in our knowledge of lipidoses; the accumulated lipids were better characterized and very often the enzymatic defect of the disease was determined. In many cases however, only the carbohydrate moiety of the accumulated glycolipids was studied, while the lipid moiety, particularly as far as the ganglioside fatty acids distribution is concerned, was poorly investigated. The data in the literature on this topic are very few; Rosemberg (1) investigated the pattern of fatty acids in Tay-Sachs' disease, but his determinations were made on the gangliosides mixture, and not on the single ganglioside. The fatty acid distribution of grey matter GM1 in GM1 gangliosidosis, type I, was reported by Suzuki (2); the fatty acid composition of nine ganglioside pure fractions in subacute sclerosing leuko-encephalitis was reported by Ledeen et al. (3); recently Eto et al. (4) gave the fatty acid composition of gangliosides GM3, GM2, GM1, GD1a in white matter in a case of Krabbe's disease.

As it was already pointed out by Rosemberg (1) "the lipid portion of gangliosides plays an important role in the formation of aggregates within the neural cell and thereby affects the mode of inclusion of these glycolipids in the architecture of the membranes"; in this respect we decided to study the fatty acids distribution of the single major gangliosides in some different neurological diseases: generalized gangliosidosis (age 15 months), Krabbe's disease (age 21 months), Niemann-Pick's disease (age 14 months) and Kufs' disease (age 46 years).

METHODS

a) Total lipids. 0.5-1 g fresh tissue were liophilized and then extracted with chloroform - methanol 2:1 (20 ml x 100 mg of liophilized material). The qualitative analyses were performed by TLC under the following experimental conditions : adsorbent : precoated plate (Merck) 20 x 20 cm, thickness 0.25 mm; solvent system: chloroform/methanol/water = 70/30/5 (v/v/v); 90 min run at room temperature; spray reagent: anisaldehyde/sulfuric acid/acetic acid - 1/2/100 (v/v/v) at 120°C x 20 min.

b) Ganglioside pattern. Gangliosides were extracted and purified according to Trams et al. (5) as modified by Tettamanti et al. (6). The quantitative analyses were performed according to Bonali et al. (7).

c) Ganglioside fatty acids distribution. Gangliosides, extracted according to Tettamanti et al. (6), were purified and fractionated by silica gel column chromatography as previously reported (8); the single gangliosides were identified by TLC with pure gangliosides of known structure prepared in our laboratory, in four different solvent systems:

1. chloroform : methanol : water = 60:35:8 (v/v/v)
2. chloroform : methanol : 2.5 N ammonia = 60:35:8 (v/v/v)
3. N-propanol : water = 70:30 (v/v)
4. N-butanol : pyridine : water = 6:2:2 (v/v/v).

Furthermore, the molar ratios of the single components (sphingosine, neutral hexose, hexosamine, sialic acid) were calculated.

The four major purified gangliosides (GM1, GD1a, GD1b, GT1b) were subjected to methanolisis in 6% sulfuric acid in methanol in screw capped teflon lined glass tube at 110°C for 2 hours. Fatty acid methyl esters were extracted with hexane and purified by TLC according to Galli et al. (9). Gas-liquid chromatographic analyses were carried out on an instrument equipped with a flame ionization detector. The following experimental conditions were used:

a) Two meter long (3 mm i. d.) spiral glass column packed with 5% polydiethylenglycolsuccinate (DEGS) coated on 80-100 mesh Diatoport S. The column temperature was programmed from 135°C to 210°C at 2.75°C/min; carrier gas N_2.

b) 80 cm (4 mm i. d.) U glass column packed with 20% WEAS coated on Chromosorb W. Programmed temperature from 150°C to 200°C at 3.5°C/min; carrier gas N_2.

Identification of peaks was obtained by comparison with pure reference standard compounds (NIH standards, Hormel Institute).

RESULTS AND DISCUSSION

a) Total lipids analyses. The pattern of total lipids in the different cases we analyzed, is shown in Fig. 1, 2, 3. It is notable that only monitoring the chloroform-methanol extract by TLC one can have sufficient indications from a qualitative point of view.

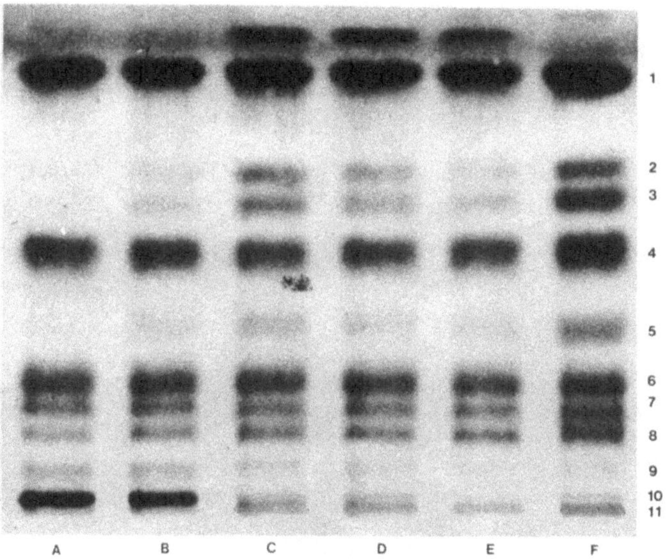

Fig. 1. TLC of total lipids. For experimental conditions see the text. Band 1: cholesterol, 2: cerebroside with normal fatty acids, 3: cerebrosides with hydroxy fatty acids, 4: phosphatidyl ethanolamine, 5: sulfatides, 6: phosphatidyl choline, 7: C_{24} sphingomyelin, 8: C_{18} sphingomyelin, 9: phosphatidyl inositol, 10: gangliosides, 11: phosphatidyl serine. A: GM1 gangliosidosis, grey matter, B: GM1 gangliosidosis, white matter, C, D, E: Krabbe's disease occipital lobe, frontal lobe and cerebellum, F: standards.

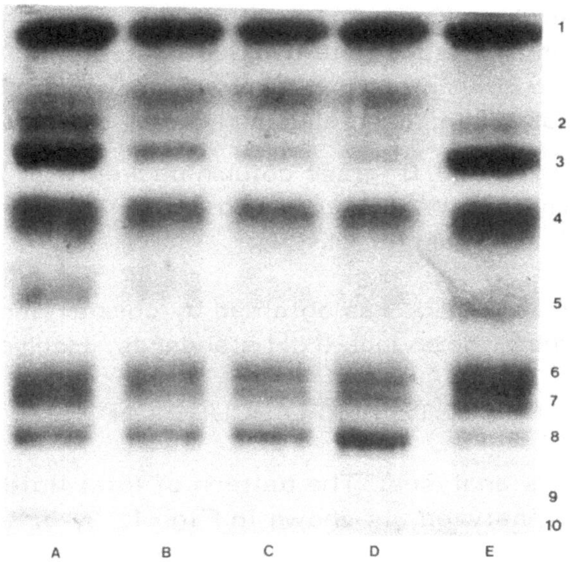

Fig. 2. TLC of total lipids (gangliosides excluded) of Niemann--Pick's disease. For experimental conditions see the text. Numbers of the bands as reported in Fig. 1. A: parietal lobe, white matter, B: parietal lobe, grey matter, C: frontal lobe, grey matter, D: temporal lobe, grey matter, E: standards.

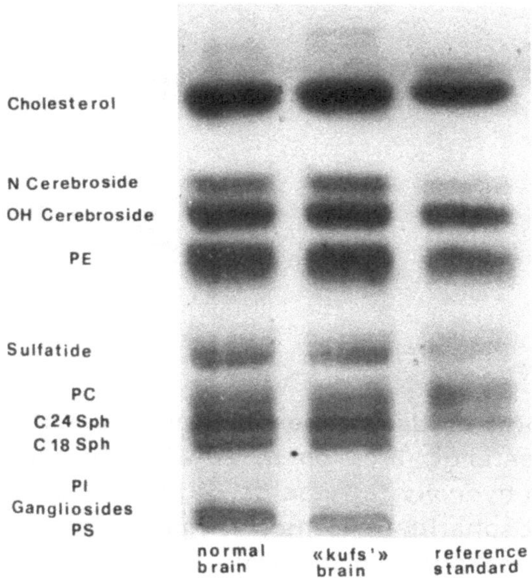

Fig. 3. TLC of total lipids of Kufs' disease. For experimental conditions see the text.

In the case of GM1 gangliosidosis (Fig. 1) the high amount of gangliosides associated with the decrease of typical myelin lipids (cerebrosides and sulfatides) is clearly significant.

In the case of Krabbe's disease (Fig. 1) even if we examined grey matter specimens, our TLC procedure shows that the ratio cerebrosides/sulfatides is higher than normal, while the absolute amounts of these glycolipids is always and significantly lower than normal.

In the case of Niemann-Pick's disease (Fig. 2) the sphingomyelin fraction, mainly the C_{18} sphingomyelin, is higher than normal, while sulfatides and cerebrosides are lower.

Finally in the case of Kufs' disease (Fig. 3) no significant differences from normal were observed.

b) Ganglioside analyses. As expected in the case of GM1 gangliosidosis, the lipid bound NANA is higher than normal either in the brain or in the examined extranervous tissues; the amount of gangliosides is particularly high in cerebellum and pons (Fig. 4, Tables I, II).

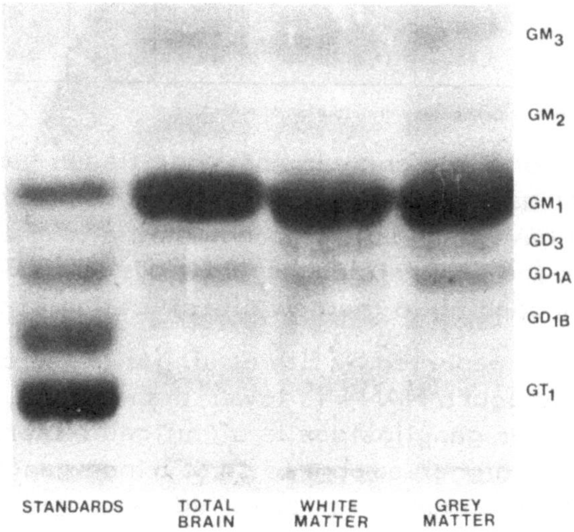

Fig. 4. TLC of brain gangliosides in GM1 gangliosidosis. Experimental conditions = adsorbent: precoated plate (Merck) 20 x 20 cm, thickness 0.25 mm; solvent system: chloroform/methanol/water = 60/35/8 (v/v/v); 12 hours run at room temperature; spray reagent: anisaldehyde/sulfuric acid/acetic acid = 1/2/100 (v/v/v) at 140°C x 10 min.

Table I - GM1 Gangliosidosis. Brain gangliosides (°).

	Grey matter	White matter	Cerebellum	Pons
Total NANA µmoles/g	4.83	3.41	8.1	8.5
GM3	3.4	2.7	0.9	1.5
GM2	5.2	4.7	0.8	0.1
GM1	72.6	73.7	78.2	82.0
GD3	2.0	1.5	1.0	1.2
GD1a	10.2	8.8	7.5	5.3
GD1b	2.6	3.5	4.5	5.1
GT1b	3.8	5.2	5.0	3.6

(°) Expressed as percentage distribution of NANA. Nomenclature of gangliosides in that of Svennerholm (J. Neurochem. 10, 613, 1963)

Table II - GM1 Gangliosidosis. Gangliosides of extra-nervous tissues (°).

Tissue	Total NANA µmoles/g	GM3	GM2	GM1	GD1a	GD1b	GT1b
thyroid	0.77	31.7	-	39.0	10.5	12.3	6.5
mediastinal lymphonode	1.2	29.2	-	39.0	11.7	9.8	10.3
myocardium	0.37	16.2	-	54.0	10.8	10.8	8.1
kidney	0.23	26.9	1.8	38.7	14.5	10.4	7.7
liver	0.63	36.6	-	41.5	7.3	4.9	9.7
spleen	0.77	40.0	-	44.0	4.7	4.7	6.6
lung	0.66	37.3	-	43.5	10.4	2.9	5.9

(°) Expressed as percentage distribution of NANA

The pattern of brain and extranervous tissue gangliosides shows that GM1 ganglioside is always present in a very high amount, so we have been able to conclude that our case can be included among the generalized gangliosidoses or GM1 gangliosidosis type I, according to the new classification of O'Brien (10).

As previously reported by Eto et al. (4) in the case of Krabbe's disease the lipid bound NANA is lower than normal, while the pattern of grey matter gangliosides is significantly abnormal owing to the presence of a higher amount of some minor gangliosides which on the basis of R_f values were identified as gangliosides GD3 and GD2 (Table III, Fig. 5).

In "Niemann-Pick's brain" the amount of lipid bound NANA is higher than in normal brain, mainly in white matter; the pattern of single gangliosides is rather similar to the normal pattern, even if minor gangliosides GM3, GM2, and GD3 are present in higher amount (Table IV, Fig. 6).

Table III - Krabbe's Disease. Brain gangliosides (°).

	Frontal lobe (")	Occipital lobe (")	Cerebellum (=)
Total NANA µmoles/g	1.95	1.87	1.44
GM3	3.1	2.3	2.8
GM2	2.1	3.0	2.5
GM1	16.8	16.4	11.4
GD3	9.5	9.1	20.0
GD1a	38.1	36.0	17.3
GD2	5.1	5.3	4.0
GD1b	11.5	14.4	21.0
GT1b	13.6	13.5	21.0

(°) Expressed as percentage distribution of NANA; (") Grey matter; (=) Total tissue.

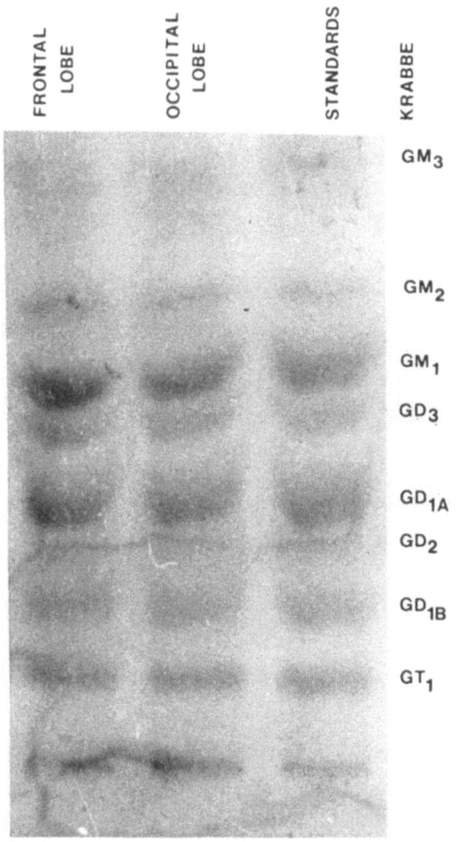

Fig. 5. TLC of brain gangliosides in Krabbe's disease. Experimental conditions as reported in Fig. 4.

Table IV - Niemann-Pick's Disease. Brain gangliosides (°).

	Temporal lobe		Frontal lobe		Parietal lobe	
	grey matter	white matter	grey matter	white matter	grey matter	white matter
Total NANA µmoles/g	3.50	2.10	3.30	1.80	3.38	2.00
GM3	5.4	< 1	5.6	< 1	6.0	< 1
GM2	7.4	4.0	8.6	5.0	8.4	4.5
GM1	15.7	22.5	16.1	23.3	17.1	23.8
GD3	4.0	2.5	5.0	1.9	5.0	1.5
GD1a	35.3	36.4	36.3	39.5	36.0	36.2
GD1b	15.6	14.5	13.6	13.4	13.5	15.3
GT1b	16.5	19.0	14.8	15.7	14.0	17.8

(°) Expressed as percentage distribution of NANA

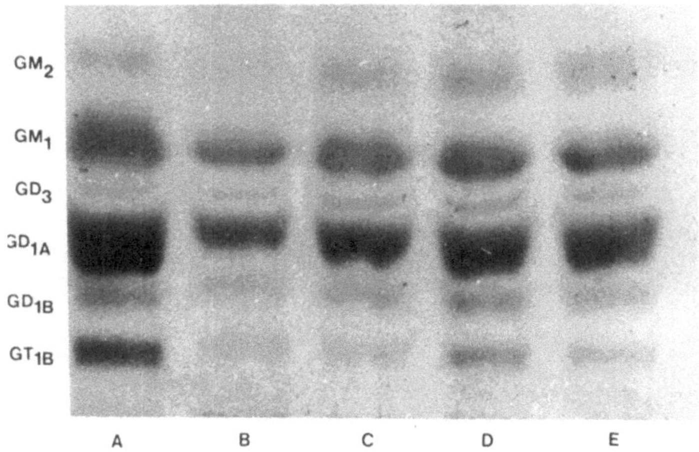

Fig. 6. TLC of brain gangliosides in Niemann-Pick's disease. Experimental conditions as reported in Fig. 4. A: standards, B: parietal lobe white matter, C: parietal lobe grey matter, D: temporal lobe grey matter, E: frontal lobe grey matter.

Finally in the case of Kufs' disease we were able to observe that the lipid bound NANA in grey matter is lower than in normal cases; on the contrary in white matter there is an increase of total gangliosides content.

From a quantitative point of view the GM2 ganglioside is present in a higher amount than in normal brain; however the main differences from normal cases could be observed when the regional pattern of different gangliosides was performed (Table V). These data agree with the histological and histochemical data reported in the literature very well (11).

c) Fatty acid analyses. In the case of GM1 gangliosidosis, Krabbe's disease and Niemann-Pick's disease the main fatty acid (90% of the total) was stearic acid (Tables VI, VII, VIII). The data reported in Table IX show considerable modifications in the fatty acid distribution of "Kufs" compared with normal gangliosides and the other pathological gangliosides analyzed. The most striking feature is the presence of very high amount of a C18 die-

Table V - Kufs' Disease. Brain gangliosides (°).

Grey Matter	Total NANA µmoles/g	GM2	GM1	GD1a	GD1b	GT1b
Lobus parietalis inf.	2.70	8.4	42.0	absent	29.6	19.9
Lobus parietalis sup.	2.6	7.1	26.7	8.9	30.2	27.2
Gyrus frontalis inf.	2.68	3.7	9.3	17.4	26.1	43.5
Gyrus cinguli	2.60	4.3	10.5	14.8	27.0	43.5
Gyrus temporalis medius	2.40	6.1	13.9	27.8	30.0	22.2
Gyrus frontalis sup.	2.70	7.4	21.8	35.4	23.4	11.7
Gyrus temporalis sup.	3.0	5.9	16.0	15.9	16.7	45.5
Nucleus leutiformis	2.1	7.4	18.5	23.2	39.4	11.6
Gyrus centralis ant.	2.72	9.2	18.4	22.2	30.0	20.1
Gyrus centralis post.	2.68	8.2	19.8	19.5	30.1	22.5
Gyrus fusiformis	1.8	13.9	18.5	22.6	35.4	9.6
Gyrus lingualis	2.26	2.3	17.9	13.1	46.0	20.6
Nucleus caudatus	2.83	2.4	11.4	17.9	40.0	28.4
Talamus	2.15	4.7	18.9	16.4	39.4	20.6
Uncus	3.40	4.8	14.7	22.0	32.4	26.0
Corpus callosus	0.54	11.2	21.8	28.1	27.9	11.9
White Matter						
Lobus parietalis inf.	1.88	5.3	56.5	absent	25.3	12.7
Lobus parietalis sup.	1.67	7.0	40.0	7.4	30.0	15.6
Gyrus frontalis inf.	1.58	7.4	12.7	23.0	32.8	24.2
Gyrus cinguli	1.51	7.9	14.1	24.8	30.8	22.2
Gyrus centralis ant.	1.06	10.6	26.9	22.8	29.7	10.0

(°) Expressed as percentage distribution of NANA.

Table VI - GM1 Gangliosidosis. Fatty acid composition of grey matter gangliosides (°).

Fatty acid	GM1	GD1a	GD1b	GT1b
14:0	trace	0.3	trace	trace
15:0	trace	trace	trace	trace
16:0	1.7	2.1	1.5	1.3
17:0	0.3	0.4	trace	trace
18:0	89.5	90.0	89.9	91.5
18:1	trace	0.2	0.3	0.5
19:0	trace	trace	trace	trace
20:0	6.1	5.2	5.8	4.3
21:0	trace	trace	trace	trace
22:0	0.2	0.3	0.2	0.4
23:0	trace	trace	trace	trace
24:0	0.2	0.2	0.2	0.2
24:1	0.5	0.4	0.5	0.3

(°) Expressed as percentage of total fatty acid

Table VII - Krabbe's Disease. Fatty acid composition of grey matter gangliosides (°).

Fatty acid	GM1	GD1a	GD1b	GT1b
14:0	0.3	0.2	0.4	0.3
15:0	0.4	0.2	0.3	0.3
16:0	4.5	4.0	3.8	3.5
16:1	0.5	0.4	0.6	0.9
17:0	trace	trace	trace	trace
18:0	82.5	82.8	82.4	83.9
18:1	3.5	4.0	2.5	2.0
18:2	0.5	0.8	0.6	0.8
19:0	0.5	0.4	trace	trace
20:0	2.8	2.7	3.5	3.0
21:0	trace	trace	trace	trace
22:0	0.8	0.5	0.5	0.4
23:0	0.4	0.2	0.5	0.4
24:0	0.9	1.0	1.5	1.4
24:1	2.8	2.6	2.5	2.7

(°) Expressed as percentage of total fatty acid

Table VIII - Niemann-Pick's Disease. Fatty acid composition of grey matter gangliosides (°).

Fatty acid	GM1	GD1a	GD1b	GT1b
14:0	trace	trace	trace	trace
15:0	trace	trace	trace	trace
16:0	1.4	1.1	1.5	1.7
17:0	trace	trace	trace	trace
18:0	88.5	87.2	90.5	89.7
18:1	1.9	1.5	0.8	1.0
19:0	trace	trace	trace	trace
20:0	4.5	5.1	3.9	4.2
21:0	trace	trace	trace	trace
22:0	1.0	1.3	0.9	0.8
23:0	0.5	0.6	trace	trace
24:0	0.4	0.6	0.2	0.4
24:1	1.2	1.8	0.8	0.9

(°) Expressed as percentage of total fatty acid

Table IX - Kufs' Disease. Fatty acid composition of grey matter gangliosides (°).

Fatty acid	GM1	GD1a	GD1b	GT1
12 : 0	2.27	1.49	trace	trace
14 : 0	2.19	trace	trace	trace
15 : 0	trace	1.49	trace	trace
16 : 0	12.39	7.47	14.38	13.61
16 : 1	trace	1.94	1.51	1.34
17 : 0	trace	trace	trace	trace
18 : 0	34.18	14.96	66.19	62.21
18 : 1	19.27	18.80	8.46	9.31
18 : 2	24.13	42.30	4.14	9.74
20 : 0	3.07	1.87	4.50	1.67
20 : 1	2.51	2.35	trace	trace
21 : 0	trace	4.98	trace	trace
22 : 0	trace	1.61	trace	trace
24 : 0	trace	trace	trace	trace

(°) Expressed as percentage of total fatty acid

noic acid. This acid was present in all the analyzed gangliosides, although in different amounts (mainly in gangliosides GD1a and GM1); it is absent from normal gangliosides and also from Kufs' brain total lipids (fraction 1 of a chloroform-methanol extract purified by Sephadex G 25 column chromatography, according to Siakotos and Rouser, 12).

The chain length, unsaturation and molecular weight of this acid were established by means of the following identification criteria:

1) The peak had the same retention time of $C_{18}:2\omega6$ (linoleic acid).

2) After a catalytic hydrogenation of the ester mixture this peak disappeared, together with the peaks of the known unsaturated acids. The disappearance of $C_{18}:2\omega6$ and $C_{18}:1\omega9$ was accompained by a corresponding increase of the saturated homologue $C_{18}:0$.

3) The analyses with combined gas chromatography-mass spectrometry provided a molecular weight and fragmentation characteristics corresponding to the methyl ester of $C_{18}:2$.

Further analysis are needed in order to clarify the following aspects:

1) The real presence of the abnormal fatty acid in the ganglioside molecule must be better supported. By now we can only ex-

clude the presence of phospholipid contaminants in the ganglioside fractions we examined; on the other hand we did not find the dienoic fatty acid in any other investigated lipids. However we are dealing with a pathological case and no sufficient informations are reported in literature about the chemical composition of brain in such a disease: we cannot exclude the presence of this acid in a lipid fraction so closely associated with gangliosides that it proved unseparable during the repeated column chromatography we made, to fractionate and purify gangliosides. But this hypothesis seems highly improbable to us.

2) The exact position of the two double bonds.

3) The cis-trans isomery of the double bond. If the IR spectra demonstrated the cis-trans or the all-trans-configuration one might easily suppose that the dienoic fatty acid would give some particular phisical-chemical features to the ganglioside molecule; and probably we could justify on this basis the presence of particular membranous bodies and lipofuscin, as observed by Zeman et al. (13) and confirmed by us in the present case (14).

In conclusion our data confirm that in GM1 gangliosidosis, Krabbe's and Niemann-Pick's diseases, the major gangliosides fatty acids pattern is normal, while they strongly show the existence of abnormalities in the lipid portion of the ganglioside molecule in case of Kufs' disease. These types of alterations had never been detected before in other neurological disorders known to affect ganglioside metabolism.

SUMMARY

Gangliosides from 4 patients died of different neurological diseases were extracted and purified. The fatty acids methyl esters of the 4 major gangliosides (GM1, GD1a, GD1b, GT1b) were analyzed by GLC. The presence of high levels of a C_{18} dienoic fatty acid in "Kufs" brain gangliosides, confirmed by combined GLC - mass spectrometry, is discussed.

ACKNOWLEDGEMENTS

The authors wish to thank Prof. H. Jatzkewitz for kindly providing pure reference standards. The skilled technical assistence of G. Cervato and M. Rancati was greatly appreciated.

REFERENCES

(1) Rosemberg, A., in Inborn Disorders of Sphingolipid Metabolism, Ed. S.M. Aronson and B. W. Volk, Pergamon Press, New York, p. 267 (1967).

(2) Suzuki, K. and Kamoshita, S., J. Neuropath. & Exptl. Neurol. 28: 25 (1969).

(3) Ledeen, R., Salsman, K. and Cabrera, M., J. Lipid Res. 9: 129 (1968).

(4) Eto, Y. and Suzuki, K., J. Neurochem. 18: 503 (1971).

(5) Trams, E.G. and Lauter, C.J., Biochim. Biophys. Acta 60: 350 (1962).

(6) Tettamanti, G., Bertona, L., Berra, B. and Zambotti, V., Ital. J. Biochem. 13: 315 (1964).

(7) Bonali, F., Tettamanti, G. and Cervato, G., Metabolismo 5: 269 (1969).

(8) Matturri, L., Coggi, G., Berra, B. and Bonali, F., Il Morgagni 1: 297 (1969).

(9) Galli, G., White, H.B. Jr. and Paoletti, R., J. Neurochem. 17: 347 (1970).

(10) O'Brien, J.S., Okada, S., Wan Ho, M., Fillerup, D.L., Veath, M. L., Adams, K., in Lipid Storage Disease, Ed. J. Bernsohn and J. H. Grossman, Academic Press, New York, p. 225 (1971).

(11) Van Bogaert, L., in Maladies nerveuses génétiques d'ordre métabolique, Rev. Méd. de Liége, Univ. de Liége, p. 17 (1962).

(12) Siakotos, A.N. and Rouser, G., J. Amer. Oil Chem. Soc. 42: 913 (1965).

(13) Zeman, W. and Donahue, S., Acta Neuropath. 3: 144 (1963).

(14) Berra, B., Coggi, G. and Matturri, L., in preparation.

SUBJECT INDEX

MIX
Papier aus verantwortungsvollen Quellen
Paper from responsible sources
FSC® C105338

If you have any concerns about our products,
you can contact us on
ProductSafety@springernature.com

In case Publisher is established outside the EU,
the EU authorized representative is:
Springer Nature Customer Service Center GmbH
Europaplatz 3, 69115 Heidelberg, Germany

Printed by Libri Plureos GmbH
in Hamburg, Germany